WIRELESS AND PERSONAL COMMUNICATIONS SYSTEMS

Feher/Prentice Hall Digital and Wireless Communications Series

Carne, E. Bryan. Telecommunications Primer: Signal, Building Blocks and Networks

Feher, Kamilo. Wireless Digital Communications: Modulation and Spread Spectrum Applications

Garg, Vijay and Joseph Wilkes. Wireless and Personal Communications Systems

Pelton, N. Joseph. Wireless Satellite Telecommunications: The Technology, the Market & the Regulations

Other Books by Dr. Kamilo Feher

Advanced Digital Communications: Systems and Signal Processing Techniques

Telecommunications Measurements, Analysis and Instrumentation

Digital Communications: Satellite/Earth Station Engineering

Digital Communications: Microwave Applications

Available from CRESTONE Engineering Books, c/o G. Breed, 5910 S. University Blvd., Bldg. C-18 #360, Littleton, CO 80121, Tel. 303-770-4709, Fax 303-721-1021, or from DIG-COM, Inc., Dr. Feher and Associates, 44685 Country Club Drive, El Macero, CA 95618, Tel. 916-753-0738, Fax 916-753-1788.

AT&T

WIRELESS AND PERSONAL COMMUNICATIONS SYSTEMS

Vijay K. Garg

Joseph E. Wilkes

AMERICAN TELEPHONE AND TELEGRAPH COMPANY

BELL LABORATORIES DIVISION

For book and bookstore information

http://www.prenhall.com

Prentice Hall PTR
Upper Saddle River, NJ 07458

Library of Congress Cataloging-in-Publication Data

Garg, Vijay Kumar, 1938–
    Wireless and personal communications systems / Vijay K. Garg,
    Joseph E. Wilkes.
      p. cm.—(Feher/Prentice Hall digital and wireless
    communication series)
    Includes bibliographical references and index.
    ISBN 0-13-234626-5 (case)
    1. Wireless communication systems. 2. Mobile communication systems.
    I. Wilkes, Joseph E. II. Title. III. Series.
    TK5103.2.G37   1996
    621.3845—dc20   95-38584
                                          CIP

Editorial/production supervision: *BooksCraft, Inc., Indianapolis, IN*
Cover design director: *Jerry Votta*
Cover design: *Design Source*
Acquisitions editor: *Karen Gettman*
Manufacturing manager: *Alexis R. Heydt*

Published by Prentice Hall PTR
Prentice-Hall, Inc.
A Simon & Schuster Company
Upper Saddle River, NJ 07458

The publisher offers discounts on this book when ordered in bulk quantities.
For more information, contact:
        Corporate Sales Department
        Prentice Hall PTR
        One Lake Street
        Upper Saddle River, NJ 07458
        Phone: 800-382-3419   Fax: 201-236-7141
        E-mail: corpsales@prenhall.com.

Printed in the United States of America

10 9 8 7 6 5 4

ISBN: 0-13-234626-5

Prentice-Hall International (UK) Limited, *London*
Prentice-Hall of Australia Pty. Limited, *Sydney*
Prentice-Hall Canada Inc., *Toronto*
Prentice-Hall Hispanoamericana, S.A., *Mexico*
Prentice-Hall of India Private Limited, *New Delhi*
Prentice-Hall of Japan, Inc., *Tokyo*
Simon & Schuster Asia Pte. Ltd., *Singapore*
Editora Prentice-Hall do Brasil, Ltda., *Rio de Janeiro*

Table of Contents

Preface

The generation that grew up watching the characters on "Star Trek" speaking to each other and sending data over handheld, pocket-sized personal communicators expects to have such devices made available to them. Personal Communications Services (PCS) promise to deliver on that expectation.

The revolution in several technologies over the last two decades makes PCS feasible.

☞ The astounding advances in low-cost, very large-scale integrated (VLSI) digital circuits—especially microprocessors and digital signal processors

☞ The advances in highly efficient solid-state radio frequency (RF) circuitry capable of operating in 1–10 gigahertz (GHz) bands

☞ The continued improvement in rechargeable batteries

☞ Advances in spread-spectrum communications

☞ The conversion of the telephone network to stored program (computer) controlled systems communicating by a special packet data network

At the same time socio-economic factors have produced the demand for PCS. These factors include

☞ An increasing demand for higher white-collar productivity, to match the gains seen on the factory floor

☞ An increasing need to work out of the office (at home or on the road)

☞ An ever increasing demand for personal mobility

☞ An increasing dependence on FAX, personal computers, electronic mail (E-mail), and centralized databases

☞ Business recognizing timeliness as a source of competitive advantage

☞ The precipitous drop in the cost of mobile communications

Technological revolutions result from the confluence of technological availability and customer needs and demands; PCS is the next such revolution.

This book describes the emerging Personal Communications Network (PCN) and PCS being envisioned. It discusses the recent history of underlying technologies that are being used to synthesize PCN and delineates the alternative approaches being considered. Although the primary focus is on the U.S. technologies, we also cover wireless technologies used in Europe and Japan.

This book can be used by the telecommunication managers engaged in managing wireless/PCS networks with little or no technical background in wireless technologies; practicing communication engineers involved in the design of wireless/PCS systems, and senior/graduate students in electrical, telecommunication, or computer engineering planning to pursue a career of a telecommunication engineer. We suggest material in chapters 1, 2, 8, 9, 10, 11, and 12 for the telecommunication managers. The practicing design engineer in telecommunications needs to cover the entire book in order to become proficient in the wireless/PCS technologies. The first seven chapters of the book deal with the history and theoretical aspects of wireless technology. Chapters 8 through 13 deal with the design aspects of the wireless/PCS system. Chapters 14 and 15 provide a general background in wireless data. If the book is used for students with a general background in electromagnetic field theory and digital systems, we suggest using the material in chapters 1, 2, 3, 4, 5, 6, and 8 in the first semester and chapters 7, 9, 10, 11, 12, 13, 14, and 15 in the second semester.

Chapter 1 presents the historical background of wireless communications starting from 1946 and examines the evolution of wireless technologies in the United States and Europe.

Chapter 2 discusses the first- and second-generation cellular systems used in the United States, Europe, and Japan. In this chapter we also examine the potential problems associated with the access technology for the second-generation-plus PCS systems and provide the vision of the third-generation PCS system.

Chapter 3 concentrates on the narrowband channelized and wideband non-channelized wireless communication systems. In this chapter, we focus on access technologies from capacity, performance, and spectral efficiency viewpoint.

Chapter 4 presents propagation and multipath characteristics of a radio wave. The concepts of delay spread and intersymbol interference are given. We also present several empirical and semiempirical models used to calculate path losses in urban, suburban, and rural environments.

Chapter 5 gives the fundamentals of cellular communications. We develop a relationship between the reuse ratio and cluster size for the hexagonal cell geometry and study the cochannel interference for the omnidirectional and sectorized cell site.

Chapter 6 deals with the digital modulation techniques and presents the modulation schemes used for cellular/wireless communications.

Chapter 7 discusses antennas and diversity. In this chapter we present different methods used to combine signals in the multipath environment.

Chapter 8 presents the analog and digital systems used in the United States. We also discuss the various air interfaces that are standardized for PCS in the United States. We provide typical call flows for origination, termination, handoff, and so on. The material in this chapter is extracted from the various standards for cellular and PCS available at the time of this writing, and it provides an end-to-end view of services. We suggest consulting the appropriate standards in order to be current.

Chapter 9 presents an overview of the Global System for Mobile Communications (GSM) system as described in the European Telecommunication Standard Institute's (ETSI's) recommendations. A brief description of Japanese Digital Cellular (JDC) system is also given. The chapter also addresses PCS 1900, a derivative of GSM, for PCS application in the United States.

Chapter 10 deals with the security issues of the wireless and PCS systems and focuses on the privacy and authentication schemes used in the U.S. and European systems.

Chapter 11 discusses the management of the PCS and cellular networks. In this chapter we present requirements for accounting management, fault management, performance management, configuration management, and security management for a PCS network.

Chapter 12 presents the interworking and interoperability issues and outlines the problems that must be addressed to achieve seamless communication.

Chapter 13 discusses the planning and engineering of a radio system. In this chapter we illustrate the process of growing a wireless system by considering a growth scenario with a frequency reuse factor of 7. We also present a traffic model of a wireless serving area for both cellular and PCS systems operating in a large metropolitan area.

In Chapters 14 and 15, we discuss Cellular Digital Packet Data (CDPD) and other packet-switched data systems such as ARDIS and RAM Mobile Data that are used for wireless data messaging services. These services use a dedicated network at the specialized mobile radio and cellular frequencies in the 800 to 900 megahertz (MHz) band.

In this book we have provided several numerical examples to illustrate the concepts. A number of problems are also given at the end of several chapters that may be assigned as homework problems to the students.

During the preparation of this book, several of our coworkers and friends have provided constructive suggestions, and we would like to thank them. In particular we thank: Dan Brown, Kamilo Feher, Reed Fisher, Larry Gitten, Jim McEowen, Bruce McNair, Ray Pickholtz, and Tippure Sundresh for their comments. We also thank V. H. MacDonald for supplying the traffic tables for Appendix C. In addition, we extend our appreciation to AT&T Bell Laboratories and AT&T Network Systems management for supporting this effort. We are grateful to Lisa Benintente, Carol Fitzgerald, Mary Klopman, and Suzanne Smith for preparing the figures. We thank the Prentice-Hall staff, in particular Karen Gettman, for providing the necessary support during the publication of this book.

Tables 4.2, 4.3, and 4.4, and 4.5 are copyrighted Pentech Press Ltd., used with permission. The material in chapter 4 Section 5 is adapted from "The Mobile Radio Propagation Channel" and copyrighted by Pentech Press Ltd, used with permission.

Figures 8.1, 8.14, 8.15, 8.16, and 8.17 are used with permission of the Telecommunications Industry Association (TIA), copyrighted by TIA.

Figures 8.2, 8.18, 8.19, 8.20, 8.21, 8.22, 8.23, 8.24, 8.25, 8.26, 8.27, 8.30, 8.31, 8.32, 8.33, 8.36, 8.37, 8.38, 8.39, 8.40, 8.41, 8.42, and 8.44 and Tables 8.1, 8.8, 8.9, 8.10, 8.13, 8.14, 8.15, 8.16, 8.20, 8.21, B.1, B.2, B.3, B.4, B.4, B.6, B.7, B.8, B.9, B.10 are copyrighted by ATIS and used with permission.

Some material in chapter 9 on GSM is adapted from an AT&T Technical Education Center Course on GSM and used with permission.

Material on Public Key cryptographic in chapter 10 is copyright 1994, AT&T, all rights reserved, reprinted with permission. Material in chapter 10, sections 1 and 2, is copyrighted by the IEEE and used with permission.

Material in chapter 14 on CDPD is adapted from the CDPD Forum and copyrighted by the CDPD Forum, used with permission.

Material in chapter 15 on RAM Mobile Data is copyrighted RAM Mobile Data and used with permission.

Vijay Garg
Joe Wilkes
June 1995

An Overview of Wireless Technologies

1.1 INTRODUCTION

Over the last decade, deployment of wireless communications in North America and Europe has been phenomenal. Wireless communications technology has evolved along a logical path, from simple first-generation analog products designed for business use to second-generation digital wireless telecommunications systems for residential and business environments.

As the industry plans and implements the second-generation digital networks in the mid-1990s, a vision of a next-generation wireless information network is emerging. Complete Personal Communications Services (PCS) will enable all users to economically transfer any form of information between any desired locations. The new network will be built on and interface with the separate first- and second-generation cordless and cellular services and will also encompass other means of wireline and wireless access such as Local Area Networks (LANs) and Specialized Mobile Radio (SMR). For at least part of the 1990s, we will see some systems that are cellular at new frequencies and some with new services. By the end of the decade, true third-generation systems offering high bandwidth multimedia applications may emerge.

1.2 HISTORICAL BACKGROUND

The first mobile telephone service was introduced in the United States in 1946 by AT&T. It was used to interconnect mobile users (usually in automobiles) to the public telephone land-line network, thus allowing telephone calls between fixed stations and mobile users. Within a year, mobile telephone service was offered in more than 25 American cities. These mobile telephone systems were based on Frequency Modulation (FM) transmission. Most of these systems used a single powerful transmitter to provide coverage of up to 50 miles or more from the base. The FM mobile telephone channels used 120 kilohertz (kHz) of spectrum to transmit a voice with an effective bandwidth of only about 3 kHz. While these systems were technically advanced for their day, modern improvements in transmitter stability, receiver noise figure, and receiver bandwidth have shown how inefficient these systems were.

Demand for mobile telephone service grew quickly and stayed ahead of the available capacity in many of the large urban cities. Offered traffic was more than the effective traffic capacity of the system. Loading factors of 50, 100, or more subscribers per radio channel were common. Service quality was terrible; blocking probabilities were as high as 65% or more. The usefulness of the mobile telephone decreased as users found that blocking often prevented them from getting a circuit during the peak periods. Users and the telephone companies, alike, realized that a handful of channels would not be enough for a true mobile telephone service to develop. Large blocks of spectrum would be needed to satisfy the demand in the urban areas.

In the mid-1960s the Bell System introduced the Improved Mobile Telephone Service (IMTS) with enhanced features, including automatic trunking, direct dialing, and full-duplex service. Improvements in transmitter and receiver design enabled a reduction in the FM channel bandwidth to 25–30 kHz. IMTS showed the fundamental viability of narrowband FM channel, automatic interconnection, trunking, and other key modern features.

In the late 1960s and the early 1970s work began on the first cellular telephone systems. The term "cellular" refers to dividing the service area into many small regions (cells) each served by a low-power transmitter with moderate antenna height. Frequencies are not reused in adjacent cells to avoid interference.

It should be recognized that the first-generation analog cellular radio was not so much a new technology as it was a new idea for organizing existing IMTS technology on a large scale. While the voice communications used the same analog FM that had been used since the end of World War II, two major technological improvements made the cellular concept a reality. In the early 1970s the microprocessor was invented; while the complex control algorithms could have been implemented in wired logic, the microprocessor made the concepts easier to implement. It also allowed more complex control algorithms to be implemented. The second improvement was in the use of a digital control link between the mobile telephone and the base station (or cell site as it was later called). In IMTS, the base station transmitted an idle tone to inform a mobile telephone that the channel was idle. The mobile telephone in turn transmitted dial pulses using tones and transmitted its own identification (ID) as a four digit number. This use of limited information between base station and mobile telephone severely limited the services available to users.

In the cellular service, channels were assigned to transmit data between the base station and the mobile telephone and from the mobile telephone to the base station using true digital data transmission. Now the mobile telephone could function more like a wireline phone, and true telephone service could be offered to people in automobiles and walking on city streets.

In the late 1980s interest emerged in a digital cellular system, where both the voice and the control were digital. The use of digital technology for reproduction of music with compact disks popularized the quality of digital audio. The idea of eliminating noise and providing clean speech to the limits of each serving area were attractive to engineers and lay people alike. By 1991 digital cellular service began to emerge to reduce the cost of wireless communications and improve the call-handling capacity of an analog cellular system. In 1993, a digital system was placed in service in some parts of the United States.

1.3 STANDARDS

In the United States, a decision by the Federal Communications Commission (FCC) to require one nationwide standard, and thus support roaming between any systems in the country, resulted in the rapid deployment of cellular systems. In Europe, several nations built their own systems that were incompatible with systems in other European nations. Similarly, Japan built its own cellular system. Thus, throughout the world, there are at least five different, incompatible, first-generation cellular standards. Each of these standards depends on frequency modulation of analog signals for speech transmission and in-band signaling to send control information between a Mobile Station (MS) and the rest of the network during a call.

Since there was already one nationwide standard in the United States, with roaming, the push for second-generation systems was not as strong in the United States as it was in Europe and Japan. Europe and Japan developed their second-generation digital mobile communications system by using new dedicated frequency bands, whereas the North American standards specify band sharing with the already successful analog cellular systems.

Current digital cellular systems use one of three standards:

1. Western Europe—Global System for Mobile Communications (GSM)
2. North American Electronic Industry Association (EIA) Standards IS-54
3. Japanese Digital Cellular (JDC).

The North American standards permit coexistence with the first-generation standard, the Advanced Mobile Phone System (AMPS), and add a digital voice transmission capability for new digital equipment. In one implementation, channels in a certain geographic area may be assigned to either a digital or analog signal, whereas in another area the two signals may share channels. Thus, the North American Standard IS-54 enhances, rather than replaces, the analog cellular technology.

Because of this history, North America and Europe appear to be moving in opposite directions. Europe is migrating from a set of incompatible analog systems to a single digital system. North America is deploying various dual-mode technologies that share a common analog standard, but with two radically different digital technologies. There is also a narrowband version of the analog standard (N-AMPS) and of the digital Time-Division Multiple Access (TDMA) standard Extended Time-Division Multiple Access (E-TDMA). We will examine these in chapter 2. Further compounding the problem is that the FCC has not mandated a single nationwide standard for PCS as they did for analog cellular and high definition television. They are letting the marketplace decide. U.S. PCS may have six or more incompatible digital standards in widespread use.

European research for digital cellular telephony began in the United Kingdom and Sweden in the early 1980s. In 1985, Conference Europeenne des Postes et Telecommunications (CEPT) started standardization of second-generation cellular telephones. Digital cellular standards were first published in the United Kingdom in 1987. These national standards specified parameters associated with operating frequency, transmitter power and spectrum, and interworking with the Public Switching Telephone Network (PSTN), but left the issue of radio interfaces open to manufacturers. There were no common air interface specifications. Debate continued over the relative merits of different technical solutions, with Sweden pushing for a Time-Division Multiple Access/Time-Division Duplex (TDMA/TDD) solution in contrast to the United Kingdom's Frequency-Division Multiple Access/Time-Division Duplex (FDMA/TDD). In January 1988, CEPT decided on the new European standards based upon a Time-Division Multiple Access/Time-Division Duplex/Multiple Carrier (TDMA/TDD/MC) approach operating just below 2 GHz frequency, subsequently known as the Digital European Cordless Telecommunications (DECT) standards. DECT was intended to support data as well as voice communications. In 1988, with the formation of the European Telecommunications Standards Institute (ETSI), responsibility for DECT standardization was moved to the ETSI Radio Equipment & System (RES) 3 subcommittee. In August 1992, DECT became a European Telecommunications Standard, ETSI 300–175. DECT has a guaranteed pan-European frequency allocation supported by the ETSI members and enforced by European Commission Directive 91/287. Figures 1.1 and 1.2 show the evolution of wireless technologies in the United States and Europe.

The details of wireless access technologies (such as FDMA, TDMA and Code-Division Multiple Access [CDMA]) will be presented in later chapters of this book.

The success of first-generation analog mobile communications systems and the enormous investment in developing second-generation digital technologies have testified to the world that PCS is in great demand. Wireless communications continue to experience rapid growth, and new applications and approaches have spawned at an unprecedented rate. The number of subscribers for cellular mobile telephones, cellular cordless phones, and radio pagers have increased manifold. Market studies continue to show large mass markets for new types of PCS, perhaps exceeding 50 to 100 million subscribers in the United States alone.

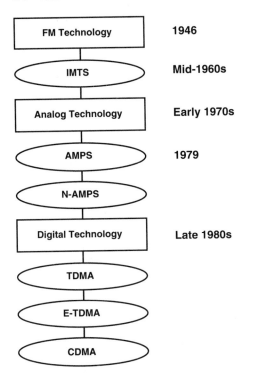

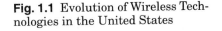

Fig. 1.1 Evolution of Wireless Technologies in the United States

Within the International Consultative Committee on Radio (CCIR), many international studies on the subject of future mobile services, radio interface design, and network architectures of PCS have been conducted under the banner of Future Public Land Mobile Telecommunications System (FPLMTS). The International Consultative Committee on Telephone and Telegraph (CCITT) is developing standards for wireline personal communications known as Universal Personal Telecommunication (UPT) communications. Also, ETSI has created a special mobile group to prepare standards for a Universal Mobile Telecommunication System (UMTS). In the U.S. standards arena, several bodies are engaged in developing standards for PCS, e.g., subcommittees T1E1, T1M1, T1S1, and T1P1 of committee T1 of the Alliance for Telecommunications Industry Solutions (ATIS), committee TR46 of the Telecommunications Industry Association (TIA), and committee 802 of IEEE.

1.4 VISION OF PCS

PCS is likely to explode worldwide in the mid to late 1990s with global revenues for services and handsets ranging from about $2 billion in 1996 to about $12 billion by the end of the decade. As the market expands, it will draw in a growing number of low-end residential users and will drive basic voice services up from half the PCS market in the mid-1990s to about three-fourths in 1999. PCS will

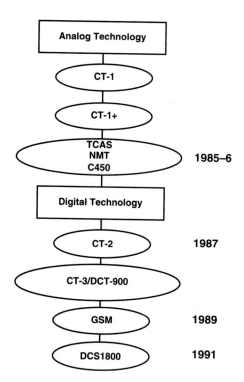

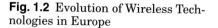

Fig. 1.2 Evolution of Wireless Technologies in Europe

use both existing and future wireline and wireless networks. Three key elements of PCS will be:

1. An easy-to-use, high-functionality handset
2. A single, personal number that can reach the subscriber anywhere
3. An individualized feature profile that follows the user and provides a customized set of services at any location

PCS will not be a single, all-encompassing wireless solution, but it will be a combination of standards, networks, and products that meet a range of user requirements at a reasonable price with a high level of support. PCS will fill gaps left by other modes of wireless and wireline telephony.

In the long run, microcellular and macrocellular services are likely to merge. This will allow seamless handoff and roaming between network types. Major data-oriented players like IBM, Apple, and DEC may also enter the market to integrate their data products with voice products. Vendors and carriers will increasingly market their offerings through mass merchants, electronics stores, and general merchandising outlets. Different price levels will be established for different end-user segments with business users—along with emergency service, health care, and public protection personnel—targeted for feature-rich portable voice and/or data units easily integrated with other types of equipment.

1.5 SUMMARY

This chapter presented the historical background of mobile communications starting from 1946 and examined the evolution of wireless technologies in the United States and Europe. The chapter also discussed American and European standards used for first- and second-generation cellular products. Finally, we presented our vision of PCS, outlining the key elements.

1.6 PROBLEMS

1. Describe five wireless services that will replace current fixed wired services. Will their estimated cost be more or less than the services they replace? In what year would the number of wireless users of the service exceed the number of wired users? What are the barriers to growth of these new services?

2. Describe two or more current wired services that should not be migrated to a wireless environment. Give reasons for your answers.

3. In the IMTS there are 11 channels at very high frequency (VHF) and 12 channels at ultra high frequency (UHF). If all 12 UHF channels are used in a service area, how many users can be loaded on the 12 channels for a blocking rate of 1%, 10%, and 50%? Assume that each user makes one call per hour in the busy hour and each call lasts for an average of three minutes. If service is sold to 100 users, what is the blocking rate in the busy hour with the given statistics?

4. Give five or more reasons why Europe, Asia, and America should each have a different standard for cellular and/or PCS systems.

5. Give five or more reasons why there should be one worldwide standard for all cellular and PCS systems.

6. Describe the main characteristics of portable wireless units for each of the following target markets: business users; residential users; emergency services users (police, fire, medical); and fleet management services (taxis, trucking, busses, and so forth).

1.7 REFERENCES

1. Calhoun, G., *Digital Cellular Radio*, Boston: Artech House, 1988.
2. EIA/TIA IS-54, "Dual-Mode Mobile Station-Base Station Compatibility Standard," December 1989, Project No. 2215.
3. Kuramoto, S., "Development of a Digital Cellular System Based on the Japanese Standard," *Wireless Communications Future Direction*, Holtzman, J. M., and D. J. Goodman, eds., Boston: Kluwer Academic Publishers, 1993.
4. Lee, W. C. Y., *Mobile Cellular Telecommunications Systems*, New York: McGraw-Hill, 1989.
5. Mehrotra, A. *Cellular Radio Analog and Digital System*, Boston: Artech House, 1994.

6. Mouly, M. and M. Pautet, *The GSM System for Mobile Communications*, Palaiseau, France, 1992.

7. NMT DOC1, "Nordic Mobile Telephone System Description," 1977.

8. O'Neill, E. F., ed., *A History of Engineering & Science in the Bell System: Transmission Tech. (1925–1975)*, Murray Hills, N.J., AT&T-Bell Labs, 1985, 408.

9. Parsons, J. D. and J. G. Gardiner, *Mobile Communication Systems*, New York: Halsted Press, 1989.

10. U.K. Total Access Communication System, Mobile Station, Land Station Compatibility Specifications, Issue 3, British Telecommunication Research Labs, March 1987.

11. Whitehead, J., "Cellular System Design: An Emerging Engineering Discipline," *IEEE Communication Magazine* 24, no. 2 (February 1986): 8–15.

12. Young, W. R., "Advanced Mobile Phone Service: Introduction, Background, and Objectives," *Bell System Technical Journal* 58, no. 1 (January 1979): 1–14.

An Overview of Cellular Systems

2.1 INTRODUCTION

In this chapter we briefly present the concept of cellular communications and discuss the first- and second-generation cellular systems used in the United States and Europe. We outline the problems associated with the second-generation-plus PCS system and provide the vision of a third-generation system.

2.2 CONCEPT OF CELLULAR COMMUNICATIONS

The idea of cellular communications is simple. During the late 1960s the Bell System proposed to alleviate the problem of spectrum congestion by restructuring the coverage areas of mobile radio systems. The traditional approach to mobile radio involved setting up a high-power transmitter located on top of the highest point in the coverage area. The mobile telephones needed to have a line of sight to the base station for adequate radio coverage. Line of sight transmission is limited to the distance to the horizon (as much as 40 or 50 miles away for a high base station antenna). The result was fair to adequate coverage over a large area. It also implied that the few available radio channels were locked up over a large area by a small number of users. In 1970 the Bell System in New York City could support just 12 simultaneous mobile conversations. The 13th caller was blocked.

The cellular concept handles the coverage problem differently. It does not use the broadcasting method; it uses a large number of low-power transmitters designed to serve only a small area. Thus, instead of an area like New York City being covered by a single transmitter, the city was divided into the smaller coverage areas called "cells." By reducing the total coverage area into small cells, it became possible to reuse the same frequencies in different cells. The problem with small cells was that not all mobile calls would now be completed within a single cell. To deal with this problem, the idea of handoff was used.

It is enormously expensive to build a system with thousands of cells right from the beginning. However, large-radius cells can evolve gracefully into small-radius cells over a period of time using cell-splitting. When the traffic reaches a point in a particular cell such that the existing allocation of channels in that cell can no longer support a good grade of service, that cell is subdivided into smaller cells with lower transmitter power to fit within the area of the former cell.

Thus, the essential elements of a cellular system are:

1. Low-power transmitter and small coverage areas or cells
2. Frequency reuse
3. Handoff and central control
4. Cell splitting to increase call capacity

2.3 FIRST-GENERATION CELLULAR SYSTEMS

As the United States was planning its cellular network in the 1970s, England, Japan, Germany, and the Scandinavian countries were also planning their systems. Each system used a different frequency band and different protocols for signaling between mobile units and base stations. They all used analog FM (with different deviations and channel spacings) for their voice communications.

During the 1970s the FCC forced the TV broadcasters off the little-used UHF channels 70–84 and made the frequencies available for two-way radio and the new cellular technology. During that time, the Bell System and Motorola actively pursued support for a 900-MHz cellular system using different designs for channel reuse and protocols. In the late 1970s, the FCC mandated that a single nationwide standard must be developed before licenses for cellular systems would be awarded. The Electronics Industry Association (EIA) formed a cellular standards committee and standardized the Advanced Mobile Phone System (AMPS) protocol for the United States.

In 1985 the Total Access Communications System (TACS) was introduced in the United Kingdom. TACS is a close relative of North America's AMPS. Other cellular systems developed in parallel were: Scandinavia's Nordic Mobile Telephone (NMT) system; West Germany's C450; Japan's Nippon Telephone & Telegraph (NTT). A comparison of various operational aspects for the five cellular systems is given in Table 2.1

The cellular approach promised virtually unlimited capacity through cell splitting. As the popularity of wireless communications escalated in the 1980s,

Table 2.1 First-Generation Cellular Systems

System Parameters	North America (AMPS)	U.K. (TACS)	Scandina-vian (NMT)	West Germany (C450)	Japan (NTT)
Transmission frequency (MHz)					
• Base station	870–890	935–960	463–467.5	461.3–465.74	870–885
• Mobile station	825–845	890–915	453–457.5	451.3–455.74	925–940
Spacing between transmitter & receiver frequency (MHz)	45	45	10	10	55
Spacing between channels (kHz)	30	25	25	20	25
Number of channels	666/832	1,000	180	222	600
Coverage radius by one base station (kilometers [km])	2–25	2–20	1.8–40.0	5–30	5 (Urban) 10 (Suburban)
Audio signal					
• Modulation	FM	FM	FM	FM	FM
• Maximum frequency deviation (kHz)	± 12	± 9.5	± 5	± 4	± 5
Control signals					
• Modulation	FSK	FSK	FSK	FSK	FSK
• Frequency deviation	± 8	± 6.4	± 3.5	± 2.5	± 4.5
Data transmission rate (kbs)	10	8	1.2	5.28	0.3
Error control on control channels	Principle of majority decision is used.	Principle of majority decision is used.	Receiving steps are predetermined according to the content of the message.	Message is sent again when an error is detected.	Transmitted signal is checked when it is sent back to the sender by the receiver.

the cellular industry faced practical limitations. For a fixed allocation of spectrum, a large increase in capacity implies corresponding reductions in cell size. For example, the U.S. AMPS design allows for cells as small as 1,600 meters (m) (1 mile). As the cells get smaller, it becomes increasingly difficult to place base stations at the locations that offer necessary radio coverage. Also, reductions in cell size demand increased signaling activity as more rapid handoffs occur; in addition, base stations are required to handle more access requests and registrations from mobile stations. The problem becomes particularly difficult in large urban areas where capacity requirements are most pressing. In addition to the capacity bottleneck, the utility of first-generation analog systems was diminished by proliferation of incompatible standards in Europe. The same mobile telephone frequencies cannot be used in different European countries. The limitations of first-generation analog systems provided motivations to the second-generation systems. The principal goals of the second-generation systems were: higher capacity and hence lower cost, and, in Europe, a continental system with full international roaming and handoff capabilities. In Europe, these goals are served by new spectrum allocations and by the formulation of a Pan-European Cellular Standard GSM.

In North America, where one standard (United States, Canada, and Mexico) existed and covered a region as large as Europe, the push for a new system was not as strong. In the new digital systems, higher capacity is derived from applications of advanced transmission techniques including efficient speech coding, error correcting channel codes, and bandwidth-efficient modulation techniques.

In Europe, the approach was to open new frequency bands for a pan-European system and not to have compatibility with existing cellular systems. In the United States, the same frequency bands were shared with the new digital systems, and the standards supported dual-mode telephones that could be used in both analog and digital systems.

2.4 TECHNOLOGIES FOR SECOND-GENERATION CELLULAR SYSTEMS

Standards and system designs exist for several new and competing technologies for the second-generation cellular systems. They are:

1. Narrowband Advanced Mobile Phone Service (N-AMPS)
2. Time-Division Multiple Access (TDMA)
3. Extended Time-Division Multiple Access (E-TDMA)
4. Spread Spectrum (SS)—Code-Division Multiple Access (CDMA)

Not all systems have seen widespread use, and some may disappear. The following sections provide a brief description and status of these technologies.

2.4.1 N-AMPS

Motorola developed N-AMPS by dividing an analog channel into three parts, thereby tripling the present analog channel capacity. Bandwidth per user is decreased from 30 kHz to 10 kHz. Each new channel is capable of handling its own calls. N-AMPS acts primarily as a bridge to digital communications that allows cellular systems to increase capacity at a low cost.

A smaller bandwidth per user in N-AMPS results in a slight degradation in speech quality that is compensated for with the addition of an interference avoidance scheme called Mobile Reported Interference (MRI). This capability, along with full call control (e.g., conference call, call waiting, call transfer, handoff, power control) is provided using a (new) continuous 100 bits per second (bps) in-band, sub-audible, signaling control channel. This scheme has the additional benefit of eliminating the audio gaps typical in the AMPS blank and burst signaling scheme. The associated control channel can also be used for sending alpha numeric characters when not actively managing call control. It has been typically used for features such as displaying calling line identification numbers, as well as features similar to those provided by an alphanumeric pager. This capability has allowed for the combining of cellular and paging applications in a single device. N-AMPS was standardized as EIA/TIA IS-88, IS-89, and IS-90 in late 1992. In 1993, IS-88 and IS-553 (AMPS) were combined to form a single analog standard called IS-91. N-AMPS has been implemented in both U.S. and international markets.

Although there has been some concern about the ultimate capacity of N-AMPS, many operators have chosen to implement N-AMPS in the identical reuse pattern as the original AMPS design. With trunking efficiencies excluded, this results in three times AMPS capacity. In a typical international market, up to 90% N-AMPS penetration has been achieved using a four-cell reuse pattern with very small (500 m) cell radii. It has been claimed by those operators that audio quality is at least as good as the AMPS systems they replaced. In addition, they report that overall dropped call performance is considerably less than AMPS due to the signaling enhancements made to the N-AMPS air interface.

2.4.2 TDMA

In TDMA, the users share the radio spectrum in time domain. An individual user is allocated a time slot. During this time slot, the user accesses the whole frequency band allocated to the system (wideband TDMA) or only part of the band (narrowband TDMA). In TDMA, transmission takes place in bursts from the mobile to the base station (uplink path), with only one user transmitting to the base station at any given time. In the downlink path (from the base station to the mobile), the base station usually transmits continuously, with the mobile listening only during the assigned time slot. TDMA channel multiplexes the bits from the number of (M) users. This requires transmission at a higher bit rate over the RF channel. Thus, TDMA is more prone to intersymbol interference because of multipath delay spread. To eliminate intersymbol interference, channel equalization techniques are required in the demodulation process of the receivers.

The cost per user of base station radio equipment is lower with TDMA since each radio is shared by M users simultaneously. In the future, the addition of new speech technology is potentially easier with TDMA because the structure of a TDMA channel can be designed to support additional time slots without changing the radio equipment. In TDMA, the spectral efficiency is reduced because of the inclusion of the guard time and synchronization sequence. Three implementations of TDMA have been used: European GSM, North American IS-54, and Japanese JDC system.

2.4.3 E-TDMA

General Motors' (GM's) effort to enter the cellular market uses E-TDMA. E-TDMA uses half-rate voice coding at 4.5 kbs that requires only one IS-54 time slot and thus allows six calls per frequency. Digital Speech Interpolation (DSI) permits deleting silence on calls. It thus reduces activity by 55–65% and allows more calls to be handled by the same number of time slots. E-TDMA is claimed to increase capacity by about 12 times that of AMPS and 4 times that of IS-54. GM plans to include E-TDMA mobile phones in a future automobile model. E-TDMA supporters must solve many technical problems, voice quality being the most important one. Furthermore, none of the cellular carriers seem to be interested in deploying it. E-TDMA does not appear to be a serious contender in the digital technology.

2.4.4 CDMA

Spread Spectrum (SS) techniques spread the bandwidth of the transmitted signal over a spectrum or band of frequencies much larger than the minimum bandwidth required to transmit the signal. Among all the SS techniques, the Direct Sequence (DS) implementation is most popular. In DS, a narrowband signal is spread by a wide bandwidth Pseudo-Noise (PN) code. A different code is assigned to each signal, and at the receiving end the signals are extracted out of the background noise by a receiver that cross-correlates the received signal with a copy of the original PN sequence. All other noise or signals with codes that do not match the receiver are filtered out. This is the direct carrier modulation by a code sequence and used in CDMA.

Most descriptions of SS assume that there is one SS transmitter and that the other transmitters are conventional (amplitude modulation [AM], FM, or single side band [SSB]) transmitters. With that description, the processing gain of SS can be considerable. Unfortunately, when the entire spectrum has all SS transmitters, the processing gain is not as high.

QUALCOMM developed the IS-95 CDMA protocol as an alternative to the TDMA cellular systems. The theoretical advantages of CDMA have generated large interest by equipment manufactures and service providers. Even with limited deployment of CDMA at cellular frequencies, CDMA has become a standard for use by PCS.

CDMA's ability to lock out conflicting signals may allow it to share a system with other radio signals without interference. CDMA can handle up to 10–20 times the number of callers as the analog AMPS. With more users on the network, costs can be distributed on a larger subscriber base, resulting in overall lower line costs. CDMA also offers fewer cell sites, better multipath resistance, superior voice quality, position location, and increased call privacy.

Capacity improvements in CDMA over other methods depend on two main technical features. The CDMA codes are generated by Walsh functions that are mathematically shown to form an orthogonal set (see chapter 8 and Appendix B). Thus, any two Walsh functions are orthogonal to each other, and two transmitters using different functions should cause no mutual interference. In practice, when the square wave Walsh functions are passed through real-world radio frequency (RF) and intermediate frequency (IF) filters, they may not stay orthogonal and the self-interference of the system may not be zero. A second technical feature of the CDMA system requires that all transmitters be under precise (less than 1 dB) power control to manage interference in the system. The power control is implemented about every millisecond (ms) but may not be fast enough to maintain precise control during deep fades of the signal.

As CDMA systems are tested and deployed, some of the large channel capacity gains may not be fully realized in practice. Although CDMA is a strong contender to the TDMA systems, only time will prove its claim to deliver on increased channel capacity.

2.5 CORDLESS PHONES AND TELEPOINT SYSTEMS

Closely related to cellular and often confused with the mobility aspects of cellular and PCS are cordless phones and Telepoint systems. Cordless phones are the low-power, low-range phones that enable an individual to move around a house or apartment and still place and receive phone calls.

First-generation cordless telephones are stand-alone consumer products. They do not require any interoperability specifications at all. Each cordless telephone comes with its own base station and needs to be compatible only with that base station. The billing, security, and privacy are achieved (to a limited degree) by preventing the phone from operating with any other base station.

Because of the popularity of cordless phones and the inability of some telephone companies to maintain public telephones in large cities, a hybrid approach was conceived, called "Telepoint." In the Telepoint system, the user owns a small low-power phone (similar in size and functionality to a cordless phone). The Telepoint phone works within 100 m of a public base station. The phone typically cannot receive calls and can place calls only when in range of the base station. The user would walk or drive within range of the base station and place the call. Roaming is not supported; the user must remain within range of the base station or the call is dropped.

This concept, based on a limited frequency allocation, was implemented as the Cordless Telephone-1 (CT-1) 900-MHz analog system. It was implemented or proposed in 13 European and Scandinavian countries.

Different incantations of the design can serve residential (wireless local loop) or public pay phone markets. CT-1 was intended to serve the residential market. An enhanced version, CT-1+, is similar to CT-1, but has added Telepoint capabilities. CT-1 uses FDMA, in which a single channel per radio carrier frequency is employed. CT-1 carries multiple narrowband carriers within a frequency band. Duplex operation, i.e., the simultaneous transmission and reception of voice signals, is implemented using separate frequencies.

Once the concept of a small low-power phone was introduced, designs were evolved to support wireless PBXs, cellular phones, PCS, and neighborhood wireless local loops. Often the system designs for second-generation cordless phones and Telepoint systems and the designs for digital cellular systems overlap.

The culmination of research in digital technologies resulted in CT-2, CT-3/DCT-900 and DECT standards for cordless telephones/Telepoint in Europe. These systems are being offered for a variety of uses—cellular, PCS, cordless phones, Telepoint, and wireless PBXs.

Cordless phones in the United States have either used FM in the 46/49-MHz band or SS in the 902–928-MHz band. Telepoint systems have not seen widespread use in the United States.

2.6 SECOND-GENERATION CELLULAR SYSTEMS

First-generation cellular systems were designed to satisfy the needs of business customers and some residential customers. With the increased demand of cellular telephones in Europe, several manufacturers began to look for new technologies that could overcome the problems of poor signals and battery performance. Poor signals resulted in poor performance for the user and a high frequency of false handoffs for the system operator. Better battery performance was needed to reduce size and cost of self-contained handheld units (handsets). Research efforts were directed toward wireless technologies to provide high-quality, interference-free speech and decent battery performance. The size of handset and better battery performance led to low-power designs and performance targets possible only with fully digital technologies. Digital cellular systems based on the GSM (TDMA) standard have emerged in Europe, while systems based on IS-54 (TDMA) and IS-95 (CDMA) are being developed in the United States. Table 2.2 provides a summary of the cellular and cordless systems.

The following sections describe the CT-2, DCT-900/DECT/CT-3, GSM, IS-54, and IS-95 systems and point out the main differences between them.

2.6.1 CT-2

These systems were developed for residential, business, and Telepoint applications in the United Kingdom. The handsets used in offices and homes

Table 2.2 Second-Generation Cellular and Cordless Systems[*]

System	IS-54	GSM	IS-95	CT-2	CT-3 DCT-900	DECT
Country	U.S.	Europe	U.S.	Europe, Asia	Sweden	Europe
Access technology	TDMA/ FDMA	TDMA/ FDMA	CDMA/ (DS) FDMA	FDMA	TDMA/ FDMA	TDMA/ FDMA
Primary use	Cellular	Cellular	Cellular	Cordless	Cordless	Cordless/ cellular
Frequency band						
• Base Station (MHz)	869–894	935–960	869–894	864–868	862–866	1800–1900
• Mobile Station (MHz)	824–849	890–915	824–849			
Duplexing	FDD	FDD	FDD	TDD	TDD	TDD
RF channel spacing (kHz)	30	200	1,250	100	1,000	1,728
Modulation	$\pi/4$ DQPSK	GMSK	BPSK/ QPSK	GFSK	GFSK	GFSK
Power, maximum/ average milli- watts (mW)	600/200	1,000/125	600	10/5	80/5	250/10
Frequency assignment	Fixed	Dynamic		Dynamic	Dynamic	Dynamic
Power control						
• Base Station	Y	Y	Y	N	N	N
• Mobile Station	Y	Y	Y	N	N	N
Speech coding	VSELP	RPE-LTP	QCELP	ADPCM	ADPCM	ADPCM
Speech rate (kbs)	7.95	13	8 (variable rate)	32	32	32
Speech channel per RF channel	3	8	—	1	8	12
Channel bit rate (kbs)	48.6	270.833	—	72	640	1152
Channel coding	1/2 rate convolu- tional	1/2 rate convolu- tional	1/2 rate forward, 1/3 rate reverse, CRC	None	CRC	CRC
Frame duration (ms)	40	4.615	20	2	16	10

[*]See later chapters for definitions of the terms used in Table 2.2.

were provided with a "value-added" public service from base stations located in railway stations, airports, and shopping centers. Although business and residential use of CT-2 offered full incoming and outgoing call facilities, Telepoint service was limited to outgoing calls only.

The United Kingdom chose FDMA for CT-2 to meet the original goal of a simple, single-user, home mobile telephone that avoids interference at call setup and supports multichannel multiplexing or handoffs. FDMA/TDD meets the needs for simple single-user channelization and simple measurement of signal power for a frequency channel from both ends of a radio link.

With the introduction of Telepoint and Wireless Private Branch Exchange (WPBX) applications in the United Kingdom, there was a need for a handset user to roam between different Telepoint operators' base stations and WPBX products. Therefore, the message protocols across the air interface needed to be well defined and common to all users. This resulted in the Common Air Interface (CAI) concept for CT-2.

CT-2 uses 4 MHz of spectrum, from 864 to 868 MHz, divided into 40 100-kHz channels. On each channel, the base station and mobile station alternate in the transmission of TDD data packets. The TDD rate, the rate at which base stations and mobile stations switch between "send" and "receive" is set at 500 Hz—1 ms each for send and receive. Power from both the mobile and base stations is restricted to 10 mW, and the mobile station selects the channel with the lowest noise. The mobile station is frequency agile during a call if the bit error rates on the selected channel reach unacceptable levels. Within this burst structure, there is a data rate of 72 kbs. In each burst, 72 bits of data are available for speech, control, signaling, and base/mobile station synchronization purposes. There is an allowance for a guard time between bursts to allow the sender to turn off its transmitter and settle into receiver mode and for the receiver to turn on its transmitter and settle at its center frequency. The guard time is nominally 4 bits long. By use of the guard period, both ends of the links are sure that the receiver is able to decode accurately the first- and later-transmitted bits in the burst.

The CT-2 modulation technique is binary frequency shift keying. With a channel spacing of 100 kHz, the bandwidth efficiency of CT-2 is 0.72 bps/Hz, about half of that of GSM. The speech coder is a standard Adaptive Differential Pulse Code Modulator (ADPCM) operating at 32 kbs.

2.6.2 DCT-900/DECT/CT-3

The application of wireless technology, particularly for a large business complex with WPBXs that support roaming and handoffs between different cells, is demanding. The system should manage the traffic in the cells in real time as the handsets move throughout the complex. This dramatically increases the complexity of the call processing software over that of a standard PBX. The software must also account for the three-dimensional environment of the system with the overlap of radio waves through different floors. Furthermore, the building environment affects the propagation of radio waves with the reflection and absorption of radio energy dependent on the construction materials. Ideally, a large building should be designed with a WPBX in mind; in practice, real buildings will have been designed before the WPBX, and compromises will be needed.

The capacity needs of a large modern office building can be met only with a high frequency reuse achieved by use of picocells with an indoor cell size of less than 50 m. Low-output power enables the handsets to be small and provides a talk time that exceeds the possibilities of other technologies. The most important requirement of the business PBX user is that the voice quality of the call is comparable to that of existing wired extensions.

The DCT-900/DECT/CT-3 choice of TDMA/TDD was dictated by the needs of multiple mobile telephones accessing multiple base units and connected to a PBX and by the shortage of paired frequency bands in Europe. The solution to this problem required the multiplexing of multiple users at a base unit and support for handoff and was readily implemented in a single-frequency TDMA/TDD. With only one frequency, TDD permitted simple rapid monitoring of power in all channels from both ends of a radio link. Dynamic time slot allocation algorithms for Dynamic Channel Allocation (DCA), with continuous transmission in at least one time slot as a "beacon" from all base units, provided a convenient mechanism for initial base-unit and time-slot selection.

The emerging standard for the large WPBX is DECT. This standard was frozen by ETSI in 1991, and the first system appeared in 1993. DECT standards do not compete directly with CT-2 because they are not oriented on the same market. DECT standards are more dedicated to PBX with large capacity, whereas CT-2 fulfills the requirements of the PBX with small capacities.

The modulation technique of DECT is Gaussian Minimum Shift Keying. The relative bandwidth of the Gaussian filter is wider (0.5 times of the bit rate) than in GSM. The bandwidth efficiency is 0.67 bps/Hz which is comparable to that of CT-2. The speech coder of DECT is ADPCM with bit rate of 32 kbs.

The DECT standard enables the development of systems specially designed to handle high capacity in a stationary environment. DECT cannot compete with cellular technology for use in vehicles, but it will be considerably cheaper in the applications it has been designed for. DECT's TDMA broadband solution may more adequately cover the businessperson's demands, such as high voice quality and data transmission capability.

2.6.3 GSM

GSM was driven by the need for a common mobile standard throughout Europe and the desire for digital transmission compatible with data and privacy. Spectrum was reallocated near 900 MHz throughout much of Europe and the surrounding region so that completely new technology could be developed by GSM. The GSM effort in the early to mid-1980s considered several system implementations including TDMA, CDMA, and FDMA technologies. A TDMA/FDMA/FDD technology was chosen with a radio link bit rate of 270 kbs.

The GSM modulation is Gaussian Minimum Shift Keying (GMSK). The bandwidth efficiency of 270-kbs signals operating with 200-kHz carrier spacing is 1.35 bps/Hz. The GSM's speech coder is referred to linear predictive coding with regular pulse excitation. The source rate is 13 kbs and transmission rate, including error detecting and correcting codes, is 22.8 kbs.

2.6.4 IS-54

In North America, where a common analog air interface was available and roaming anywhere in Canada, the United States or Mexico was possible, there was no need to replace the existing analog systems. Therefore, the Cellular Tele-communication Industry Association (CTIA) requested the TIA to specify a system that could be retrofitted into the existing AMPS system. The high cost of the cell sites was the major driving force. Thus, the important factor in the IS-54 was to maximize the number of voice channels that can be supported by a cell site within the available cellular spectrum. Several TDMA/FDMA and pure FDMA system proposals were considered before the IS-54 standard was selected. IS-54 fits three TDMA 8-kbs encoded speech channels into each 30 kHz AMPS channel.

IS-54 uses a linear modulation technique, Differential Quadrature Phase Shift Keying (DQPSK) to provide a better bandwidth efficiency. The transmission rate is 48.6 kbs with a channel spacing of 30 kHz. This gives bandwidth efficiency of 1.62 bps/Hz, a 20% improvement over GSM. The main penalty of linear modulation is power efficiency that affects the weight of handsets and time between battery charging. The IS-54 speech coder is a type of code book excited linear predictive coding referred to as Vector Sum Excited Linear Prediction (VSELP). The source rate is 7.95 kbs and the transmission rate is 13 kbs.

2.6.5 IS-95

Recently a CDMA protocol has been proposed by QUALCOMM and standardized in the United States as IS-95. IS-95 is aimed at the dual-mode operation with the existing analog cellular system. The basic idea behind increased capacity in IS-95 is the use of a wideband channel (the proposed channel width is 1.25 MHz in each cell or 42 30-kHz channels) where many subscribers can talk together without interfering with each other. Each channel is shared by many users with different codes. IS-95 proposes soft handoff to improve voice quality and RAKE receiver to take advantage of multipath fading and to lower signal-to-interference (S/I) ratio. Other factors that affect the channel capacity include use of variable rate vocoder, voice activity factor, and power control in the forward and reverse channel.

2.6.6 JDC

Japanese Digital Cellular standards are aimed to replace the three incompatible analog cellular systems in Japan. The basic radio channel design defined in the JDC standard is comparable with the North American IS-54 (TDMA) digital and European GSM system. The JDC systems use three-channel TDMA. Two frequency bands have been reserved: 800-MHz band with 130 MHz of duplex separation and 1.5-GHz band with 48 MHz of duplex separation. The 800-MHz band will be used first, whereas the 1.5-GHz band is for future use. The modulation scheme is $\pi/4$-QPSK with interleaved carrier spacing of 25 kHz. The speech coder uses 11.2-kbs VSELP including channel coding.

2.7 SECOND-GENERATION-PLUS PCS SYSTEMS

Although many people describe PCS as a third-generation system, the U.S. implementation uses modified cellular protocols. The opening of the 2-GHz band by the FCC has generated a flurry of activity to develop new systems. Unfortunately, in the race to deploy systems, most work has been to up band the existing cellular systems to the new 2-GHz band. Whether the protocol is GSM, IS-54, or IS-95, each proponent wants to make minimum changes in its protocol to win the PCS race.

It may not be until later in the 1990s before true third-generation systems offering wireless multimedia access emerge. The initial offering may be tailored to the environment and the need for rapid entry into the marketplace.

A further factor is the need to support wireless residential service (wireless CENTREX), cordless phones, Telepoint, wireless PBXs, low-mobility (on-street) portable phones, and high-mobility (in-vehicle) mobile phones. Although there is a desire for one protocol to support all needs, cost constraints may result in several solutions, each one optimized to a particular need. This need and demand for wireless communications in several environments has been shown by the rapid growth of different technologies that are optimized for particular applications and environments. Examples are:

1. Residential mobile telephones and their evolution to digital technology in CT-2 and to DCT-900/DECT/CT-3 for in-building PBX environments
2. Analog cellular telephones for widespread mobile service and their digital evolutions to GSM, E-TDMA, IS-54, and CDMA
3. Wireless data networks both for low-rate wide-area coverage and higher-rate Wide Local Area Networks (WLANs)

Basic needs for PCS include standardized low-power technology to provide voice and moderate-rate data to small, lightweight, economical, pocket-size personal handsets that can be used for tens of hours without attention to batteries and to be able to provide such communications economically over wide areas, including in homes and other buildings, outdoors for pedestrians in neighborhoods and urban areas, and anywhere there are reasonable densities of people.

The CT-2 and DCT-900/DECT/CT-3 technologies look attractive for providing low-priced personal communications services with volume penetration. To permit widespread use of these technologies in outside environments where base stations have less attenuation between themselves than between mobile stations and base stations, time synchronization of base station transmissions is required to achieve good performance with TDD. While the DCT-900/DECT/CT-3 technology was appropriate for WPBXs, it needs modification for more widespread PCS applications, for which it also incurs synchronization requirements and additional complexity.

The DCS1800 is a standard for PCN that has been developed by ETSI. It is a derivative of the GSM 900 MHz cellular standard. In Europe DCS has been allocated frequencies from 1710 to 1785 MHz and 1805 to 1880 MHz to provide a maximum theoretical capacity of 375 radio carriers, each with 8 or 16 (half-rate) voice/data channels. In DCS1800 there are provisions for national roaming between operators with overlapping coverage. These modifications have enabled the GSM cellular standard to be enhanced to provide a high-capacity, quality PCN system that can be optimized for handheld operation. The 1800-MHz operating band in the DCS results in a small cell structure that is compatible with the PCN concept. The 1800-MHz band is occupied by fixed radio links for which alternative technologies exist, and clearance of the band can be more readily effected than attempting to manage coexistence and transition between the first- and second-generation cellular systems at 800/900 MHz.

The initial implementation of European PCN is based on the provision of a high-quality small cell network (cell radius less than 1 km in a dense urban environment to 5 km in the rural environment). Radio coverage and system parameters are optimized for low-power handsets, and emphasis is placed on providing a significantly higher statistical call success and quality level for the handheld portable than current cellular networks provide.

The future evolution of DCS1800 may include microcell structure for coverage and capacity enhancement into buildings such as airport terminals, railway stations, and shopping centers, where large numbers of people gather. A further development would then be in "private" cells within offices to provide business communications. Ubiquitous deployment of microcells in a PCN environment will require a very fast handoff processing capability that is not currently available on DCS1800. How successfully the DCS1800 technology can be implemented in office environments to replace DCT-900/DECT/CT-3 is in question.

2.8 VISION OF THE THIRD-GENERATION SYSTEMS

First-generation analog and second-generation digital systems are designed to support voice communication with limited data communication capabilities. Third-generation systems are targeted to offer a wide variety of services listed in Table 2.3. Most of the services are wireless extensions of Integrated Services Digital Network (ISDN), whereas services such as navigation and location information are mobile specific. Wireless network users will expect a quality of service similar to that provided by the wireline networks such as ISDN. Service providers will require higher-complexity protocols in the physical link layer because of the unpredictable nature of the radio propagation environment and the inherent terminal mobility in a wireless network. These protocols will use powerful for-

Table 2.3 Proposed Teleservices for a Third-Generation System

Teleservices	Throughput (kbs)	Target Bit Error Rate
Telephony	8–32	10^{-3}
Teleconference	32	10^{-3}
Voice mail	32	10^{-3}
Program sound	128	10^{-6}
Video telephony	64	10^{-7}
Video conference	384–768	10^{-7}
Remote terminal	1.2–9.6	10^{-6}
User profile editing	1.2–9.6	10^{-6}
Telefax (Group 4)	64	10^{-6}
Voiceband data	64	10^{-6}
Database access	2.4–768.0	10^{-6}
Message broadcast	2.4	10^{-6}
Unrestricted digital information	64–1,920	10^{-6}
Navigation	2.4–64.0	10^{-6}
Location	2.4–64.0	10^{-6}

ward error correction and digital speech interpolation techniques to match the quality of service of the fixed network.

Because of the multitude of teleservices offered in different operating scenarios, the teletraffic density generated will depend on the environment, the mix of terminal types, and the terminal density. Teletraffic density will vary substantially for high-bit-rate services provided in business areas, whereas basic services such as speech and video telephony will be offered in all other environments.

The third-generation network will concentrate on the service quality, system capacity, and personal and terminal mobility issues. The system capacity will be improved by using smaller cells and the reuse of frequency channels in a geographically ordered fashion. A third-generation network will use different cell structures according to the operational environment. Cell structures will range from conventional macrocells to indoor picocells. In particular, microcells with low transmission power will be widely deployed in urban areas, while other cell structures will be used according to the environment to provide ubiquitous coverage. It is expected that the cost of base station equipment for microcells will be significantly reduced because of the elimination of costly high-power amplifiers and the economies of scale in microcell base station manufacturing. Nevertheless, the system's cost will still play a dominating role in the design of the network infrastructure because more microcellular base stations will be required to provide adequate radio coverage. Microcells with a radius less than 1,000 m will be used extensively to provide coverage in metropolitan areas. Microcell base stations will be mounted on lamp posts or on buildings where electric supply is readily available. For high-user-density areas such as airport terminals, railway stations, and shopping malls, picocells with coverage of tens of meters will be used. To facilitate efficient handoff when the vehicle-based user crosses microcells at high speed, these calls will be handled by umbrella cells (or overlay macrocells) whose coverage areas will contain several to tens of microcells.

The planning of third-generation systems will be more complicated than the design of present speech-oriented, macrocell-based mobile systems and will require a more advanced and intelligent network planning tool.

2.9 SUMMARY

This chapter discussed the first-generation cellular systems used in Europe, Japan, and the United States. A brief description of wireless technologies used for the second-generation cellular systems was also given. The second-generation cellular systems used in Europe and the United States were discussed. The chapter concluded by examining the possible problems associated with the access technology for the second-generation-plus PCS system and providing a vision of the third-generation PCS system.

2.10 PROBLEMS

1. Describe some of the problems that occur as cells are split from a nominal 20-km cell radius to 10, 5, 2.5, and 1.25 km and smaller. What are the practical limits for an on-street cellular system? How does your answer change for solutions in cities and suburban and rural areas?

2. Describe the advantages of a digital system over an analog system when used for wireless communications. Describe some disadvantages of a digital system over an analog system and how they might be overcome.

3. Describe the advantages and disadvantages of a TDMA system compared to a CDMA system. Pick one system and make a convincing argument, as a seller of systems, why a system provider should buy your system. Now as a service provider, convince the user of the service why he/she should buy your digital service as opposed to the analog system and as opposed to the other digital system that your competitors have deployed.

4. Describe the advantages and disadvantages to a user of a Telepoint system. How does your answer change if the same handset can be used at home as a cordless telephone?

5. Describe some third-generation services and compare them for modulation type needed, error rates, and bandwidths.

2.11 REFERENCES

1. Bellcore Technical Advisories, "Generic Framework Criteria for Universal Digital Personal Communications Systems (PCS)," FA-TSY–001013, Issue 1 (March 1990), and FA-NWT–001013, Issue 2 (December 1990); "Generic Criteria for Version 0.1 Wireless Access Communications Systems (WACS)," TA-NWT–001313, Issue 1 (July 1992).

2. Cheung et al., "Network Planning for Third-Generation Mobile Radio Systems," *IEEE Communications Magazine* 32, no. 11 (November 1994): 54–69.

3. Chia, "Beyond the Second-Generation Mobile Radio Systems," *British Telecom Engineering* 10 (January 1992): 326–335.

4. Cox, D. C., "Personal Communications—A Viewpoint," *IEEE Communications Magazine* (November 1990): 8–20.

5. European Telecommunications Standards Institution (ETSI), "Recommendations for GSM900/DCS1800" (ETSI, Cedex, France).

6. Gardiner, J. D. "Second Generation Cordless Telephony in U.K: Telepoint Services and the Common Air Interface," *IEE Electronics and Communications Engineering Journal* (April 1992).

7. Gilhousen, K. S., et al., "On the Capacity of a Cellular CDMA System," *IEEE Transaction on Vehicular Technology* VT–40, no. 2 (May 1991): 303–12.

8. Goodman, D. J., "Trends in Cellular and Cordless Communications," *IEEE Communications Magazine* (June 1991): 31–40.

9. Grillo and MacNamee, "European Perspective on Third Generation Personal Communication Systems," Proceedings of the IEEE VTC Conference, Orlando, Florida, May 1990.

10. Mehrotra, A. *Cellular Radio—Analog & Digital Systems*, Boston-London: Artech House, 1994.

11. Owen and C. Pudney, "DECT-Integrated services for cordless communications," Proceedings of the Fifth International Conference on Mobile Radio and Personal Communications, Institution of Electrical Engineers, Warwick, United Kingdom, December 1989.

12. Tuttlebee, *Cordless Telecommunication in Europe*, London-Heidelberg-New York: Springer-Verlag, 1990.

13. Viterbi, A. J., and Roberto Padovani, "Implications of Mobile Cellular CDMA," *IEEE Communications Magazine* (December 1992): 30 no. 12 38–41.

CHAPTER 3

Access Technologies

3.1 INTRODUCTION

In this chapter, we discuss the narrowband channelized and wideband non-channelized systems for wireless communications and focus on access technologies including the Frequency-Division Multiple Access (FDMA), Time-Division Multiple Access (TDMA), and Code-Division Multiple Access (CDMA). We examine these access technologies from a capacity, performance, and spectral efficiency viewpoint.

3.2 NARROWBAND CHANNELIZED SYSTEMS

Traditional architectures for the analog and digital wireless systems are channelized. In a channelized system, the total spectrum is divided into a large number of relatively narrow radio channels that are defined by carrier frequency. Each radio channel consists of a pair of frequencies. The frequency used for transmission from the base station to the mobile unit is called the forward channel, and the frequency used for transmission from the mobile unit to the base station is referred to as the reverse channel. A user is assigned both frequencies for the duration of the call. The forward and reverse channels are assigned widely separated frequencies to keep the interference between transmission and reception to a minimum.

A narrowband channelized system demands precise control of output frequencies for an individual transmitter. In this system, the transmission by a given mobile unit is confined within the specified narrow bandwidth to avoid interference with adjacent channels. The tightness of bandwidth limitations plays a dominant role in the evaluation and selection of modulation technique. It also influences the design of transmitter and receiver elements, particularly the filters, which can greatly affect the cost of a mobile unit.

3.2.1 North American Narrowband Analog Channelized System

In North America, the spectrum originally allocated by the FCC for cellular telephone service consists of two blocks of frequencies, each 20 MHz wide. The frequency block between 825 and 845 MHz is used for transmissions from mobile to base station whereas the frequency block between 870 and 890 MHz is used for transmissions from base station to mobile. Each frequency block is divided into 30-kHz channels. Thus, 666 channels are available for transmission in each direction. Each two-way channel uses a separate frequency for each direction of transmission. The frequencies are always separated by 45 MHz.

For convenience of discussion and to signal the channel information from base stations to mobile stations, each channel has a designated channel number. The numbers start at the lowest frequency and increase by one for each 30 kHz of increment. Thus, channel 1 is centered at 825.03 MHz (for mobile to base station direction), channel 2 at 825.06 MHz, and so on.

Initial FCC regulations specified that each Cellular Geographical Service Area (CGSA) would be served by two competing cellular systems: by the local telephone company providing land service in the area, referred to as the "wireline" carrier; by some other company referred to as the "nonwireline" carrier. Each of the competing systems are allocated half of the available channels. Nonwireline systems operate on channels 1 through 333, and wireline systems on channels 334 through 666.

In cellular telephone systems, calls are initially set up using "setup channels," also known as control channels. The voice communications are carried on "voice channels" that are separate from the setup channels. Each system uses 21 of its 333 channels as dedicated control channels. This leaves 312 channels to serve as voice channels. In nonwireline systems the dedicated control channels are channels 313 through 333, whereas a wireline system uses channels 334 through 354 as the control channels.

In 1987 the FCC allocated an extra 5 MHz of spectrum for each system, 2.5 MHz for mobile to base station transmissions, and 2.5 MHz for base station to mobile transmissions. Each carrier received 83 new 30-kHz channels. All the new channels are used as voice channels. Therefore, each system now has 21 control channels and 395 voice channels. Figure 3.1 shows the allocated spectrum for the two systems. The original spectrum channels are the "A band" (nonwireline) and "B band" (wireline) channels. To these the A', A", and B' bands were added. The first 1.5 MHz above the B band is the A' band that is followed by 2.5 MHz of B' band. The additional 1-MHz A" band allocated to nonwireline carriers was placed immediately below the original A band. The frequencies shown in Figure 3.1 are for mobile to base station transmissions. The corresponding frequencies 45 MHz higher are allocated for base station to mobile transmissions. The channels in the A', A", and B' bands are referred to as the "expanded spectrum" channels. In chapter 8, we discuss the band plans in more detail and show that the expanded spectrum channels are numbered sequentially with channel numbers described in Tables 8.5 and 8.7. Channel numbers start at channel 1 at the bottom of the A band and run sequentially through channel 799 at the top of the B' band. The channel numbers for the A" band start with 991 and continue through 1023. Channel 1023 would be 30 kHz below channel 1.

3.2.2 Narrowband Digital Channelized Systems

As mentioned in the previous chapter, first-generation analog cellular systems showed signs of capacity saturation in major urban areas, even with a modest total user population. A major capacity increase was needed to meet future demand. Several digital techniques were deployed to solve the capacity problem of the analog cellular systems. There are two basic digital strategies whereby a fixed-spectrum resource can be allocated to different users.

☞ Using different frequencies—FDMA

☞ Using different time slots—TDMA

3.2.2.1 Frequency-Division Multiple Access In FDMA, signals from various users are assigned different frequencies, just as in the analog system described previously. Guard bands are maintained between adjacent signal spectra to minimize cross talk between channels (see Figure 3.2).

The advantage of FDMA is that a capacity increase can be obtained by reducing the information bit rate and using efficient digital codes. Technological advances required for implementation are simple. A system can be configured so

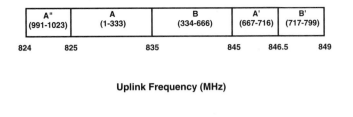

Uplink Frequency (MHz)

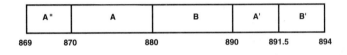

Downlink frequency (MHz)

Fig. 3.1 AMPS Frequency Band and Channel Assignments

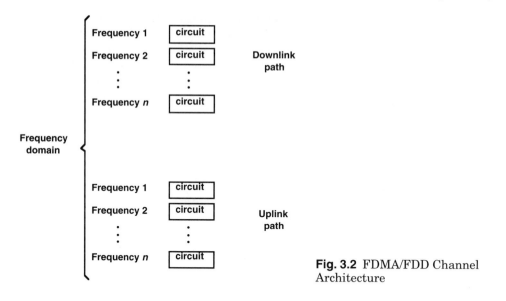

Fig. 3.2 FDMA/FDD Channel Architecture

that improvements in terms of speech coder bit rate reduction could be readily incorporated.

The disadvantages of FDMA include:

1. Since the system architecture, based on FDMA, does not differ significantly from the analog system, the improvement available in capacity depends on operation at a reduced S/I ratio. But the narrowband digital approach gives only limited advantages in this regard so that modest capacity improvements could be expected from a given spectrum allocation.

2. Narrowband technology involves narrowband filters, and, because these are not realized in very large-scale integrated (VLSI) digital circuits, this may set a high cost "floor" for terminals even under volume production conditions.

3. The maximum bit rate per channel is fixed and small, inhibiting the flexibility in bit-rate capability that may be a requirement for computer file transfer in some applications in the future.

Traditional systems use Frequency-Division Duplex (FDD) in which the transmitter and receiver operate simultaneously on different frequencies. Separation is provided between the downlink and uplink channels to prevent the transmitter from interfering with or desensing the receiver. Other precautions are also needed to prevent desensing, such as the use of two antennas or, alternatively, one antenna with a duplexer (a special arrangement of RF filters protecting the receiver from the transmit frequency). A duplexer adds weight, size, and cost to a radio transceiver and can limit the minimum size of a subscriber unit.

3.2.2.2 Time-Division Multiple Access

In a TDMA system, data from each user is conveyed in time intervals called slots (see Figure 3.3). Several slots

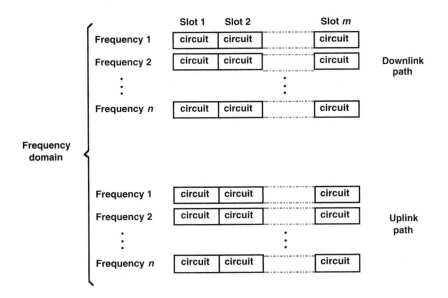

Fig. 3.3 TDMA/FDD Channel Architecture

make up a frame. Each slot is made up of a preamble plus information bits addressed to various stations as shown in Figure 3.4. The function of the preamble is to provide identification and incidental information and to allow synchronization of the slot at the intended receiver. Guard times are used between each user's transmission to minimize cross talk between channels.

Most TDMA systems time divide a frame into multiple slots used by different transmitters. This approach, called Time-Division Multiplex (TDM), uses

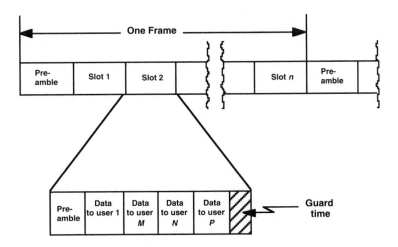

Fig. 3.4 TDMA Frame

several fixed-rate bit streams time-division multiplexed onto a TDMA bit stream. The data is transmitted via a radio carrier from a base station to several active mobiles in the downlink. In the reverse direction (uplink), transmission from mobiles to base stations is time sequenced and synchronized on a common frequency for TDMA.

A TDMA system using multiple slots can support a wide range of user bit rates by selecting the lowest multiplexing rate or a multiple of it. This enables supporting a variety of voice-coding techniques at different bit rates with different voice qualities. Data communications customers could make the same kinds of decisions, choosing and paying for digital data rate as required. This would allow customers to request and pay for bandwidth on demand.

The advantages of TDMA are:

1. TDMA permits a flexible bit rate, not only for multiples of a basic single-channel rate but also submultiples for low-bit-rate broadcast-type traffic.

2. TDMA potentially integrates into VLSI without narrowband filters, giving a low cost floor in volume production.

3. TDMA offers the opportunity for frame-by-frame monitoring of signal strength/bit error rates to enable either mobiles or base stations to initiate and execute handoffs.

4. TDMA utilizes bandwidth more efficiently because no frequency guard band is required between channels.

5. TDMA transmits each signal with sufficient guard time between time slots to accommodate:

   ✗ Time inaccuracies because of clock instability
   ✗ Delay spread
   ✗ Transmission time delay because of propagation distance
   ✗ The "tails" of signal pulses in TDMA because of transient responses

The disadvantages of TDMA include:

1. For mobiles and particularly handsets, TDMA on the uplink demands high peak power in transmit mode, which shortens battery life.

2. TDMA requires a substantial amount of signal processing for matched filtering and correlation detection for synchronizing with a time slot.

A TDMA system can be designed to use one frequency band by using TDD. In TDD, a bidirectional flow of information is achieved using a simplex-type scheme by automatically alternating in time the direction of transmission on a single frequency. At best, TDD can provide only a quasi-simultaneous bidirectional flow, since one direction must be off while the other is using the frequency. However, with a high enough transmission rate on the channel, the off time is not noticeable during conversations, and, with a digital speech system, the only effect is a very short delay.

The amount of spectrum required for both FDD and TDD is similar. The difference lies in the use of two bands of spectrum separated by a certain minimum bandwidth for FDD, whereas TDD requires only one band of frequencies but of twice the bandwidth. TDD's strength lies in the designer's ability to find a single band of unassigned frequencies being easier than finding two bands separated by the required bandwidth.

With TDD systems, the transmit time slot and the receiver time slot of the subscriber unit occur at different times. With the use of a simple RF switch in the subscriber unit, the antenna can be connected to the transmitter when a transmit burst is required (thus disconnecting the receiver from the antenna) and to the receiver for the incoming signal. The RF switch thus performs the function of the duplexer, but is less complex, smaller in size and less costly. TDD uses a burst mode scheme like TDMA and therefore also does not require a duplexer.

Since the bandwidth of the transmitter and receiver in a TDD system is twice that of a transmitter and receiver in a FDD system, RF filters in all the transmitters and receivers for TDD systems must be designed to cover twice the bandwidth of FDD system filters.

Depending on the data rate used and the number of slots per frame, a TDMA system can use the entire bandwidth of a system or can use FDD. The resultant multiplexing is a mixture of frequency division and time division. Thus, the entire frequency band is divided into a number of duplex channels (spaced about 350 to 400 kHz apart). These channels are deployed in a frequency reuse pattern, where radio-port frequencies are assigned using an autonomous adaptive frequency assignment algorithm. Each channel is configured in a Time-Division Multiplexing (TDM) mode for downlink (base station to mobile) direction and a TDMA mode for the uplink (mobile to base station) direction.

3.3 SPECTRAL EFFICIENCY

An efficient use of the frequency spectrum is the most desirable feature of a mobile communications system. To realize an efficient use of a spectrum, a number of techniques have been proposed or already implemented in mobile communications systems. Some of these techniques employed to improve spectral efficiency include reduction of the channel bandwidth, information compression, variable bit-rate control, and improved channel assignment algorithms. Spectral efficiency of a mobile communications system also depends on the choice of a multiple access scheme. A precise measure of spectral efficiency enables one to estimate the capacity of a mobile communications system and allows one to set up a minimum standard as a reference of measure.

The overall efficiency of a mobile communications system can be estimated by knowing the modulation and the multiple-access spectral efficiencies separately.

3.3.1 Spectral Efficiency of Modulation

The spectral efficiency with respect to modulation is defined as:

$$\eta_m = \frac{(\text{Total number of channels available in the system})}{(\text{Bandwidth})\ (\text{Total coverage area})} \qquad (3.1a)$$

$$\eta_m = \frac{\dfrac{B_w}{B_c} \times \dfrac{N_c}{N}}{B_w \times N_c \times A_c} \qquad (3.1b)$$

$$\eta_m = \frac{1}{B_c \times N \times A_c} \qquad (3.1c)$$

where:

η_m = modulation efficiency (channels/MHz/ km^2),
B_w = bandwidth of the system (MHz),
B_c = channel spacing (MHz),
N_c = number of cells in a cluster,
N = frequency reuse factor of system, and
A_c = area covered by a cell (km^2).

Eq. (3.1c) shows that the spectral efficiency of modulation does not depend on the bandwidth of the system. It depends only on the channel spacing, the total coverage area, and the frequency reuse factor. By reducing the channel spacing, the spectral efficiency of modulation for the system can be increased, provided the total coverage area remains unchanged. If a radio plan or modulation scheme can be designed to reduce N, then more channels are available in a cell and efficiency is improved.

Another definition of spectral efficiency of modulation is Erlangs/MHz/ km^2.

$$\eta_m = \frac{(\text{Total traffic carried by the system})}{(\text{Bandwidth})\ (\text{Total coverage area})} \qquad (3.2a)$$

$$\eta_m = \frac{\text{Total traffic carried by } \left(\dfrac{B_w/B_c}{N} \right) \text{channels}}{B_w A_c} \qquad (3.2b)$$

By introducing the trunking efficiency factor, η_t, in Eq. (3.2b), the total traffic carried through the system is given as:

$$\eta_m = \frac{\eta_t \left(\dfrac{B_w/B_c}{N} \right)}{B_w A_c} \qquad (3.2c)$$

$$\eta_m = \frac{\eta_t}{B_c N A_c} \qquad (3.2d)$$

where:

η_t is a function of the blocking probability and $\dfrac{B_w}{B_c}$

Several observations can be made from Eq. (3.2d):

1. The voice quality depends on the frequency reuse factor N, which is a function of the S/I ratio of the modulation scheme used in the mobile communications system.

2. The relationship between system bandwidth B_w and the amount of traffic carried by the system is non-linear, i.e., for a given percentage increase in B_w, the increase in the traffic carried by the system is more than the increase in B_w.

3. From the average traffic per user (Erlang/user) during the busy hour and erlang/km^2/MHz, the capacity of the system in terms of users/ km^2/ MHz can be obtained.

4. The spectral efficiency depends on the blocking probability.

E X A M P L E   3 – 1

Problem Statement

In the North American Narrowband analog channelized cellular system, the one-way bandwidth of the system is 12.5 MHz. The channel spacing is 30 kHz. Calculate the spectral efficiency (for a dense metropolitan area with small cells) using the following parameters:

(a) Area of a cell = 8 km^2

(b) Total coverage area = 4,000 km^2

(c) Average number of calls per user during the busy hour = 1.2

(d) Average holding time of a call = 100 seconds

(e) Call blocking probability = 2%

(f) Frequency reuse factor = 7

Solution

Number of 30-kHz channels = $(12.5 \times 1,000)/30 = 416$

Number of signaling channels = 21

Number of voice channels = $416 - 21 = 395$

Number of channels per cell = $395/7 = 56$

Number of cells = $4,000/8 = 500$

With 2% blocking for omnidirectional case, the total traffic carried by 56 channels (using Erlang-B formula) = 45.9 Erlangs/cell = 5.74 Erlangs/ km^2

Number of calls per hour = $(45.9 \times 3,600)/100 = 1,652.4$ calls/hour/cell = $1,652.4/8 = 206.6$ calls/hour/ km^2

Number of users/cell = $1,652.4/1.2 = 1,377$ users/hour/cell = $1,377/56 = 24.6$ users/hour/channel

$\eta_m = (45.9 \times 500)/(4,000 \times 12.5) = 0.459$ Erlangs/MHz/km^2

3.3.2 Multiple Access Spectral Efficiency

In FDMA, users share the radio spectrum in the frequency domain. In FDMA, the multiple access for speech efficiency is reduced because of guard bands between channels and also because of signaling channels. In TDMA, the efficiency is reduced because of guard time and synchronization sequence.

Multiple access spectral efficiency is defined as the ratio of the total time-frequency domain dedicated for voice transmission to the total time-frequency domain available to the system. Thus, the multiple access spectral efficiency is a dimensionless number with an upper limit of unity.

3.3.2.1 FDMA Spectral Efficiency   For FDMA, multiple access spectral efficiency is given as:

$$\eta_a = \frac{B_c N_T}{B_w} \leq 1 \tag{3.3}$$

where:
η_a = multiple access spectral efficiency and
N_T = total number of voice channels in the covered area.

E X A M P L E   3 – 2

Problem Statement
Refer to Example 3–1 and calculate multiple access spectral efficiency for FDMA.

Solution

$$\eta_a = \frac{30 \times 395}{12.5 \times 1,000} = 0.948$$

3.3.2.2 TDMA Spectral Efficiency   For the wideband TDMA, multiple access spectral efficiency is given as:

$$\eta_a = \frac{\tau M_t}{T_f} \tag{3.4}$$

where:
τ = duration of a time slot,
T_f = frame duration, and
M_t = number of time slots per frame.

In Eq. (3.4) it is assumed that the total available bandwidth is shared by all users. For the narrowband TDMA schemes, the total band is divided into a number of subbands, each using the TDMA technique. For the narrowband TDMA system, frequency domain efficiency is not unity, as the individual user channel does not use the whole frequency band available to the system. The efficiency of the narrowband TDMA system is given as:

$$\eta_a = \left(\frac{(\tau M_t)}{T_f} \right)\left(\frac{(B_u N_u)}{B_w} \right)$$

(3.5)

where:
B_u = bandwidth of an individual user during his or her time slot and
N_u = number of users sharing the same time slot in the system, but having
       access to different frequency sub bands.

3.3.2.3 Overall Spectral Efficiency of FDMA and TDMA System   The overall spectral efficiency η of a mobile communications system can be obtained by considering both the modulation and multiple access efficiencies

$$\eta = \eta_m \eta_a$$

E X A M P L E   3 – 3

Problem Statement
In the North American Narrowband TDMA cellular system, the one-way bandwidth of the system is 12.5 MHz. The channel spacing is 30 kHz, and there are 395 total voice channels in the system. The frame duration is 40 ms, with 6 time slots per frame. The system has an individual user data rate of 16.2 kbs in which the speech with error protection has a rate of 13 kbs. Calculate the efficiency of the TDMA system.

Solution
The time slot duration, $\tau = (13/16.2) \times (40/6) = 5.35$ ms

$T_f = 40$ ms, $M_t = 6$, $N_u = 395$, $B_u = 30$ kHz, and $B_w = 12.5$ MHz

$$\eta_a = \frac{5.35 \times 6}{40} \times \frac{30 \times 395}{12,500} = 0.76$$

The overhead portion of the frame = $1.0 - 0.76 = 24\%$

3.3.2.4 Capacity and Frame Efficiency of a TDMA System
Capacity
The capacity of a TDMA system is given by:

$$N_u = \frac{\eta_b \mu}{v_f} \times \frac{B_w}{RN}$$

(3.6)

where:
N_u = number of channels (mobile users) per cell,
η_b = bandwidth efficiency factor,
μ = bit efficiency (= 2 for Quadrature Phase Shift Keying [QPSK]),
v_f = voice activity factor (equal to one for TDMA),
B_w = one-way bandwidth of the system,
R = information bit rate plus overhead, and
N = frequency reuse factor.

$$\text{Spectral Efficiency } \eta = \frac{N_u \times R}{B_w} \text{ bits/sec/Hz} \tag{3.7}$$

E X A M P L E   3 – 4

Problem Statement

Calculate the capacity and spectral efficiency of a TDMA system using the following parameters: bandwidth efficiency factor $\eta_b = 0.9$, bit efficiency (with QPSK) $\mu = 2$, voice activity factor $v_f = 1.0$, one-way system bandwidth $B_w = 12.5$ MHz, information bit rate $R = 16.2$ kbs, and frequency reuse factor $N = 19$.

Solution

$$N_u = \frac{0.9 \times 2}{1.0} \times \frac{12.5 \times 10^6}{16.2 \times 10^3 \times 19}$$

$N_u = 73.1$ (say 73 mobile users per cell)

Spectral efficiency $\eta = (73 \times 16.2)/(12.5 \times 1{,}000) = 0.094$ bit/sec/Hz

Efficiency of a TDMA Frame

The number of overhead bits per frame (see Figure 3.4) is:

$$b_o = N_r b_r + N_t b_p + (N_t + N_r) b_g \tag{3.8}$$

where:
N_r = number of reference bursts per frame,
N_t = number of traffic bursts (slots) per frame,
b_r = number of overhead bits per reference burst,
b_p = number of overhead bits per preamble per slot, and
b_g = number of equivalent bits in each guard time interval.

The total number of bits per frame is:

$$b_T = T_f \times R_{rf} \tag{3.9a}$$

where:
T_f = frame duration and
R_{rf} = bit rate of the radio-frequency channel.

$$\text{Frame efficiency } \eta = (1 - b_o/b_T) \times 100\% \tag{3.9b}$$

It is desirable to keep at least 90% efficiency.

The number of bits per data channel (user) per frame $b_c = R T_f$, where $R =$ bit rate of each channel.

$$\text{Number of channels/frame } N_{CF} = \frac{(\text{Total data bits}) / (frame)}{(\text{Bits per channel}) / (frame)}$$

$$N_{CF} = \frac{\eta R_{rf} T_f}{R T_f} \tag{3.10a}$$

$$N_{CF} = \frac{\eta R_{rf}}{R} \tag{3.10b}$$

E X A M P L E   3 – 5

Problem Statement

Consider a TDMA system with the following parameters: $N_r = 1$, $N_t = 4$, $b_r = 48$ bits, $b_p = 12$ bits, $b_g = 20$ bits, $T_f = 5$ ms, $R_{rf} = 500$ kbs, and $R = 4.8$ kbs. Calculate the frame efficiency and the number of channels per frame.

Solution

$b_o = 1 \times 48 + 4 \times 12 + (1 + 4) \times 20 = 196$ bits per frame

$b_T = 5 \times 10^{-3} \times 500 \times 10^3 = 2{,}500$ bits per frame

$\eta = \left(1 - \frac{196'}{2{,}500}\right) \times 100 = 92.16\%$

Number of channels/frame $= \dfrac{0.9216 \times 500}{4.8} = 96$

Number of channels/traffic burst (slot) $= \dfrac{96}{4} = 24$

3.4 WIDEBAND SYSTEMS

In wideband systems, the entire system bandwidth is made available to each user and is many times larger than the bandwidth required to transmit information. Such systems are known as "Spread Spectrum" (SS) systems. There are two fundamental types of SS: Direct Sequence Spread Spectrum (DSSS) (see Figure 3.5) and Frequency Hop Spread Spectrum (FHSS).

One advantage of DSSS systems is that the transmission bandwidth exceeds the coherence bandwidth. The received signal, after despreading, resolves into multiple signals with different time delays. A RAKE receiver (See chapter 7) can recover the multiple time-delayed signals and combine them into one signal, providing an inherent time diversity receiver with lower frequency of deep fades. Thus, the DSSS systems provide an inherent robustness against mobile channel degradations. Another potential benefit of DSSS systems is the greater resistance to interference effects in a frequency reuse situation. Also, there may be no hard limit on the number of mobile users who can simultaneously gain access.

Frequency hopping (FH) is the periodic changing of frequency or frequency set associated with transmission. If the modulation is multiple frequency-shift keying (FSK), two or more frequencies are in the set that change at each hop. For other modulations, a single center or carrier frequency is changed at each hop.

An FH signal may be considered as a sequence of modulated pulses with pseudorandom carrier frequencies. The set of possible carrier frequencies is called the hop set. Hopping occurs over a frequency band that includes a number of frequency channels. The bandwidth of a frequency channel is called the

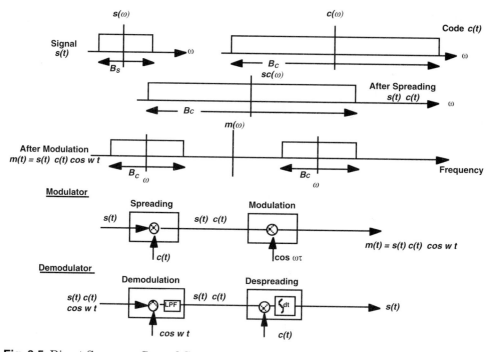

Fig. 3.5 Direct Sequence Spread Spectrum

instantaneous bandwidth (B_I). The bandwidth of the frequency band over which the hopping occurs is called the total hopping bandwidth (B_H). The time duration between hops is called the hop duration or hopping period (T_H).

FH may be classified as fast or slow. Fast FH occurs if there is a frequency hop for each transmitted symbol. Thus, fast FH implies that the hopping rate equals or exceeds the information symbol rate. Slow FH occurs if two or more symbols are transmitted in the time interval between frequency hops.

FH allows communicators to hop out of frequency channels with interference. To exploit this capability, error-correcting codes, appropriate interleaving, and disjointed frequency channels are nearly always used.

A frequency synthesizer is used for an FH system to convert a stable reference frequency into the various frequency of hop sets.

FH communicators do not often operate in isolation. Instead, they are usually elements of a network of FH systems that cause mutual multiple-access interference. This network is called a Frequency-Hopping Multiple-Access (FHMA) network.

If the hoppers of a FHMA network all use the same M frequency channels, but coordinate their frequency transitions and their hopping sequence, then the multiple-access interference for a lightly loaded system can be greatly reduced over a nonhopped system. For the number of hopped signals (M_h) less than the number of channels (N_c), a coordinated hopping pattern can eliminate interference. As the number of hopped signals increases beyond N_c, then the interference will increase in proportion to the ratio of the number of signals to the number of channels. Since there is no interference suppression system in frequency hopping, for high channel loadings, the performance of an FH system is no better than that of a nonhopped system. FH systems are best for light channel loadings in the presence of conventional nonhopped systems.

Network coordination for FH systems are simpler to implement than for DS-CDMA systems because the timing alignments must be within a fraction of a hop duration rather than a fraction of a sequence chip. Because of operational complications of coordination, asynchronous FHMA networks are usually preferable.

FH systems work best when a limited number of signals are sent in the presence of nonhopped signals where mutual interference can be avoided. In general, FH systems reject interference by trying to avoid it, whereas DS systems reject interference by spreading it. The interleaving and error-correcting codes that are effective with FH systems, are effective with DS systems. Error-correcting codes are more essential for FH systems than for DS systems because partial-band interference is a more pervasive threat than high-power pulsed interference.

The major problems with FH systems with increasing hopping rates are: the cost of a frequency synthesizer increases, and its reliability decreases; synchronization becomes more difficult.

In theory, a wideband system can be overlaid on existing, fully loaded, narrowband channelized systems. Thus, it may be possible to create a wideband network right on top of the narrowband cellular system using the same spectrum. In practice, the noninfinite processing gain of wideband systems may limit their deployment in the same spectrum as narrowband systems.

3.5 COMPARISONS OF FDMA, TDMA, AND DS-CDMA

The primary advantage of DS-CDMA is its ability to tolerate a fair amount of interfering signals compared to FDMA and TDMA, which typically cannot tolerate any such interference (Figure 3.6). As a result of the interference tolerance of CDMA, the problems of frequency band assignment and adjacent cell interference are greatly simplified. Also, flexibility in system design and deployment are significantly improved since interference with others is not a problem. On the other hand, FDMA and TDMA radios must be carefully assigned a frequency or time slot to assure that there is no interference with other similar radios. Therefore, sophisticated filtering and guard-band protection is needed with FDMA and TDMA technologies.

Capacity improvements with DS-CDMA also result from voice activity patterns during two-way conversation (i.e., times when a party is not talking) that cannot be cost-effectively exploited in FDMA or TDMA systems. DS-CDMA radios can, therefore, accommodate more mobile users than FDMA/TDMA radios on the same bandwidth.

With DS-CDMA, adjacent microcells share the same frequencies, whereas with FDMA/TDMA it is not feasible for adjacent microcells to share the same frequencies because of interference.

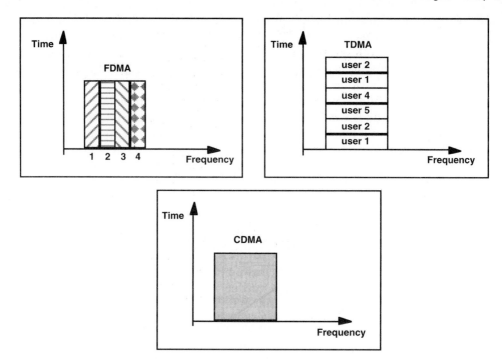

Fig. 3.6 Comparison of Multiple Access Methods

Further capacity gains can also result from antenna technology advancement by using direction antennas, which allows the microcell area to be divided into sectors.

Table 3.1 provides a summary of access technologies for wireless systems.

3.6 CAPACITY OF A DS-CDMA SYSTEM

The capacity of a DS-CDMA system depends on the processing gain, B_w/R, the energy per bit-to-interference ratio, E_b/I_o, the voice duty cycle, the DS-CDMA omnidirectional frequency reuse efficiency, η_f, and the number of sectors, G in the cell-site antenna.

The received signal power at the cell, from a mobile unit, is $S = R \times E_b$.

The S/I ratio is

$$\frac{S}{I} = \frac{R}{B_w} \times \frac{E_b}{I_o} \tag{3.11}$$

In a cell containing N_u mobile transmitters (users), the number of effective interferers is $N_u - 1$ regardless of how they are distributed within the cell since Automatic Power Control (APC) is used in the mobiles. APC operates such that the incident power at the center of the cell from each mobile is the same as for

Table 3.1 Access Technologies for Wireless Systems

System	Access Technology	Mode of Operation	Frame Rate (kbs)
AMPS	FDMA/FDD	Analog FM	—
North American IS-54 (Dual Mode)	TDMA/FDD FDMA/FDD	Digital/ Analog FM	48.6 —
North American IS-95 (Dual Mode)	DS-CDMA/FDD FDMA/FDD	Digital/ Analog FM	1,288 —
Wideband CDMA	DS-CDMA/FDD	Digital	4,096, 8,192, or 12,288
PACS	TDMA	Digital	384
PCS-2000	CDMA/TDMA	Digital	5,000
GSM Pan-European	TDMA/FDD	Digital	270.833
CT-2 Cordless	FDMA/TDD	Digital	72.0
DECT Cordless	TDMA/TDD	Digital	1,152.0

every other mobile in the cell, regardless of the distance from the center of the cell. APC conserves battery power in the mobiles, minimizes interference to other users, and helps to overcome fading.

In a hexagonal cell structure, the S/I ratio is given as:

$$\frac{S}{I} = \frac{1}{(N_u - 1) \times [1 + 6 \times k_1 + 12 \times k_2 + 18 \times k_3 + \ldots]} \tag{3.12}$$

where: N_u = number of mobile users in the band, B_w, and k_i, $i = 1, 2, 3, \ldots$ is the interference contribution from all terminals in individual cells in tiers 1, 2, 3, and so on, relative to the interference from the center cell. This loss contribution is a function of both the path loss to the center cell and the power reduction because of power control to an interfering mobile's own cell center.

If we define a frequency reuse efficiency, η_f, as

$$\eta_f = \frac{1}{[1 + 6 \times k_1 + 12 \times k_2 + 18 \times k_3 + \ldots]} \tag{3.13a}$$

$$\frac{S}{I} = \frac{\eta_f}{(N_u - 1)} \tag{3.13b}$$

$$\frac{E_b}{I_o} = \frac{B_w}{R} \times \frac{\eta_f}{(N_u - 1)} \tag{3.14}$$

Equation (3.14) does not include the effect of background thermal and spurious noise (ρ/S) in the spread bandwidth B_w. Including this as an additive degradation term in the denominator results in a bit energy-to-interference ratio of:

$$\frac{E_b}{I_o} = \frac{B_w}{R} \times \frac{\eta_f}{(N_u - 1) + \rho/S} \qquad (3.15)$$

The capacity of the DS-CDMA system is reduced by ρ/S which is the ratio of background thermal plus spurious noise-to-power level.

For a fixed B_w/R ratio, one way to increase the capacity of the DS-CDMA system is to reduce the E_b/I_o ratio, which depends upon modulation and coding scheme. By using a powerful coding scheme, the E_b/I_o ratio can be reduced, but this increases system complexity. Also, it is not possible to reduce the E_b/I_o ratio indefinitely. The only other way to increase the system capacity is to reduce the interference ρ/S. Two approaches are used: one is based on the natural behavior of human speech, and the other is based on the application of sectorized antennas. From experimental studies it has been found that, typically, in a full-duplex two-way voice conversation, the duty cycle of each voice is, on the average, less than 40%. Thus, for the remaining period of the time, the interference induced by the speaker is eliminated. Since the channel is shared between all the users, noise induced in the desired channel is reduced due to the silent interval of other interfering channels. It is not cost-effective to exploit the voice activity factor in the FDMA or TDMA systems because of the time delay associated with reassigning the channel resource during the speech pauses. If we define v_f as the voice activity factor, then Eq. (3.15) can be written as:

$$\frac{E_b}{I_o} = \frac{\eta_f}{v_f} \times \frac{B_w}{R} \times \frac{1}{(N_u - 1) + \rho/S} \qquad (3.16a)$$

$$(N_u - 1) + \frac{\rho}{S} = \left[\frac{\eta_f}{v_f}\right] \times \left[\frac{B_w}{R}\right] \times \left[\frac{I_o}{E_b}\right] \qquad (3.16b)$$

The equation for determination of capacity in the DS-CDMA system should also include additional parameters to reflect the bandwidth efficiency factor, capacity degradation factor to account for imperfect APC, and the number of sectors in the cell-site antenna. Eq. (3.16b) is augmented by the additional factors to provide the following equation for DS-CDMA capacity at one cell:

$$N_u = \frac{\eta_f \eta_b c_d G}{v_f} \times \frac{B_w}{R \times (E_b/I_o)} + 1 - \frac{\rho}{S} \qquad (3.17a)$$

Eq. (3.17a) can be rewritten as Eq. (3.17b) by neglecting the last two terms.

$$N_u = \frac{\eta_f \eta_b c_d G}{v_f} \times \frac{B_w}{R \times (E_b/I_o)} \qquad (3.17b)$$

where:

η_f = frequency reuse efficiency,
η_b = bandwidth efficiency factor < 1,
c_d = capacity degradation factor to account for imperfect APC,
v_f = voice activity factor,
B_w = one-way bandwidth of the system,
R = information bit rate plus overhead,
E_b = energy per bit of the desired signal,
E_b / I_o = bit energy-to-interference ratio, and
G = number of sectors in cell-site antenna.

For digital voice transmission, E_b / I_o is the required value for a bit error rate of 10^{-3} or better, and η_f depends on the quality of macroscopic diversity. Under the most optimistic assumption, $\eta_f < 0.5$. The voice activity factor, v_f, is generally assumed to be less than or equal to 0.5. E_b / I_o for a bit error rate of 10^{-3} can be as high as 63 (18 dB) if no coding is used and as low as 5 (7 dB) for a system using a powerful coding scheme. The capacity degradation factor, c_d will depend on the implementation but will always be less than 1.

E X A M P L E   3 – 6

Problem Statement

Calculate the capacity and spectral efficiency of a single sector DS-CDMA system using the following data:

(a) Bandwidth efficiency $\eta_b = 0.9$

(b) Frequency reuse efficiency $\eta_f = 0.45$

(c) Capacity degradation factor $c_d = 0.8$

(d) Voice activity factor $v_f = 0.4$

(e) Information bit rate $R = 16.2$ kbs

(f) $E_b / I_o = 7$ dB

(g) One-way system bandwidth $B_w = 12.5$ MHz

Neglect other sources of interference.

Solution

$E_b / N_o = 5.02$ (7 dB)

$$N_u = 1 + \frac{0.45 \times 0.9 \times 0.8 \times 1}{0.4} \times \frac{12.5 \times 10^6}{16.2 \times 10^3 \times 5.02}$$

$N_u = 125.5$ (say 126)

Spectral efficiency, $\eta = (126 \times 16.2)/(12.5 \times 10^3) = 0.163$ bits/sec/Hz

In these calculations, an omnidirectional antenna is assumed. If a three-sector antenna ($G = 3$) is used at cell site, the capacity will be increased to 378 mobile users per cell, and spectral efficiency be 0.49 bits/sec/Hz.

3.7 COMPARISON OF DS-CDMA AND FDMA/TDMA SYSTEM CAPACITY

Using Eqs. (3.6) and (3.17b), the ratio of the capacity for DS-CDMA and TDMA systems is given as:

$$\frac{N_{CDMA}}{N_{TDMA}} = \frac{c_d N \eta_f}{E_b/I_o} \times \frac{1}{v_f} \times \frac{G}{\mu} \tag{3.18}$$

E X A M P L E   3 – 7

Problem Statement

Using the data given in Example 3-4 and 3-6, compare the capacity of the DS-CDMA and TDMA systems.

Solution

$$\frac{N_{CDMA}}{N_{TDMA}} = \frac{0.8 \times 19 \times 0.45}{5.02} \times \frac{1}{0.4} \times \frac{1}{2}$$

$$\frac{N_{CDMA}}{N_{TDMA}} = 1.70$$

3.8 SUMMARY

This chapter described the access technologies used for wireless communications. We discussed FDMA, TDMA, and CDMA technologies, along with their advantages and disadvantages. Illustrated examples showed calculations for determination of capacity of TDMA and CDMA systems. We also presented brief descriptions of the FDD, TDD, TDM/TDMA, and TDM/TDMA/FDD approaches.

3.9 PROBLEMS

1. In a proposed TDMA cellular system, the one-way bandwidth of the system is 40 MHz. The channel spacing is 30 kHz, and there are 1,333 total voice channels in the system. The frame duration is 40 ms divided equally between 6 time slots. The system has an individual user data rate of 16.2 kbs in which the speech with error protection has a rate of 13 kbs. Calculate the efficiency of the TDMA system. Repeat the calculations for 20, 60, 80, and 100 MHz.

2. In problem 1, what is the optimum bandwidth for a TDMA system? What are the practical limits that prevent one from using the optimum bandwidth?

3. Calculate the access efficiency (η_a) of the FDMA cellular system with a frequency reuse factor equal to 7. The total number of voice channels available is 395, the total allocated bandwidth is 12.5 MHz, and the channel spacing is 30 kHz.

4. In the GSM (TDMA/FDD), the multiframe duration is 120 ms. The multiframe consists of 26 frames of 8 time slots each. The transmit bit rate is 270.833 kbs. The system contains 125 channels of 200 kHz bandwidth each. The total system bandwidth is 25 MHz. Calculate the access efficiency of the GSM given that only two groups of 57 bits carry user information per slot.

5. In the IS-54 (TDMA/FDD), the frame duration is 40 ms. The frame contains 6 time slots. The transmit bit rate is 48.6 kbs. Each time slot carries 260 bits of user information. The total number of 30-kHz voice channels available is 395, and the total system bandwidth is 12.5 MHz. Calculate the access efficiency of the system.

6. Calculate the capacity and spectral efficiency (η) of the GSM using the following parameters: bandwidth efficiency factor $\eta_b = 0.96$; bit efficiency (i.e., modulation efficiency with GMSK) $\mu = 1.354$; voice activity factor $v_f = 1.0$; information bit rate = 22.8 kbs; frequency reuse factor $N = 4$; and system bandwidth $B_w = 12.5$ MHz.

7. Calculate the capacity and spectral efficiency (η) of an IS-54 system using the following parameters: $\eta_b = 0.96$; $\mu = 1.62$ (i.e., modulation efficiency with π/4-DQPSK); voice activity factor $v_f = 1.0$; information bit rate = 19.5 kbs; frequency reuse factor = 7; system bandwidth = 12.5 MHz.

3.10 REFERENCES

1. Bellamy, J. C., *Digital Telephony*, 2d ed., New York: John Wiley & Sons, Inc., 1990.

2. Bellcore Research, "Generic Framework Criteria for Version 1.0 Wireless Access Communication System (WACS)," FA-NWT-001318 (June 1992).

3. Calhoun, G., *Digital Cellular Radio*, Boston: Artech House, 1988.

4. Gilhousen, K. S., et al., "Increased Capacity Using CDMA for Mobile Satellite Communication," *IEEE Journal on Selected Areas in Communications* 8, no. 4 (May 1990).

5. Jacobs, I. M., et al., "Comparison of CDMA and FDMA for the MobileStar System," Proceedings of the Mobile Satellite Conference, Pasadena, California, May 3–5, 1988, 283–90.

6. Lee, W. C. Y., "Spectrum Efficiency in Cellular," *IEEE Transactions on Vehicular Technology* 38 (May 1989); 69–75.

7. Lee, W. C. Y., "Overview of Cellular CDMA," *IEEE Transactions on Vehicular Technology* 40, no. 2 (May 1991); 291–302.

8. Lee, W. C. Y., *Mobile Communications Design Fundamentals*, 2d ed., New York: John Wiley & Sons, Inc., 1993.

9. Mehrotra, A., *Cellular Radio—Analog and Digital Systems*, Boston: Artech House, 1994.

10. Parsons, D., and J. G. Gardiner, *Mobile Communication Systems*, New York: Halsted Press, 1989.

11. Torrieri, J., *Principles of Secure Communication Systems*, 2d ed., Boston: Artech House, 1992.

12. Viterbi, A. J., "When Not to Spread Spectrum—A Sequel," *IEEE Communications Magazine* 23, (April 1985); 12–17.

13. Ziemer, E., and R. L. Peterson, *Introduction to Digital Communications*, New York: Macmillan Publishing Company, 1992.

CHAPTER 4

Fundamentals of Radio Communications

4.1 INTRODUCTION

In this chapter we discuss propagation and multipath characteristics of a radio wave. The concepts of delay spread, which causes channel dispersion, and intersymbol interference are also presented.

Since the mathematical modeling of the propagation of radio waves in a real-world environment is complicated, several authors have developed empirical models.

We present these empirical and semiempirical models used for calculating the path losses in the urban, suburban, and rural environments and compare the results obtained with each model.

4.2 RADIO-WAVE PROPAGATION

The portion of the electromagnetic spectrum that constitutes radio waves extends from a frequency of 30 kHz (wavelength 10 km) to 300 GHz (wavelength 0.1 cm).

Electromagnetic energy in the form of radio waves propagates outward from a transmitting antenna. In free space, radio waves propagate in straight lines and are reflected off objects using rules similar to those of light waves. Thus, in space, the transmitter and receiver must "see" each other, or the radio wave must be reflected off a conducting object.

Radio waves on the earth are affected by the terrain of the ground, the atmosphere, and the natural and artificial objects on the terrain. For radio waves on the earth, depending on the frequency of transmission, there are three main propagation means: ground waves, tropospheric waves, and ionospheric or sky waves.

The **ground wave** is that portion of the radiation that is directly affected by terrain and objects on the terrain. The ground wave travels in contact with the earth's surface by scattering off buildings, vegetation, hills, mountains, and other irregularities in the earth's surface. Unless the transmitter and receiver can "see" each other, the ground wave propagation provides the dominant local signal at a receiver.

The ground wave effects all frequencies from very low frequencies to microwave frequencies. For medium-wave (0.3–3 MHz) and shortwave (3–30 MHz) signals, it provides the local daytime propagation means. At very high frequency (VHF) and higher, it provides the most reliable propagation means. The signal dies off rapidly as the distance from the transmitter increases.

The **tropospheric wave** is that portion of the radiation that is kept close to the earth's surface as a result of refraction (bending) in the lower atmosphere. This tropospheric refraction can sometimes make VHF communication possible over distances far greater than can be covered with the ordinary ground wave. The amount of bending increases with frequency, so tropospheric communication improves as frequency is raised. It is relatively inconsequential below 30 MHz.

For VHF and ultra high frequency (UHF) propagation used for cellular and PCS communications, the tropospheric wave becomes an annoyance that causes interference and is not a dependable form of communications. Certain communications systems depend on the tropospheric wave, but they are outside the scope of this book.

The **ionospheric or sky wave** arises from radio waves that leave the antenna at angles somewhat above the horizontal. Under certain conditions, these waves can be sufficiently reflected by the ionospheric layers high in the earth's atmosphere to reach the ground again at distances ranging from 0 to about 4,000 km from the transmitter. By successive reflections at the earth's surface and in the upper atmosphere, a communication can be established over distances of thousands of miles, and on occasion radio signals can travel completely around the earth. Propagation by the regular layers of the ionosphere takes place mainly at frequencies below 30 MHz. Ionospheric propagation is dependent on the sunspot cycle; at times of high sunspot activity, the Maximum Usable Frequency (MUF) between two points on the earth can exceed 50 MHz. At periods of low sunspot activity, the MUF can be as low as 2 or 3 MHz.

Sunspots come and go on about an 11-year cycle. Since ionospheric propagation does not occur for the UHF signals that are used for cellular and PCS communications, further discussions on this are outside the scope of this book. However, when UHF signals pass through the ionosphere, as in satellite communications, the distorting effects of the ionosphere must be accounted for.

For two-way mobile radio, the vehicle or portable antenna seldom has a direct line-of-sight path to the base station. Radio waves penetrate into buildings to a limited extent and, because of diffraction, appear to bend sightly over minor variations in the ground. Due to multiple scattering and reflection, radio waves also propagate into built-up areas, although the signal strength is substantially reduced by all these various effects. A sensitive receiver is able to detect signals even in heavily built-up areas and within buildings.

Because of the short-range propagation characteristics of VHF and UHF, it is possible to allocate the same frequency to different mobile users in areas separated by distances of 50 to 100 km (31–62 miles) with a high degree of confidence that the signals will not interfere with each other, except under anomalous propagation conditions (such as tropospheric propagation), as long as signal power is low and antenna height is restricted.

4.3 MULTIPATH CHARACTERISTICS OF A RADIO WAVE

Radio waves propagate through space as traveling electromagnetic (EM) waves. The energy of signals exists in the form of electrical (E) and magnetic (H) fields. Both electrical and magnetic fields vary sinusoidally with time. The two fields always exist together because a change in electrical field generates a magnetic field and a change in magnetic field generates an electrical field. Thus, there is a continuous flow of energy from one field to the other.

Radio waves arrive at a mobile receiver from different directions with different time delays. They combine via vector addition at the receiver antenna to give a resultant signal with a large or small amplitude depending upon whether the incoming waves combine to reinforce each other or cancel each other. As a result, a receiver at one location may experience a signal strength several tens of dB different from a similar receiver located only a short distance away. As a mobile moves from one location to another, the phase relationship between the various incoming waves also changes. Thus, there are substantial amplitude and phase fluctuations, and the signal is subjected to fading. It should also be noted that whenever relative motion exists there is a **Doppler shift** in the received signal. In the mobile radio case, the fading and Doppler shift occur as a result of motion of the receiver through a spatially varying field.

It also results from the motion of the scatterers of the radio waves (e.g., cars, trucks, vegetation). The effect of the multipath propagation is to produce a received signal with an amplitude that varies quite substantially with location. At UHF and higher frequencies, the motion of the scatterers also causes fading to occur even if the mobile set or handset is not in motion.

Figure 4.1 illustrates the fading characteristics of a mobile radio signal. The rapid fluctuations caused by the local multipath are known as fast fading (Rayleigh fading). The long-term variation in the mean level is known as slow fading (log-normal fading).The slow fading is caused by movement over distances large enough to produce gross variations in the overall path between the base station and the mobile. Fast fading is usually observed over distances of about half a wavelength. For VHF and UHF, a vehicle traveling at 30 miles per hour can pass through several fast fades in a second. Therefore, the mobile radio signal, as shown in Figure 4.1, consists of a short-term fast-fading signal superim-

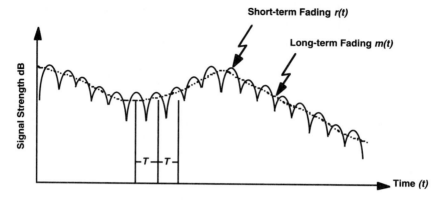

Fig. 4.1 A Mobile Radio Signal Fading Representation

posed on a local mean value (which remains constant over a small area but varies slowly as the receiver moves). As noted above, even a stationary handset will see fading, though the rate is lower.

The received signal $s(t)$ can be expressed as the product of two parts: the signal subject to long-term fading $m(t)$ and the signal subject to short-term fading $r(t)$ as:

$$s(t) = m(t)r(t) \tag{4.1}$$

We refer to Figure 4.2 and assume that the z-axis is perpendicular to the surface of the earth and x-y plane lies on the surface of the earth. At every receiv-

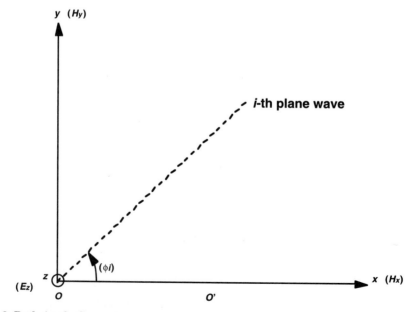

Fig. 4.2 Path Angle Geometry

ing point, we assume the existence of N plane waves of equal amplitude. The path-angle geometry for the i-th scattered plane wave is shown in Figure 4.2. If the transmitted signal is vertically polarized (i.e., the electrical field vector is aligned along the z-axis), the field components at mobile are the electrical field E_z, the magnetic field component H_x, and the magnetic field component H_y. These components at the receiving point, O, are expressed in the complex equivalent baseband form using Clarke's model as:

$$E_z = E_o \sum_{i=1}^{N} e^{j\phi_i} \tag{4.2}$$

$$H_x = -\frac{E_o}{\eta} \sum_{i=1}^{N} \sin\alpha_i e^{j\phi_i} \tag{4.3}$$

$$H_y = \frac{E_o}{\eta} \sum_{i=1}^{N} \cos\alpha_i e^{j\phi_i} \tag{4.4}$$

where:
ϕ_i = phase angle relative to the carrier phase,
E_o = amplitude of the N plane wave, and
η = intrinsic wave impedance which is given as:

$$\eta = \sqrt{\frac{\mu_o}{\varepsilon_o}} \approx 377 \text{ ohms} \tag{4.5}$$

in which μ_o = free-space magnetic permeability $(4\pi \times 10^{-7})$ H/m and ε_o = free-space electric permittivity (8.854×10^{-12}) F/m.

The speed of light, c, is given as:

$$c = \frac{1}{\sqrt{\mu_o \varepsilon_o}} = 3 \times 10^8 \text{ m/s} \tag{4.6}$$

For a vehicle moving in the x direction with a constant velocity, v, the received carrier is Doppler shifted by

$$f_d = f_m \cos\phi = \frac{v \cos\phi}{\lambda} = \frac{v_{eff}}{\lambda} = \frac{v_{eff} f_c}{c} \tag{4.7}$$

where:
$f_m = v/\lambda$ = maximum value of Doppler frequency, f_d at $\phi = 0$,
v_{eff} is the effective velocity of the vehicle,
f_c is the carrier frequency, and
ϕ is the path angle.

The Doppler frequency f_m is directly related to the phase change $\Delta\phi$ caused by the change in the path length. The Doppler shift is bounded to $\pm f_m$, which in general is much smaller than the carrier frequency f_c. The waves arriving from ahead of the vehicle experience a positive Doppler shift, whereas those coming from behind the vehicle have a negative Doppler shift. Thus, each component of the received signal is shifted by different values of Doppler frequency. For example, a vehicle traveling at 50 mph and receiving signals at a carrier frequency of 880 MHz will introduce a maximum Doppler shift of $v/\lambda = (50 \times 0.447 \times 880 \times 10^6)/(3 \times 10^8) = 65.56$ Hz.

4.3.1 Short-Term Fading

By applying the central limit theorem and observing that α_i and ϕ_i are independent, it follows that E_z, H_x, and H_y are complex Gaussian random variables for large N.

We consider the RF version of Eq. (4.2) for the field intensity E_z:

$$E_z = E_o \sum_{i=1}^{N} e^{j(\omega_c t + \phi_i)} \tag{4.8}$$

The real part of E_z is given as:

$$Re\,[E_z] = E_o \sum_{i=1}^{N} \cos \omega_c t \cos \phi_i - E_o \sum_{i=1}^{N} \sin \omega_c t \sin \phi_i \tag{4.9}$$

Let $A_c = E_o \sum_{i=1}^{N} \cos \phi_i$ and $A_s = E_o \sum_{i=1}^{N} \sin \phi_i$, then Eq. (4.9) can be written as:

$$Re\,[E_z] = A_c \cos \omega_c t - A_s \sin \omega_c t \tag{4.10}$$

Since ϕ_i is uniformly distributed between 0 to 2π, therefore the mean values of A_c and A_s are zero. The mean square values of A_c and A_s are $E\left(A_c^2\right) = E\left(A_s^2\right) = \dfrac{E_o^2 N}{2} = P_o$, the mean received power at the mobile.

Since A_c and A_s are uncorrelated, and therefore independent, we can write $E[A_c A_s] = 0$. Thus, the density of A_c and A_s follows a normal distribution, and the envelope of A_c and A_s is given by:

$$r = \left(A_c^2 + A_s^2\right)^{1/2} \tag{4.11}$$

and the phase, θ, is given as:

$$\theta = a\tan \frac{A_s}{A_c} \tag{4.12}$$

The square root of the sum of the square of two Gaussian functions is a Rayleigh distribution. Therefore, the probability density function for short-term or multipath fading is given by the Rayleigh distribution (refer to Figure 4.3):

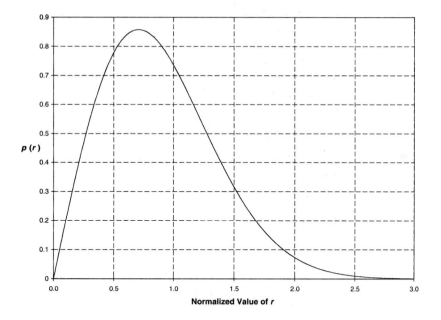

Fig. 4.3 Rayleigh Distribution—Short-Term Fading

$$p(r) = \frac{r}{P_o} e^{(-r^2)/(2P_o)} \tag{4.13}$$

where:
$2P_o = 2\sigma^2$ is the mean square power of the component subject to short-term
fading and
r^2 is the instantaneous power.

The corresponding cumulative distribution function is

$$prob(r \le R) = P(R) = \int_o^R \frac{r}{P_o} e^{(-r^2)/(2P_o)} dr \tag{4.14a}$$

$$P(R) = -e^{(-r^2)/(2P_o)} \Big|_0^R \tag{4.14b}$$

$$P(R) = 1 - e^{-R^2/2P_o} \tag{4.14c}$$

$$r_{mean} = E[r] = \int_0^\infty r p(r) dr = 1.2533\sqrt{P_o} = 1.2533\sigma \tag{4.15}$$

$$\text{Mean Square:} = E[r^2] = \int_0^\infty r^2 p(r) dr = 2P_o = 2\sigma^2 \tag{4.16}$$

$$\text{Variance:} \ \sigma_r^2 = E[r^2] - (E[r])^2 = 0.4292 P_o = 0.4292\sigma^2 \tag{4.17}$$

The mean value r_M, is defined as that for which $P(r_M) = 0.5$,

$$\therefore 1 - e^{\left(-r_m^2\right)/(2P_o)} = 0.5 \tag{4.18a}$$

and

$$r_M = 1.1774\sqrt{P_o} = 1.774\sigma \tag{4.18b}$$

It is often convenient to write Eqs (4.13) and (4.14b) in terms of mean, mean-square value, or median rather than in terms of P_o.

Let $E[r] = \bar{r}$ and $E[r^2] = \overline{r^2}$.

In terms of the mean-square value

$$p(r) = \frac{2r}{\overline{r^2}}e^{\left[-\frac{r^2}{\overline{r^2}}\right]} \tag{4.19a}$$

$$P(R) = 1 - e^{\left[-\frac{R^2}{\overline{r^2}}\right]} \tag{4.19b}$$

In terms of the mean

$$p(r) = \frac{\pi r}{2\bar{r}^2}e^{\left[-\frac{\pi r^2}{(2\bar{r})^2}\right]} \tag{4.20a}$$

$$P(R) = 1 - e^{\left[-\frac{\pi R^2}{(2\bar{r})^2}\right]} \tag{4.20b}$$

In terms of the median

$$p(r) = \frac{2r\ln 2}{r_M^2}e^{\left(-\frac{r^2\ln 2}{2r_M^2}\right)} \tag{4.21a}$$

$$P(R) = 1 - 2^{-\left(\frac{R}{r_M}\right)^2} \tag{4.21b}$$

The Rayleigh probability density function describes the first-order statistics of the signal envelope over distances short enough for the mean level to be regarded as constant. First-order statistics are those for which distance is not a factor, and the Rayleigh distribution gives information such as the overall percentage of locations (or time) for which the envelope lies below a specified value.

System engineers are interested in a quantitative description of the rate at which fades of any depth occur and the average duration of a fade below any given depth. This provides a valuable aid in selecting transmission bit rates, word lengths, and coding schemes in digital radio systems and allows an assessment of system performance. The required information is provided in terms of level crossing rate and average fade duration below a specified level.

4.3.2 Level Crossing Rate

The level-crossing rate, $N(R)$, at a specified signal level R is defined as the average number of times per second that the signal envelope crosses the level in a positive-going direction $(\dot{r} > 0)$.

$$N(R) = \int_0^\infty \dot{r} p(R, \dot{r}) \, d\dot{r} \tag{4.22}$$

where $p(R, \dot{r})$ is the joint Probability Density Function (PDF) of R and $\dot{r}$, and a dot indicates the time derivative.

Using derivations given in references [5] and [6], the average level crossing rate at a level R can be shown to be:

$$N(R) = \sqrt{\frac{\pi}{\sigma^2}} R f_m e^{\left(\frac{-R^2}{2\sigma^2}\right)} \tag{4.23}$$

Since $2\sigma^2$ = mean-square value, therefore $\sqrt{2}\sigma$ is the root mean square (RMS) value. The level crossing rate for a vertical monopole antenna can then be given as:

$$N(R) = \sqrt{2\pi} f_m \rho e^{-\rho^2} = n_o n_R \tag{4.24}$$

where:

$\rho = \dfrac{R}{\sqrt{2}\sigma} = \dfrac{R}{R_{RMS}}$ = the ratio between the specified level and the RMS amplitude of the fading envelope,

$f_m = \dfrac{v}{\lambda}$,

$n_o = \sqrt{2\pi} f_m$,

$n_R = \rho e^{-\rho^2}$,

n_R is the normalized level crossing that is independent of wavelength and vehicle speed,

v = speed of vehicle, and

λ = carrier wavelength.

Figure 4.4 gives the n_R at a level R.

An approximate expression for $N(R)$ below a given level, $\rho = R/R_{RMS}$, can be given as:

$$N(R) \approx \sqrt{2\pi} \frac{v}{\lambda} \rho \tag{4.25}$$

The above equations neglect the effect of the motion of the scatterers. When their motion is taken into account, the effect is to increase the fade rate.

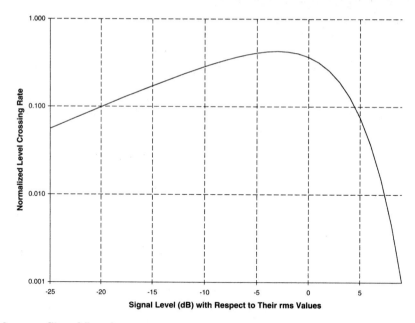

Fig. 4.4 n_R vs. Signal Level

4.3.3 Average Fade Duration

The average fade duration is the average of $\tau_1, \tau_2, ..., \tau_n$. The average duration of fades below the specified level R can be found from

$$E\,[\tau_R] \;=\; \tau\,(R) \;=\; \frac{prob\,[r \leq R]}{N\,(R)} \tag{4.26}$$

$$\tau\,(R) \;=\; \frac{e^{\rho^2} - 1}{\sqrt{2\pi} f_m \rho} \;=\; \frac{e^{\rho^2} - 1}{n_o \rho} \tag{4.27}$$

An approximate expression for $\tau(R)$ can be given as:

$$\tau\,(R) \;\approx\; \frac{\lambda}{v}\,\frac{\rho}{\sqrt{2\pi}} \tag{4.28}$$

E X A M P L E   4 – 1

Problem Statement
Calculate the level-crossing rate at a level of –10 dB and average duration of fade for a cellular system of 900 MHz and a vehicle speed of 24 km/h. Assume the free-space speed of propagation for electromagnetic waves = 3×10^8 m/s. Neglect the effects of the motion of the scatterers. Compare the results obtained using the approximate expressions.

Solution

At 900 MHz, $\lambda = \dfrac{3 \times 10^8}{900 \times 10^6} = \dfrac{1}{3}$ m, $v = 6.67$ m/s, $f_m = \dfrac{6.67}{\dfrac{1}{3}} = 20$ Hz,

$$n_o = \sqrt{2\pi}f_m = 50 .$$

From Fig. 4.3, $n_R = 0.32$ at –10 dB.

$$N(R) = 0.32 \times 50 = 16.0 \text{ fades/sec}$$

$$\rho e^{\rho^2} = n_R = 0.32$$

$$\rho = 0.294$$

$$\tau(R) = \dfrac{(1.09 - 1)}{50 \times 0.294} = 0.0061 \text{ sec} = 6.1 \text{ ms}$$

Using the approximate expressions we get:

$$\text{Fading level} = \rho = -10 \text{ dB}$$

$$20 \log \rho = -10$$

$$\rho = 10^{-10/20} = 0.3162$$

$$N(R) \approx \sqrt{2\pi} \times \dfrac{6.67}{\dfrac{1}{3}} \times 0.3162 = 15.85 \text{ fades/sec}$$

$$\tau(R) \approx \dfrac{1}{3 \times 6.67} \dfrac{0.3162}{\sqrt{2\pi}} = 0.0063 = 6.3 \text{ ms}$$

These results are quite close to those obtained using the exact expressions.

4.3.4 Long-Term Fading

The probability density function for long-term fading is given by the log-normal distribution (refer to Figure 4.5).

$$p(m) = \dfrac{1}{m\sigma_m\sqrt{2\pi}} e^{\left[-(\log m - \overline{m})^2 / \left(2\sigma_m{}^2\right)\right]}, \quad m > 0 \tag{4.29}$$

where:
$\overline{m}$ is the mean of log m and
σ_m is the standard deviation.

Using $z = (\log m - \overline{m})/\sigma_m$, the cumulative distribution function is given as:

$$prob(z \le Z) = P(z \le Z) = \dfrac{1}{2} + \dfrac{1}{2}erf\left(\dfrac{Z}{\sqrt{2}}\right) \approx 1 - \dfrac{1}{\sqrt{2\pi}Z} e^{(-Z^2/2)} \tag{4.30}$$

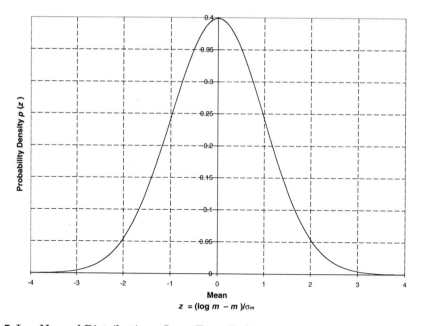

Fig. 4.5 Log-Normal Distribution—Long-Term Fading

4.3.5 Delay Spread

The radio signal follows different paths because of multipath reflection. Each path has a different path length, so the time of arrival for each path is different. The effect, which smears or spreads out the signal, is called "delay spread." As an example, if an impulse is transmitted by the transmitter, by the time this impulse is received at the receiver, it is no longer an impulse but rather a pulse that is spread (refer to Figure 4.6). In a digital system, the delay spread causes intersymbol interference, thereby limiting the maximum symbol rate of a digital multipath channel.

The mean delay spread τ_d is:

$$\tau_d = \frac{\int_0^\infty tD(t)\,dt}{\int_0^\infty D(t)\,dt} \tag{4.31}$$

where:
$D(t)$ is the delay probability density function and
$\int_0^\infty D(t)\,dt = 1$.

Some representative delay functions are:

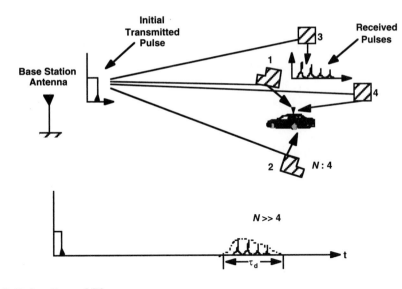

Fig. 4.6 Delay Spread Phenomenon

☞ Exponential:

$$D\,(t)\ =\ \frac{1}{\tau_d}e^{\frac{t}{\tau_d}}$$

☞ Uniform:

$$D\,(t)\ =\ \frac{\tau_d}{2},\ 0 \le t \le 2\tau_d$$

$$D(t) = 0, \text{ elsewhere}$$

The measured data suggest that the mean delay spreads are different in different environments (refer to Table 4.1).

4.3.6 Coherence Bandwidth

The coherence bandwidth B_c is the bandwidth for which either the amplitudes or phases of two received signals have a high degree of similarity. Specifi-

Table 4.1 Measured Data for Delay Spread

Type of Environment	Delay Spread τ_d (μs)
Open area	< 0.2
Suburban area	0.5
Urban area	3

cally, the coherence bandwidth is defined as that bandwidth where the correlation function $|R_T(\Omega)| = 0.5$, for two fading signal envelopes at two frequencies f_1 and f_2, respectively, where

$$\Delta f = |f_1 - f_2| \tag{4.32}$$

Two frequencies that are further apart than the coherence bandwidth, B_c, will fade independently. This concept is also useful for diversity reception (see chapter 7).

 The coherence bandwidth for two fading amplitudes of two received signals is

$$\Delta f > B_c = \frac{1}{2\pi\tau_d} \tag{4.33}$$

The coherence bandwidth for two random phases of two received signals is

$$\Delta f > \overline{B}_c = \frac{1}{4\pi\tau_d} \tag{4.34}$$

4.3.7 Intersymbol Interference

 In a time-dispersive medium, the transmission rate R for a digital transmission is limited by delay spread. If a low bit-error-rate performance is required, then

$$R < \frac{1}{2\tau_d} \tag{4.35}$$

In a real situation, R is determined based upon the required bit error rate, which may be limited by the delay spread.

4.4 CAPACITY OF A COMMUNICATION CHANNEL

Consider a digital communication channel that transmits S watts of power. The majority of the transmitted power is contained in a bandwidth B_w. We assume that the only effect of the channel is to add thermal noise to the transmitted signal. The bandwidth of this noise is very large compared to the signal bandwidth B_w. If the statistics of this noise are assumed to be Gaussian, the channel is called **Additive White Gaussian Noise (AWGN)** channel. Given these constraints, there exists a maximum rate at which information can be transmitted over the channel with arbitrarily high reliability. This transmission rate is called as the **error-free capacity** of a communication channel. The work of C. E. Shannon suggested that signaling schemes exist such that error-free transmission can be achieved at any rate lower than channel capacity. The normalized error-free capacity of a communication channel based on Shannon's work is given as:

$$\frac{C}{B_w} = log_2\left[1 + \frac{S}{N_o B_w}\right] = log_2\left[1 + \frac{E_b}{N_o}\left(\frac{R}{B_w}\right)\right] \tag{4.36}$$

where:
C = channel capacity (bits/s),
B_w = one-way transmission bandwidth (Hz),
E_b = energy per bit of the received signal (Joule),
R = information rate (bits/s),
S = $E_b R$ = signal power, and
N_o = single-sided noise power spectral density (W/Hz).

An ideal communication channel can be defined as one in which information is transmitted at the maximum rate $R = C$ bits/second(s).

$$\therefore \frac{C}{B_w} = log_2\left[1 + \frac{E_b}{N_o}\left(\frac{C}{B_w}\right)\right] \tag{4.37}$$

Solving Eq. (4.37) for E_b/N_o, we get

$$\frac{E_b}{N_o} = \frac{2^\alpha - 1}{\alpha}$$

where $\alpha = \frac{C}{B_w}$

If the information rate R is less than C, Shannon proved that it is theoretically possible, by coding, to achieve error-free transmission through the channel. Eq. (4.36) is plotted in Figure 4.7. This plot can be separated into a bandwidth-limited region, where $R/B_w > 1$ and a power-limited region, where $R/B_w < 1$. Thus, when the number of bits/s/Hz is more than 1, one has an efficient scheme in terms of utilizing bandwidth. If the number of bits/s/Hz is less than 1, one has an efficient scheme in terms of power utilization. When $R/B_w \to 0$, the limiting signal-to-noise ratio (SNR), E_b/N_o, is about −1.6 dB. Thus, for any given rate-to-bandwidth, an SNR exists above which error-free transmission is possible and

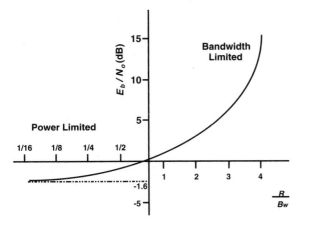

Fig. 4.7 Power-Bandwidth Trade-off for Error-free Transmission

below which it is not. Quite often, practical communication channels are compared with this ideal channel by choosing a probability of error 10^{-6} and finding the SNR ratio necessary to achieve it.

E X A M P L E   4 – 2

Problem Statement

The data transmission rate through an RF channel with a bandwidth of 30 kHz is 19.2 kbs. What E_b/N_o is required to achieve a communication reliability of one error or less in 10^6 transmitted bits?

Solution

$$\frac{R}{B_w} = \frac{19.2}{30} = 0.64$$

$$0.64 = log_2\left[1 + \left(\frac{E_b}{N_o}\right)(0.64)\right]$$

$$\frac{E_b}{N_o} = 0.8724 = -0.593 \text{ dB}$$

The channel is power controlled; if the communicator uses enough power to obtain $E_b/N_o = -0.593$ dB or more, error correction can be used to obtain an arbitrarily lower error rate.

For cellular and PCS systems with small cells, the dominant effect is from interference not noise. However, the effects of interference can be converted to an effective noise level to calculate error rates.

4.5 PROPAGATION LOSSES IN BUILT-UP AREAS

Since the mathematical modeling of the propagation of radio waves in a real-world environment is complicated, several empirical models to predict propagation losses have been developed by various authors. In the following sections, we present the empirical and semiempirical models used to calculate the propagation path losses in the urban, suburban, and rural environments and compare the results obtained with each model.

4.5.1 Classification of Built-up Areas

The propagation of radio waves in the built-up areas is strongly influenced by the environment, in particular the size, type, and density of building. For propagation studies for mobile radio, the environment is described as rural, suburban, and urban. Urban areas are dominated by tall buildings, offices, and other commercial structures. Suburban areas contain residential houses and parks, whereas rural areas contain open farmland with farm buildings, wooded areas, and forests. These qualitative descriptions, however, are not precise and are open to different interpretations by different users.

Various experimenters have found that natural and man-made objects affect radio propagation, and they use the following characteristics to classify land usage types:

☞ Building characteristics: density, height, location, and size
☞ Vegetation density
☞ Terrain variations

Using some or all of these characteristics, various experimenters have determined propagation loss graphics and formulas.

With a computer simulation and using the following parameters, Kafaru [9] proposed a method to classify land usage types. His method included:

☞ Building location with respect to some reference point
☞ Building size or base area
☞ Total area occupied by buildings
☞ Number of buildings in the area concerned
☞ Building height
☞ Terrain variations
☞ Parks and/or gardens with trees and vegetation

With this information, Kafaru developed the following data:

1. **Building size distribution (BSD):** A PDF defined by a mean and standard deviation. The standard deviation indicates homogeneity. A small value shows an area where the buildings are of fairly uniform size; a large value implies a more diverse range.
2. **Building area index (BAI):** An area factor of occupied buildings, it is defined as the percentage of the 500 m × 500 m test square that is covered by buildings, regardless of their heights.
3. **Building height distribution (BHD):** A PDF of the heights of all buildings within the area concerned.
4. **Building location distribution (BLD):** A PDF describing the location of buildings within the area.
5. **Vegetation index (VI):** The percentage of the area covered by trees, and so on.
6. **Terrain variation index (TVI)**

He then proposed the following classifications and subclassifications of the environment:

Class 1 (rural)
     A—flat
     B—hilly
     C—mountainous

Class 2 (suburban)
    A—residential with some open spaces
    B—residential with little or no open space
    C—high-rise residential
Class 3 (urban and dense urban)
    A—shopping area
    B—commercial area
    C—industrial area

The parameters associated with the subclasses in class 2 and 3 environments are given in Table 4.2.

4.5.2 Models to Predict Propagation Losses

Before describing the models to predict propagation losses, it is worth re-emphasizing that there is no single model that is universally applicable in all situations. The accuracy of a particular model in a given environment depends on the fit between the parameters required by the model and those available for the area concerned. Generally, we are concerned with predicting the mean signal strength in a small area and with the variation in signal strength as the mobile moves.

Several models have been proposed. In the following section, we focus only upon the five most widely used empirical or semiempirical models in Japan, the United Kingdom, and the United States.

4.5.2.1 Okumura's Model   Based on extensive measurements in and around Tokyo at frequencies up to 1,920 MHz, Y. Okumura et al. [7] developed an empirical model to predict signal strength. The model is based on the free-space path loss between the points of interest. The value of $A_{mU}(f,d)$ obtained from Figure 4.8 is added to the free-space loss. A_{mU} is the median attenuation relative to free space in an urban area over quasismooth terrain with a base station effec-

Table 4.2 Parameters for Class 2 and 3 Environments[a]

Class	BAI (%)	VI (%)	BSD (m^2)		BHD (No. of Stories)	
			μ_s	σ_s	μ_h	σ_h
2A	12–20	≥ 2.5	95–115	55–70	2	1
2B	20–30	< 5	100–120	70–90	2–3	1
2C	≥ 12	≤ 2	≥ 500	> 90	≥ 4	1
3A	≥ 45	0	200–250	≥ 180	≥ 4	1
3B	30–40	0	150–200	≥ 160	3	1
3C	35–45	≤ 1	≥ 250	≥ 200	2–3	1

a.where μ = the mean and σ = the standard deviation

tive antenna height $h_{Te} = 200$ m and mobile antenna height $h_R = 3$ m. A_{mU} is a function of frequency (in the range 100–3,000 MHz) and distance from base station (1–100 km). Correction factors are applied to account for antennas not at the reference heights. The basic formulation for the model is:

$$L_{50} = L_f + A_{mU} + G_{Tu} + G_{Ru} \ \text{dB} \qquad (4.38)$$

where:
L_{50} is the median path loss,
$A_{mU}(f,d)$ = median attenuation relative to free space in an urban area (refer to
          Figure 4.8),
L_f = free-space loss,
G_{Tu} = base station antenna height gain factor (refer to Figure 4.9), and
G_{Ru} = mobile antenna height gain factor (refer to Figure 4.10).

Additional correction factors, in graphical form, are used to account for street orientation and transmission in suburban and rural areas and over irregular terrain. These corrections are added or subtracted as necessary. Irregular terrain is further classified as rolling hilly terrain, isolated mountain, general sloping terrain, and mixed land-sea path. The terrain-related parameters, that are evaluated to determine the various correction factors, include:

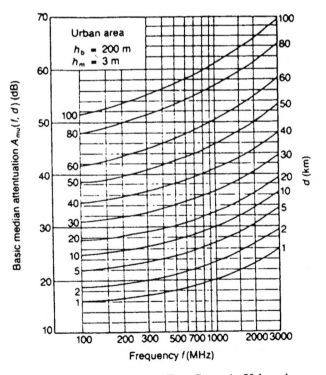

Fig. 4.8 Basic Median Path Loss Relative to Free Space in Urban Areas over Quasi-Smooth Terrain (after Okumura)

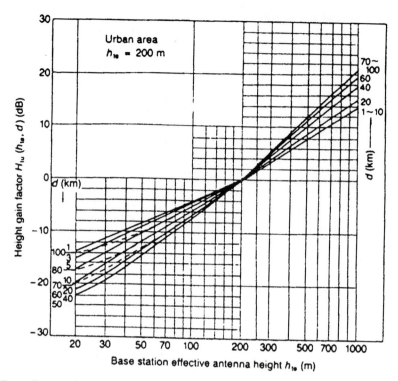

Fig. 4.9 Base Station Antenna Height/Gain Factor in Urban Areas as a Function of Range (Reference Height = 200 m)

1. Effective base station antenna height (H_{Te}): height of the base station antenna above the average ground level calculated over the range interval 3–15 km in the direction toward the receiver (refer to Figure 4.11).

2. Terrain variation height (Δh): terrain irregularity parameter that defines the average height taken over a distance of 10 km from the receiver in a direction toward the transmitter.

3. Isolated ridge height: When the propagation path has a single obstructing mountain, its height is measured relative to the average ground level between it and base station.

4. Average ground slope: For sloping ground, the angle θ (+ or −) is measured over 5 to 10 km.

5. Mixed land-sea path parameter: Percentage of the total path length covered by water.

Okumura's model is the most widely quoted of the available models.

4.5.2.2 Sakagmi and Kuboi Model

Shuji Sakagmi and Kiyoshi Kuboi [12] extended Okumura's model to develop a more general empirical model using a multiple regression analysis of the experimental data collected at the Chiyoda

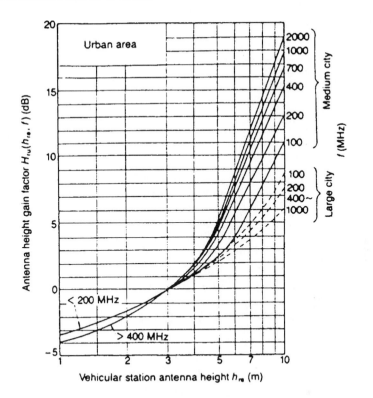

Fig. 4.10 Mobile Station Antenna Height/Gain Factor in Urban Areas as a Function of Frequency and Urbanization (Reference Height = 3 m)

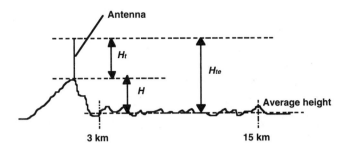

Fig. 4.11 Method of Calculating the Effective Base Station Antenna Height

and Shibuya station test sites in Tokyo. The Chiyoda test site involved mobile movement around the Otemachi, Ginza, and Nihonbashi areas and represented the built-up areas. The Shibuya test site included mobile movement around the Shibuya and Aoyama areas and also in the residential area of Azabu. The test parameters used in the experimental setups are shown in Table 4.3.

Table 4.3 Test Parameters

	Chiyoda	Shibuya
Transmitting antenna height from ground	37 m, 86 m	20 m, 41 m
Receiving antenna height from ground	1.5 m	1.5 m
Transmitting and receiving antennas	λ/2 dipole	λ/2 dipole
Power and frequency	14.1 W, 813.275 MHz 13.2 W, 1,432.9 MHz	14.1 W, 813.275 MHz 13.2 W, 1,432.9 MHz

The following empirical formula was developed:

$$L_{50} = 100 - 7.1 \log W + 0.023\theta + 1.4 \log h_s + 6.1 \log \langle H \rangle \tag{4.39}$$

$$- \left[24.37 - 3.7 \left[\frac{h}{h_{To}} \right]^2 \right] \log h_T + [43.42 - 3.1 \log h_T] \log d$$

$$+ 20 \log f_c + e^{[13 \log f_c - 3.23]} \text{ dB}$$

where:
W = street width (5–50 m),
θ = street angle (0–90 degrees),
h_S = building height along the street (5–80 m),
$\langle H \rangle$ = average building height (5–50 m: height above the mobile station ground),
h_T = base station antenna height (20–100 m: height above the mobile station ground),
h_{To} = base station antenna height from the ground (m): height from the base station ground,
f_c = frequency (450–2,200 MHz),
h = building height near base station (m),
$h \le h_{To}$: height from the base station ground, and
d = distance from base station ground (0.5–10 km).

4.5.2.3 Hata's Model   M. Hata [3] developed empirical formulas to describe the graphical data given by Okumura. Hata's formulation applies to certain ranges of input parameters and is valid only for quasismooth terrain. The following are the expressions:
Urban area:

$$L_{50} = 69.55 + 26.16 \log f_c - 13.83 \log h_T - a(h_R)$$
$$+ [44.9 - 6.55 \log h_T] \log d \qquad\qquad \text{dB} \tag{4.40}$$

where:
$150 \le f_c \le 1500$ MHz,
$30 \le h_T \le 200$ m.,
$1 \le d \le 20$ km., and
$a(h_R)$ is the correction factor for the mobile antenna height and is computed as:

For a small- or medium-sized city:

$$a(h_R) = (1.1 \log f_c - 0.7) h_R - (1.56 \log f_c - 0.8) \ \text{dB} \tag{4.40a}$$

where $1 \le h_R \le 10$ m.
For a large city:

$$a(h_R) = 8.29 (\log 1.54 h_R)^2 - 1.1 \ \text{dB}, f_c \le 200 \ \text{MHz} \tag{4.40b}$$

$$a(h_R) = 3.2 (\log 11.75 h_R)^2 - 4.97 \ \text{dB}, f_c \le 400 \ \text{MHz} \tag{4.40c}$$

Suburban area:

$$L_{50} = L_{50}(urban) - 2 \left[\log \left[\frac{f}{28} \right]^2 - 5.4 \right] \ \text{dB} \tag{4.41}$$

Open area:

$$L_{50} = L_{50}(urban) - 4.78 (\log f_c)^2 + 18.33 \log f_c - 40.94 \ \text{dB} \tag{4.42}$$

Hata's model does not account for any of the path-specific corrections used in Okumura's model. Okumura's model gives predictions that correlate reasonably well with measured data for most urban and suburban areas. However, it tends to average over some of the extreme situations and does not respond sufficiently quickly to rapid changes in the radio path profile. The distance-dependent behavior of Okumura's model is in agreement with the measured values.

4.5.2.4 M. F. Ibrahim and J. D. Parsons Model
A series of field trials were conducted in London at frequencies in the range between 168 and 900 MHz [4,10]. Data were collected in 500-m squares, and a median path loss between isotropic antennas was extracted from the data for each square. The range from the transmitter varied from 2 to 10 km. The various measured path losses were used to determine the effects of range, transmission frequency, terrain characteristics, and degree of urbanization.

The following equation was developed as the best fit for London data:

$$L_{50} = -20 \log (0.7 h_T) - 8 \log h_R + \frac{f_c}{40} + 26 \log \frac{f_c}{40} - 86 \log \left(\frac{f_c + 100}{156} \right)$$

$$\left\{ 40 - 14.15 \log \left[\frac{f_c + 100}{156} \right] \right\} \log d + 0.256 L - 0.37 H + K \tag{4.43}$$

where:
K = $0.087 U - 5.5$ for the city center, otherwise $K = 0$,
L_{50} = median path loss between two isotropic antennas (dB),
h_T and h_R = effective heights of the transmitting and receiving antennas (m),
        and $h_R \le 3$ m,
f_c = transmission frequency (MHz),
d = range in meter($d \le 10{,}000$ m),
L = land usage factor,
H = difference in height between the squares containing the transmitter and receiver, and
U = degree of urbanization factor.

The semiempirical model based on the plane-earth equation suggests that the relationship between the received and transmitted powers is expressed by

$$\frac{P_R}{P_T} = G_T G_R \left[\frac{h_T h_R}{d^2}\right] \times (\text{clutter factor}) \tag{4.44}$$

$$L_{50} = 40\log d - 20\log (h_T h_R) + \beta \ \text{dB} \tag{4.45}$$

where β = clutter factor in dB

$$\beta = 20 + \frac{f_c}{40} + 0.18L - 0.34H + K$$

with $K = 0.094 \ U - 5.9$. K is applicable only in the highly urbanized city center, otherwise $K = 0$.

The RMS errors produced by two models are summarized in the Table 4.4, the RMS error is defined as:

$$RMS_{error} = \sqrt{\frac{\sum (Y_p - Y_m)^2}{N}}$$

where:
Y_p = predicted value (dB),
Y_m = measured value (dB), and
N = number of values considered.

4.5.2.5 W. C. Y. Lee's Model   Lee's model [5] is intended for use at 900 MHz, and it is applicable in two modes, an area-to-area mode and a point-to-point mode. In the area-to-area mode, the prediction is based on three parameters: (1) median transmission loss at a range of 1 km (L_o); (2) slope of the path-loss curve(γ) dB/decade, and (3) adjustment factor (F_o).

$$L_{50} = L_o + \gamma \log d + F_o \tag{4.46}$$

The values of L_o and γ were derived from experimental data and are given in Table 4.5. The following test parameters were used to obtain the result given in Table 4.5.

☞ f_c = 900 MHz
☞ h_T = 30.5 m
☞ Transmitter power = 10 W
☞ G_T with respect to λ / 2 dipole = 6 dB

Table 4.4 RMS Prediction Errors Produced by Two Models

Frequency (MHz)	168	455	900
Empirical Model	2.1	3.2	4.19
Semiempirical Model	2.0	3.3	5.8

Table 4.5 Lee's Model Parameters

Environment	L_o (dB)	γ
Free space	91.3	20.0
Open (rural)	91.3	43.5
Suburban	104.0	38.0
Urban:		
Tokyo	128.0	30.0
Philadelphia	112.8	36.8
Newark	106.3	43.1

☞ $h_R = 3$ m

$$F_o = F_1 F_2 F_3 F_4 F_5 \qquad (4.46a)$$

where:

$F_1 = \left[\dfrac{\text{Actual base station antenna height (m)}}{30.5}\right]^2$

$F_2 = \dfrac{\text{Actual transmitter power (W)}}{10}$

$F_3 = \dfrac{\text{Actual gain of base station antenna}}{4}$

$F_4 = \left[\dfrac{\text{Actual mobile antenna height (m)}}{3}\right]^2, h_R > 3$ (m)

or

$F_4 = \left[\dfrac{\text{Actual mobile antenna height (m)}}{3}\right]^2, h_R < 3$ (m)

$F_5 = \left[\dfrac{f_c}{f_o}\right]^n, n=2\text{–}3$

To predict propagation loss in the point-to-point mode, Lee's model takes some account of the terrain by using effective base station antenna height (refer to Figure 4.12). Eq. (4.46) becomes:

$$\overline{L_{50}} = L_{50} + 20 \log\left[\frac{H_e}{30}\right] \qquad (4.47)$$

E X A M P L E   4 – 3

Problem Statement

The measured path loss at a distance of 10 km in a large metropolitan area is 160 dB. The test parameters used in the experiment were the following:

(a) Base station antenna height: 30 m

(b) Mobile station antenna height: 3 m

(c) Carrier frequency: 1,000 MHz

(d) Antenna Type: $\gamma / 2$ Dipole

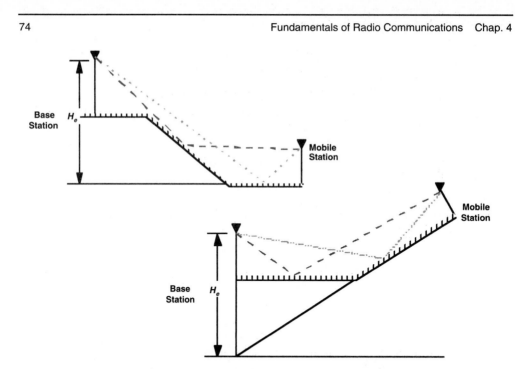

Fig. 4.12 Method of Calculating the Effective Base Station Antenna Height in Lee's Model

Compare the measured path loss with Hata's and Lee's model.

Solution
Hata's Model

$$L_{50} = 69.55 + 26.16 \log 1{,}000 - 13.82 \log 30 - a(h_R) + (44.9 - 6.55 \log 30) \log 10 \text{ dB}$$

$$a(h_R) = 3.2 (\log 11.75 \times 3)^2 - 4.97 \text{ dB}$$

$$a(h_R) = 2.69 \text{ dB}$$

$$L_{50} = 69.55 + 78.48 - 20.41 - 2.69 + 35.22 = 160.15 \text{ dB}$$

Lee's Model

For large urban area, $L_o \approx 118$ dB and $\gamma \approx 40$ dB

$$F_1 = \left(\frac{30}{30.5}\right)^2 = 0.9675,$$

$$F_2 = 1,$$

$$F_3 = 1,$$

$$F_4 = 1, \text{ and}$$

$$F_s = \left(\frac{1,000}{900}\right)^2 = 1.23$$

$$F_o = 0.9675 \times 1 \times 1 \times 1 \times 1.23 = 1.19 \text{ dB}$$

$$L_{50} = 118 + 40 \log 10 + 1.19 = 159.19 \text{ dB}$$

E X A M P L E   4 – 4

Problem Statement

Estimate path loss at a distance of 10 km in a large metropolitan area using the following data:

(a) Street width (W): 20 m

(b) Street angle(θ): 40 degrees

(c) Building height along the street (h_s): 40 m

(d) Average building height $<H>$: 30 m

(e) Building height near base station (h): 25 m

(f) Base station antenna height from ground (h_{T0}): 30 m

(g) Carrier frequency (f_c): 850 MHz

(h) Base station antenna height (h_T): 30 m

(i) Mobile station antenna height (h_R): 3 m

(j) Land usage factor (L): 40%

(k) Difference in heights between the ground containing transmitter and receiver: 10 m

Solution

We will use Sakagmi/Kuboi and Ibrahim/Parsons models to estimate path loss.

Sakagmi and Kuboi Model

$$L_{50} = 100 - 7.1 \log 20 + 0.023 \times 40 + 1.4 \log 40$$
$$+ 6.1 \log 30 - \left[24.37 - 3.7\left(\frac{25}{30}\right)^2\right] \log 30$$
$$+ [43.42 - 3.1 \log 30] \log 10 + 20 \log 850 + e^{13 (\log 850 - 3.23)}$$

$$L_{50} = 100 - 9.24 + 0.92 + 2.24 + 9.01 - 32.2 + 38.84 + 58.59 + 0.02$$

$$L_{50} = 168.18 \text{ dB}$$

Ibrahim and Parsons Model

• Empirical model:

$$L_{50} = -20 \log (0.7 \times 30) - 8 \log 3 + \frac{850}{40} + 26 \log \frac{850}{40} - 86 \log \frac{950}{156}$$
$$+ \left(40 + 14.15 \log \frac{950}{156}\right) \log 10,000 + 0.256 \times 40 - 0.37 \times 10$$

$$L_{50} = -26.4 - 3.82 + 21.25 + 34.51 - 67.48 + 204.41 + 10.24 - 3.7 = 169 \text{ dB}$$

• Semiempirical model:

$$\beta = 20 + \frac{850}{40} + 0.18 \times 40 - (0.34 \times 10) = 45.05 \text{ dB}$$

$$L_{50} = 40 \log 10,000 - 20 \log (30 \times 3) + 45.05 = 166 \text{ dB}$$

You may notice that the results obtained from the two models are fairly close and are in agreement with Sakagmi and Kuboi result.

4.6 SUMMARY

In this chapter we discussed the short-term and long-term fading characteristics of a radio signal. A numerical example was given for calculating level crossing rate and average duration of the short-term fade. We discussed several empirical and semiempirical models to calculate the path losses in different environments.

4.7 PROBLEMS

1. Tropospheric propagation can cause interference between cellular/PCS systems in different cities. It can occur during certain weather patterns (e.g., temperature inversions) and is most severe when there are no obstacles between the cities. From a map of North America, identify five city pairs that would likely see interference during times of strong tropospheric propagation. Describe some methods for reducing the interference between these cities.

2. If the earth was a smooth sphere with no atmosphere, then the ground wave propagation would be limited to "line of sight," i.e., the transmitter and receiver would have to "see" each other. Calculate the maximum range for a car-mounted mobile telephone with an antenna mounted 6 feet off of the ground and a cell-site (base station) height of 100 feet. Repeat the calculation for base station heights of 50 feet, 150 feet, 500 feet, and 1,000 feet. **Hint:** For maximum range, the path will just graze the earth at right angles to the radius to the center of the earth (this is called the horizon) and two right-angle triangles are formed. One side of each triangle is R, the radius of the earth; the hypotenuse is $R + H$, where H is the height of the antenna; the remaining side is the distance from one antenna to the horizon. When sides for both triangles are found, the sum is the range.

3. Find the years from 1900 to 2000 when sunspots were a maximum and a minimum. What was the Maximum Usable Frequency (MUF) during the peaks and valleys of sunspot activities? Considering the peak MUF during a sunspot maximum, explain why the 30–50-MHz band is not usable for a cellular or PCS system.

4. Calculate the level-crossing rate at a level of –20 dB and average duration of fade for a PCS system at 2,200 MHz and a vehicle speed of 100 km/hour (h). Assume the free-space speed of propagation for electromagnetic waves $= 3 \times 10^8$ m/second (s).

Neglect the effects of the motion of the scatterers. Compare the results obtained using the approximate expressions.

5. Repeat the previous problem for a pedestrian walking at 2.5 km/h and a level of −10 dB.

6. Characterize the terrain in the vicinity of your school. Which propagation model would you use to calculate the path loss for a system that covers the neighborhood of your school?

4.8 REFERENCES

1. Clarke, R. H., "A Statistical Theory of Mobile Radio Reception," *Bell System Technical Journal* 47 (July–August 1968): 957–1000.

2. "An Environment-dependent Approach to Wideband Modeling and Computer Simulation of UHF Mobile Propagation in Built-up Areas," Ph.D. thesis, University of Liverpool, U.K, 1989.

3. Hata, M., "Empirical Formula for Propagation Loss in Land Mobile Radio Services" *IEEE Transactions on Vehicular Technology* 29, no. 3 (1980).

4. Ibrahim, M. F., and J. D. Parsons, "Signal Strength Prediction in Built-up Area, Part 1: Median Signal Strength," IEE Proceedings 130, Part F, no. 5, 1983.

5. Lee, W. C. Y., *Mobile Communications Engineering,* New York: McGraw-Hill, 1982.

6. Mehrotra, A., *Cellular Radio Performance Engineering,* Boston: Artech House, 1994.

7. Okumura, Y., et al., "Field Strength and its Variability in the VHF and UHF Land Mobile Radio Service," Review Electronic Communication Lab 16, no. 9–10, 1968.

8. Parsons, J. D., *The Mobile Propagation Channel,* New York: Halsted Press—John Wiley and Sons, 1992.

9. Parsons, J. D., and J. G. Gardiner, *Mobile Communication Systems,* New York: Halsted Press—John Wiley and Sons, 1989.

10. Parsons, J. D., and M. F. Ibrahim, "Signal Strength Prediction in Built-up Area, Part 2: Signal Variability," IEE Proceedings 130, Part F, no. 5, 1983.

11. Rice, S. O., "Statistical Properties of a Sine Wave plus Random Noise," *Bell System Technical Journal* 27 (1948): 109–57.

12. Sakagmi, Shuji, and Kiyoshi Kuboi, "Mobile Propagation Loss Prediction for Arbitrary Urban Environment," *Electronics and Communications in Japan, Part 1* 74, no. 10 (1991).

Fundamentals of Cellular Communications

5.1 INTRODUCTION

In this chapter, we present the fundamentals of the cellular communications and develop a relationship between the reuse ratio (q) and cluster size (N) for hexagonal cell geometry. The chapter also covers cochannel interference for the omnidirectional and sectorized cell site. Cell splitting, registration, terminal authentication, and handoff procedures used in the cellular communications are also discussed.

5.2 CELLULAR SYSTEM

Most commercial radio and television systems are designed to cover as much area as possible. These systems typically operate at the maximum power and with the highest antennas allowed by the FCC. The frequency used by the transmitter cannot be reused until there is enough geographical separation so that one station does not interfere significantly with another station assigned to that frequency.

The cellular system takes the opposite approach. It seeks to make an efficient use of available channels by using low-power transmitters to allow frequency reuse at much smaller distances (see Figure 5.1). Maximizing the number of times each channel may be reused in a given geographic area is the key to an efficient cellular system design.

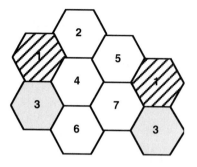

Fig. 5.1 Frequency Reuse

Cellular systems are designed to operate with groups of low-power radios spread out over the geographical service area. Each group of radios serve mobile units presently located near them. The area served by each group of radios is called a "cell." Each cell has an appropriate number of low-power radios for communications within itself. The power transmitted is chosen to be large enough to communicate with mobile units located near the edges of the cell. The radius of each cell may be chosen to be perhaps 26 km (about 16 miles) in a start-up system with relatively few subscribers, down to less than 2 km (about 1 mile) for a mature system requiring considerable frequency reuse.

As the traffic grows, new cells and channels are added to the system. If an irregular cell pattern is selected, it would lead to an inefficient use of the spectrum due to its inability to reuse frequencies on account of cochannel interference. In addition, it would also result in an uneconomical deployment of equipment, requiring relocation from one cell site to another. Therefore, a great deal of engineering effort would be required to readjust the transmission, switching, and control resources every time the system goes through its development phase. All these difficulties lead to the use of a regular cell pattern in a cellular system design.

In reality, the cell coverage is an irregularly shaped circle. The exact coverage of the cell will depend on the terrain and other factors, as described in the previous chapter. For design convenience and as a first-order approximation, we assume that the coverage areas are regular polygons. For example, for an omnidirectional antenna with constant signal power, each cell-site coverage area would be circular. To achieve full coverage without dead spots, a series of regular polygons for cell sites are required. Any regular polygon, such as an equilateral triangle, a square, or a hexagon, can be used for cell design. The hexagon is used for two reasons: first, a hexagonal layout requires fewer cells and therefore, fewer transmitter sites and second, a hexagonal cell layout is less expensive compared to square and triangular cells. In practice, after the polygons are drawn on a map of the coverage area, radial lines are drawn and the SNR ratio calculated for various directions using propagation models from the previous chapter or computer programs. For the remainder of this chapter, we will assume regular polygons for the coverage areas even though in practice that is only an approximation.

5.3 Geometry of a Hexagonal Cell

We use the u–v axes to calculate the distance D between points C_1 and C_2 (refer to Figure 5.2). C_1 and C_2 are the centers of the hexagonal cells with coordinates (u_1, v_1) and (u_2, v_2).

$$D = \{ (u_2 - u_1)^2 (\cos 30°)^2 + [(v_2 - v_1) + (u_2 - u_1) \sin 30°]^2 \}^{1/2} \tag{5.1}$$

$$D = \{ (u_2 - u_1)^2 + (v_2 - v_1)^2 + (v_2 - v_1)(u_2 - u_1) \}^{1/2} \tag{5.1a}$$

If we assume $(u_1, v_1) = (0, 0)$ or the origin of the coordinate system is the center of a hexagonal cell and restrict (u_2, v_2) to be integer value (i, j) then Eq. (5.1a) can be written as:

$$D = [i^2 + j^2 + ij]^{1/2} \tag{5.2}$$

The normalized distance between two adjacent cells is unity $(i = 1, j = 0)$ or $(i = 0, j = 1)$. The actual center-to-center distance between two adjacent hexagonal cells is $2R \cos 30°$ or $\sqrt{3} R$ where R is the center-to-vertex distance.

We assume the size of all the cells is roughly the same. As long as the cell size is fixed, cochannel interference will be independent of the transmitted power of each cell. The cochannel interference is a function of q where $q = D/R$. Furthermore, D is a function of N_I and S/I in which N_I = number of cochannel interfering

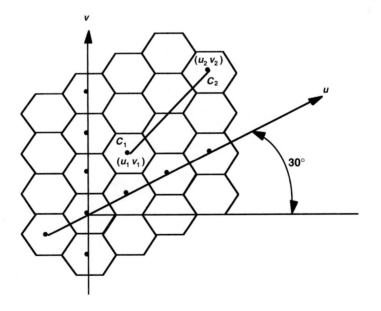

Fig. 5.2 Coordinate System

cells in the first tier (refer to Figure 5.3) and S/I = received signal-to-interference ratio at the desired mobile receiver.

From Figure 5.4, we see the radius of the large cell D (cochannel separation) is given as:

$$D^2 = 3R^2 (i^2 + j^2 + ij) \tag{5.3}$$

Since the area of a hexagon is proportional to the square of distance between center and vertex, the area of the large hexagon is:

$$A_{large} = k\,[\,3R^2 (i^2 + j^2 + ij)\,] \tag{5.4}$$

where:
k is a constant.

Similarly the area of the small hexagon is given as:

$$A_{small} = k\,(R^2) \tag{5.5}$$

Comparing Eqs. (5.4) and (5.5) and using Eq. (5.3), we can write

$$\frac{A_{large}}{A_{small}} = 3\,(i^2 + j^2 + ij) = \frac{D^2}{R^2} \tag{5.6}$$

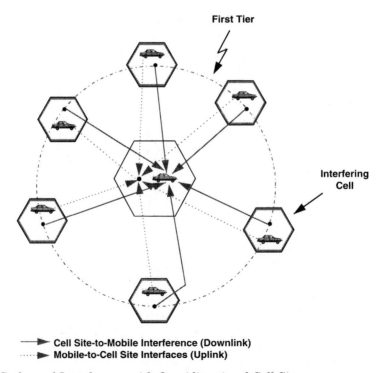

Fig. 5.3 Cochannel Interference with Omnidirectional Cell Site

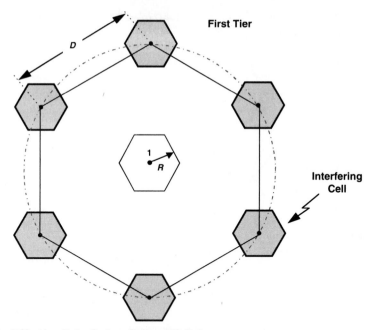

Fig. 5.4 Six Effective Interfering Cells of Cell 1

From symmetry, we can see that the large hexagon encloses the center cluster of N cells plus one-third the number of the cells associated with six other peripheral hexagons. Thus, the total number of cells enclosed is equal to $N + 6 (1/3\ N) = 3N$. Since the area is proportional to the number of cells, $A_{large} = 3N$, and $A_{small} = 1$.

$$\frac{A_{large}}{A_{small}} = 3N \tag{5.7}$$

Substituting Eq. (5.7) into (5.6), we get:

$$3N = 3\,(i^2 + j^2 + ij) \tag{5.8}$$

$$\therefore \frac{D^2}{R^2} = 3N \tag{5.9}$$

$$\frac{D}{R} = q = \sqrt{3N} \tag{5.9a}$$

where:
q = reuse ratio (refer to Figure 5.5).

Table 5.1 lists the values of q for different values of N.

Eq. (5.9a) is important because it affects both the traffic-carrying capacity of a cellular system and the cochannel interference. By reducing q the number of

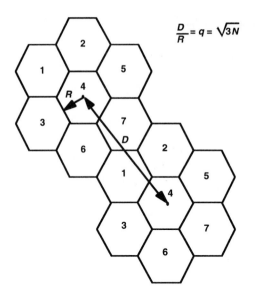

$$\frac{D}{R} = q = \sqrt{3N}$$

Fig. 5.5 Relationship between q and N

Table 5.1 Cochannel Reuse Ratio vs. Frequency Reuse Pattern

i	j	N	q = D/R
1	0	1	1.73
1	1	3	3.00
2	0	4	3.46
2	1	7	4.58
3	0	9	5.20
2	2	12	6.0
3	1	13	6.24
4	0	16	6.93
3	2	19	7.55
4	1	21	7.94
4	2	28	9.17

cells per cluster is reduced. If total RF channels are constant, then the number of channels per cell is increased, thereby increasing the system traffic capacity. On the other hand, the cochannel interference is increased with small q. The reverse is true when q is increased—i.e., an increase in q reduces cochannel interference and also the traffic capacity of the cellular system.

5.4 Cochannel Interference Ratio

The S/I ratio at the desired mobile receiver is given as:

$$\frac{S}{I} = \frac{S}{\sum\limits_{k=1}^{N_I} (I_k)} \tag{5.10}$$

In a fully equipped hexagonal-shaped cellular system, there are always six cochannel interfering cells in the first tier (i.e., N_I = 6, Figure 5.3). Cochannel interference can be experienced both at the cell site and at the mobile units in the center cell. In a small cell system, interference will be the dominating factor and thermal noise can be neglected. Thus S/I ratio can be given as:

$$\frac{S}{I} = \frac{1}{\sum\limits_{k=1}^{6} \left(\dfrac{D_k}{R}\right)^{-\gamma}} \tag{5.11}$$

where:
$2 \leq \gamma \leq 5$ is the propagation path-loss slope and
γ depends upon the terrain environment.

If we assume D_k is the same for the six interfering cells for simplification, i.e., $D = D_k$, then Eq. (5.11) becomes:

$$\frac{S}{I} = \frac{1}{6\,(q)^{-\gamma}} = \frac{q^{\gamma}}{6} \tag{5.12}$$

$$\therefore q = \left[6\left(\frac{S}{I}\right)\right]^{\frac{1}{\gamma}} \tag{5.13}$$

For analog systems using FM, normal cellular practice is to specify an S/I ratio to be 18 dB or higher based on subjective tests. An S/I of 18 dB is the measured value for the accepted voice quality from the present-day cellular mobile receivers.

Using an S/I ratio equal to 18 dB (i.e., 63.1) and $\gamma = 4$ in Eq. (5.13), then

$$q = [6 \times 63.1]^{0.25} = 4.41 \tag{5.14}$$

Substituting q from Eq. (5.14) into Eq. (5.9a) yields

$$N = \frac{(4.41)^2}{3} = 6.49 \approx 7 \tag{5.15}$$

Eq. (5.15) indicates that a seven-cell reuse pattern is needed for an S/I ratio of 18 dB. Based on $q = D/R$, we can select D by choosing the cell radius R.

E X A M P L E   5 – 1

Problem Statement

We consider a cellular system with 395 total allocated voice channel frequencies. If the traffic is uniform with an average call holding time of 120 seconds and the call blocking during the system busy hour is 2%, calculate:

(a) The number of calls per cell site per hour

(b) The mean S/I ratio for cell reuse factors equal to 4, 7, and 12.

Assume omnidirectional antennas with six interferers in the first tier and a slope for the path loss of 40 dB/decade ($\gamma = 4$).

Solution

For a reuse factor $N = 4$, the number of voice channels per cell site = 395/4 ≈ 99 and $q = \sqrt{3 \times 4} = 3.5$. Using the Erlang-B traffic table for 99 channels with 2% blocking, we find a traffic load of 87 Erlangs.

$$\therefore \frac{\text{No. of calls per cell site per hour} \times 120}{3,600} = 87$$

$$\therefore \text{No. of calls per cell site per hour} = 87 \times 30 = 2,610$$

Using Eq. (5.12) we can calculate the mean S/I ratio as S/I = $(3.5)^4/6$ = 25 = 14 dB.

The results for $N = 7$ and $N = 12$ are given in Table 5.2.

It is evident from the results in the table that, by increasing the reuse the factor from $N = 4$ to $N = 12$, the mean S/I ratio is increased from 14 dB to 23.3 dB (a 66.4% improvement). However, the call capacity of the cell site is reduced from 2,610 to 739 calls per hour (a 72% reduction).

Table 5.2 Cell Reuse Factor vs. Mean S/I Ratio and Call Capacity

N	q	Voice Channels per Cell	Calls per Cell per Hour	Mean S/I dB
4	3.5	99	2,610	14.0
7	4.6	56	1,376	18.7
12	6.0	33	739	23.3

5.5 CELLULAR SYSTEM DESIGN IN WORST-CASE SCENARIO WITH AN OMNIDIRECTIONAL ANTENNA

In the previous section we showed that the value of $q \approx 4.6$ is adequate for the normal interference case with a seven-cell reuse pattern. We reexamine the seven-cell reuse pattern and consider the worst case in which the mobile unit is located at the cell boundary (refer to Figure 5.6) where it receives the weakest signal from its own cell but is subjected to strong interference from all the interfering cells in the first tier. The distances from the six interfering cells are shown in Figure 5.6.

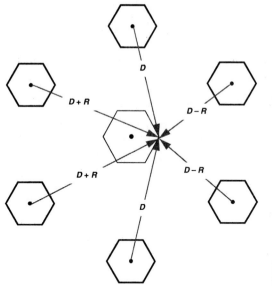

Fig. 5.6 Worst-Case Scenario for Cochannel Interference

The S/I ratio can be expressed as:

$$\frac{S}{I} = \frac{R^{-\gamma}}{2\,(D-R)^{-\gamma} + 2D^{-\gamma} + 2\,(D+R)^{-\gamma}} \tag{5.16}$$

Using $\gamma = 4$ and $D/R = q$, we rewrite Eq. (5.16) as:

$$\frac{S}{I} = \frac{1}{2\,(q-1)^{-4} + 2q^{-4} + 2\,(q+1)^{-4}} \tag{5.17}$$

where:
$q = 4.6$ for a normal seven-cell reuse pattern.

Substituting $q = 4.6$ in Eq. (5.17), we get S/I = 54.3 or 17.3 dB. For a conservative estimate, if we use the shortest distance (D-R) then

$$\frac{S}{I} = \frac{1}{6\,(q-1)^{-4}} = \frac{1}{6\,(3.6)^{-4}} = 28 \text{ or } 14.47 \text{ dB} \tag{5.18}$$

In the real situation, because of imperfect cell-site locations and the rolling nature of the terrain configuration, the S/I ratio is often less than 17.3 dB. It could be 14 dB or lower. Such a condition can easily occur in a heavy traffic. Therefore, the cellular system should be designed around the S/I ratio of the worst case. If we consider the worst case for a seven-cell reuse pattern, we conclude that a cochannel interference reduction factor of $q = 4.6$ is not enough in an omnidirectional cell system. In an omnidirectional cell system, $N = 9$ ($q = 5.2$) or $N = 12$ ($q = 6.0$) cell reuse pattern would be a better choice. These cell reuse patterns would provide the S/I ratio of 19.78 dB and 22.54 dB, respectively.

5.6 COCHANNEL INTERFERENCE REDUCTION WITH THE USE OF DIRECTIONAL ANTENNAS

In case of increased call traffic, the frequency spectrum should be used efficiently. We should avoid increasing the number of cells N in a frequency reuse pattern. As N increases, the number of frequency channels assigned to a cell becomes smaller, thereby decreasing the efficiency of the frequency reuse pattern.

Instead of increasing N in a set of cells, we use a directional antenna arrangement to reduce the cochannel interference. In this scheme, each cell is divided into three or six sectors and uses three or six directional antennas at the base station (refer to Figures 5.7 and 5.8). Each sector is assigned a set of channels (frequencies). The cochannel interference decreases as discussed below.

5.7 DIRECTIONAL ANTENNAS IN SEVEN-CELL REUSE PATTERN

5.7.1 Three-Sector Case

The three-sector case for the seven-cell reuse pattern is shown in Figure 5.7. We consider the worst case where the mobile unit is at position M (see Figure 5.9). In this situation, the mobile receives the weakest signal from its own cell and fairly strong interference from two interfering cells 1 and 2. Because of

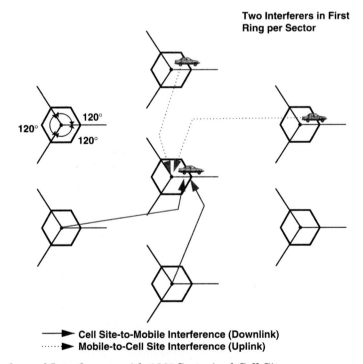

Fig. 5.7 Cochannel Interference with 120° Sectorized Cell Sites

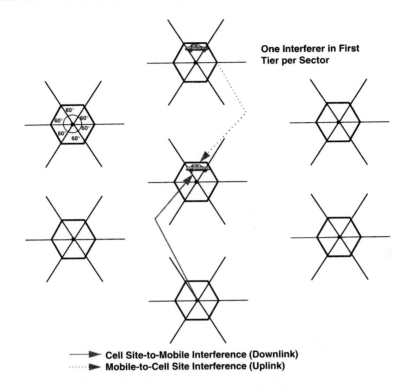

Fig. 5.8 Cochannel Interference with 60° Sectorized Cell Sites

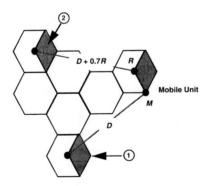

Fig. 5.9 Worst-Case Scenario in a 120° Sectorized Cell Site

the use of directional antennas, the number of interfering cells is reduced from six to two. At the point M, the distances between the mobile unit and the two interfering antennas are D and $(D + 0.7\,R)$, respectively. The S/I ratio in the worst case is given as:

$$\frac{S}{I} = \frac{R^{-4}}{D^{-4} + (D + 0.7R)^{-4}} \tag{5.19}$$

$$\frac{S}{I} = \frac{1}{q^{-4} + (q + 0.7)^{-4}} \tag{5.19a}$$

Using $q = 4.6$ in Eq. (5.19a), we get S/I = 285 or 24.5 dB. The S/I for a mobile unit served by a cell site with a 120° directional antenna exceeds 18 dB in the worst case. It is evident from Eq. (5.19a) that the use of a directional antenna is helpful in reducing the cochannel interference. In real situations, under heavy traffic, the S/I could be 6 dB weaker than in Eq. (5.19a) due to irregular terrain configurations and imperfect site locations. The resulting 18.5 dB S/I is still adequate.

5.7.2 Six-Sector Case

In this case the cell is divided into six sectors by using six 60° beamwidth directional antennas as shown in Figure 5.8. In this case, only one interference can occur. The worst case S/I ratio will be (see Figure 5.10):

$$\frac{S}{I} = \frac{R^{-4}}{(D + 0.7R)^{-4}} = (q + 0.7)^4 \tag{5.20}$$

For $q = 4.6$, Eq. (5.20) gives S/I = 789 or 29 dB. This indicates a further reduction of cochannel interference. Using the argument as was used for the three-sector case and subtracting 6 dB from 29 dB, the remaining 23 dB is still more than adequate. For heavy traffic, the 60° sector configuration can be used to reduce the cochannel interference. However, with the six-sector configuration the trunking efficiency is decreased.

E X A M P L E   5 – 2

Problem Statement

Fig. 5.11 shows the traffic in Erlangs for a seven-cell cellular system located in a busy metropolitan area. The total available channels in the system are 395. Assuming each subscriber in the system generates 0.03 Erlangs of traffic with an average call holding time of 120 seconds and the system covers an area of 1,200 square miles with cells designed for Grade-of-Service (GOS) of 2%, compute:

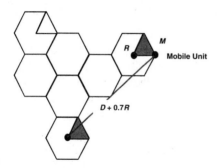

Fig. 5.10 Worst-Case Scenario in a 60° Sectorized Cell Site

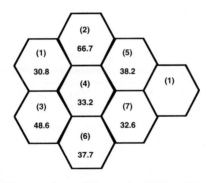

Fig. 5.11 Traffic in Erlangs for a Seven-Cell Cellular System

(a) The number of channels required in each cell
(b) The number of subscribers served by the system
(c) The average number of subscribers per channel
(d) The number of calls supported by the system
(e) The subscriber density per square mile
(f) The call density per square mile
(g) The cell radius in miles
(h) The channel reuse factor

Solution

$$\frac{\text{No. of calls per hour per subscriber} \times 120}{3,600} = 0.03$$

$$\text{No. of calls per hour per subscriber} = 0.9$$

Refer to Table 5.3 for the following calculations:

(a) Average number of subscribers/channel = 9,597/358 = 26.8
(b) Erlangs per mile^2 = 287.9/1,200 = 0.24
(c) Subscriber density = 9,597/1,200 = 8.0 subscribers per mile^2
(d) Call density = 8,637/1,200 = 7.2 calls per mile^2
(e) Area of each cell = 1,200/7 = 171.4 miles^2
(f) $A_{\text{hex}} = 2.6\,R^2 = 171.4$
(g) Radius of the cell, $R = 8.12$ miles

Since there are a total of 395 voice channels, the channel reuse factor is 358/395 = 0.906.

E X A M P L E   5 – 3

Problem Statement
The offered traffic density E in Erlangs per square mile for a cellular system can be expressed as

$$E = CW^{0.68}$$

where:
C = estimated market penetration (in percent) for the service and
W = number of working channels per square mile.

Table 5.3 Required Number of Channels, Subscribers, and Calls per Cell

Cell Number	Traffic (Erlangs)	No. of Subscribers per Cell	No. of Calls per Cell	No. of Channels Required	Channel Utilization
1	30.8	1,026.7	924	40	0.77
2	66.7	2,223.3	2,001	78	0.86
3	48.6	1,620.0	1,458	59	0.82
4	33.2	1,106.7	996	43	0.77
5	38.2	1,273.3	1,146	48	0.80
6	37.8	1,260.0	1,134	48	0.79
7	32.6	1,086.7	978	42	0.78
Total	287.9	9,597.0	8,637	358	

In a particular cellular system in a suburban environment, the average size of a cell is 200 square miles. The system has 30 cells designed to provide a 2% GOS. The 395 voice channels in the system are divided into 30 frequency groups, with a frequency reuse factor of 1.52. Calculate the estimated market penetration for the service, the total number of subscribers the system can serve, and the number of subscribers per square mile. Assume the average traffic generated by each subscriber in the system is 0.025 Erlangs.

Solution

(a) The number of channels per cell = $(395 \times 1.52)/30 = 20$

(b) Erlangs per cell with 2% blocking = 13.2

(c) Total traffic supported by the system = $13.2 \times 30 = 396$ Erlangs

(d) $E = 13.2/200 = 0.066$ Erlangs per mile^2

(e) $W = 20/200 = 0.10$ channels per mile^2

(f) $C(0.1)^{0.68} = 0.066$; $C = 31.6\%$

(g) The number of subscribers served by the system = $396/0.025 = 15,840$

(h) Number of subscribers per mile^2 = $15,840/(30 \times 200) = 2.64$

E X A M P L E   5 – 4

Problem Statement

Compare the spectral efficiency of the digital system with respect to the present analog system using the following data:

(a) The total number of channels in the U.S. cellular system = 416

(b) The number of control channels = 21

(c) The number of voice channels = 395

(d) The channel band width = 30 kHz

(e) The reuse factor $N = 7$

(f) The total available band width in each direction = 12.5 MHz

(g) The total coverage area = 10,000 km^2

(h) The required S/I ratio for analog system = 18 dB (63.1)

(i) The required S/I ratio for digital system = 14 dB (25.1)

(j) The call blocking = 2%

Solution
Analog System

(a) The number of voice channels per cell site 395/7 = 56

(b) The offered traffic load (using Erlang-B traffic tables) = 45.9 Erlangs per cell site

(c) The carried traffic load = $(1.0 - 0.02) \times 45.9 = 44.98$ Erlangs per cell site

$$\text{Spectral efficiency} = \frac{44.98 \times \left(\dfrac{10{,}000}{2.6R^2}\right)}{12.5 \times 10{,}000} = \frac{1.384}{R^2} \text{ Erlangs/\{km}^2\}/\text{MHz}$$

$$q = \left(6\,\frac{S}{I}\right)^{\frac{1}{4}}$$

$$q^2 = \left(6\,\frac{S}{I}\right)^{\frac{1}{2}}$$

$$\therefore \frac{q_{digital}^2}{q_{analog}^2} = \sqrt{\frac{25.1}{63.1}} = 0.6307$$

Digital System

(a) Number of channels per 30 kHz = 3

(b) Number of voice channels per cell site = $56 \times 3 = 168$

(c) Offered traffic load = 154.5 Erlangs per cell site

(d) Carried traffic load = 151.4 Erlangs per cell site

$$\text{Spectral efficiency} = \frac{151.4}{12.5 \times 2.6R^2 \times 0.6307} = \frac{7.386}{R^2} \text{ Erlangs/\{km}^2\}/\text{MHz}$$

Thus, the relative spectral efficiency of a digital system with respect to the present analog system is 7.386/1.384 = 5.34.

E X A M P L E   5 – 5

Problem Statement
Consider a cellular system with 395 total allocated voice channels of 30 kHz each. The total available bandwidth in each direction is 12.5 MHz. The traffic is uniform with average call holding time of 120 seconds, and call blocking during the system busy hour is 2%. Calculate:

(a) The calls per cell site per hour

(b) The mean S/I ratio

(c) The spectral efficiency in Erlangs/km²/MHz

For a cell reuse factor N equal to 4, 7, and 12, respectively, and for omnidirectional, 120° and 60° systems, calculate the call capacity.

Plot spectral efficiency versus cell radius for $N = 7$ and comment on the results. Assume that there are 10 mobiles/km² with each mobile generating traffic of 0.02 Erlangs. The slope of path loss is $\gamma = 40$ dB per decade.

Solution

We consider only the first-tier interferers and neglect the effects of cochannel interference from the second and other higher tiers. The mean S/I ratio can then be given as:

$$\text{Mean } \frac{S}{I} = \gamma \log (\sqrt{3N}) - 10 \log m$$

where:
γ = slope of path loss (dB/decade),
m = number of interferers in the first tier,
m = 6 for omnidirectional system,
m = 2 for 120° sectorized system, and
m = 1 for 60° sectorized system.

The traffic carried per cell site $= V \times t \times A_c = V \times t \times 2.6R^2$.

where:
V = number of mobiles per km²,
t = traffic in Erlangs per mobile, and
A_c = area of hexagonal cell (i.e., $A_c = 2.6\ R^2$).

The traffic carried per cell site $= 10 \times 0.02 \times 2.6R^2 = 0.52R^2$.

$$\text{The spectral efficiency} = \frac{\text{Traffic carried per cell} \times N_c}{B_w \times A}$$

where:
N_c = number of cells in the system (i.e., $A/2.6R^2$) and
A = area of the system.

$$\therefore \text{The spectral efficiency} = \frac{\text{Traffic carried per cell}}{2.6R^2 \times B_w}$$

We will demonstrate the procedure for calculating the results in one row in Tables 5.4 and 5.5; the remaining calculations can be made without any difficulty.

120° Sectorized Cell Site

(a) $N = 7$

(b) Number of voice channels per sector $= 395/(7 \times 3) \approx 19$

(c) Offered traffic load per sector from Erlang-B tables $= 12.3$ Erlangs

(d) Offered traffic load per cell site $= 3 \times 12.3 = 36.9$ Erlangs

(e) Carried traffic load per cell site $= (1 - 0.02) \times 36.9 = 36.2$ Erlangs

$$\therefore \frac{\text{No. of calls per cell site per hour} \times 120}{3,600} = 36.2$$

Thus, the number of calls per cell site per hour = 1,086

The spectral efficiency for a cell radius = 2 km will be:

$$\text{Spectral efficiency} = \frac{36.2}{2.6 \times 12.5 \times 2^2} = 0.278$$

$$\therefore R = \sqrt{\frac{36.2}{0.52}} = 8.3 \text{ km, and}$$

$$\text{mean } \frac{S}{I} = 40 \log \sqrt{21} - 10 \log 2 = 26.44 - 30.1 = 23.43 \text{ dB}$$

From the results in Tables 5.4 and 5.5, and Figure 5.12, we can draw the following conclusions:

1. Sectorization reduces cochannel interference and improves the mean S/I ratio for a given cell reuse factor. However, it reduces trunking efficiency since the channel resource is distributed more thinly among the various sectors. As a result, spectrum efficiency of a sectorized system is reduced if the cluster size is kept constant.

2. Since a sectorized cellular system has fewer cochannel interferers, it is possible to reduce the cluster size, hence increasing the spectrum efficiency of the overall system.

Table 5.4 Omni vs. Sectorized Cellular System Performance

System	N	Channels per Sector	Offered Load (E) per Cell	Carried Load (E) per Cell	Calls/Cell per Hour	Cell Radius (km)	Mean S/I dB
	4	99	87.0	85.3	2,559	12.8	13.8
Omni	7	56	45.9	45.0	1,350	9.3	18.7
	12	33	24.6	24.1	723	6.8	23.3
	4	33	73.8	72.3	2,169	11.8	18.6
120° Sector	7	19	36.9	36.2	1,086	8.3	23.4
	12	11	17.5	17.2	516	5.8	28.1
	4	17	64.2	62.9	1,887	11.0	21.6
60° Sector	7	9	26.0	25.5	765	7.0	26.4
	12	6	13.7	13.4	402	5.1	31.1

Table 5.5 Spectral Efficiency in Erlangs/km²/MHz vs. Cell Radius (km)

System	N	q = D/R	Cell Radius (km)				
			2	4	6	8	10
Omni	4	3.5	0.656	0.164	0.073	0.041	0.026
	7	4.6	0.346	0.087	0.038	0.022	0.014
	12	6.0	0.185	0.046	0.021	0.012	0.007
120° Sector	4	3.5	0.556	0.139	0.062	0.035	0.022
	7	4.6	0.278	0.070	0.031	0.017	0.011
	12	6.0	0.132	0.033	0.015	0.008	0.005
60° Sector	4	3.5	0.484	0.121	0.054	0.030	0.019
	7	4.6	0.196	0.049	0.022	0.012	0.008
	12	6.0	0.103	0.026	0.012	0.006	0.004

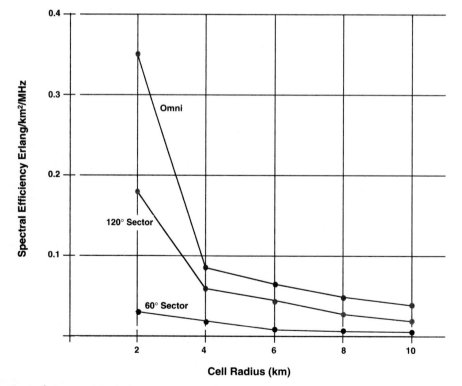

Fig. 5.12 Spectral Efficiency vs. Cell Radius

3. An omnidirectional cellular system requires a cluster size of 7, while a 120° sectorized system requires a cluster size of 4 and a 60° sectorized system requires a cluster size of 3 for desired mean S/I ratio of approximately 18 dB.

5.8 CELL SPLITTING

As the traffic within a particular cell increases, the cell is split into smaller cells. This is done in such a way that cell areas, or the individual component coverage areas of the cellular system, are further divided to yield yet more cell areas (refer to Figure 5.13). The splitting of cell areas by adding new cells provides for an increasing amount of channel reuse and, hence, increasing subscriber serving capacity.

Decreasing cell radii imply that cell boundaries will be crossed more often. This will result in more handoffs per call and a higher processing load per subscriber. Simple calculations show that a reduction in a cell radius by a factor of four will produce about a tenfold increase in the handoff rate per subscriber. Since the call processing load tends to increase geometrically with the increase in the number of subscribers, with cell splitting the handoff rate will increase exponentially. Therefore, it is essential to perform a cost-benefit study to compare the overall cost of cell splitting versus other available alternatives to handle increased traffic load.

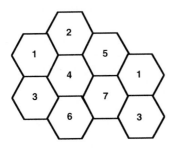

Growing by Splitting Cell 4 into Cells of Small Size

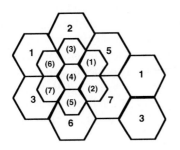

Fig. 5.13 Cell Splitting

5.9 REGISTRATION

Mobile station (MS) registration is a process where MS characteristics such as location or status are provided to the network. Registration may be initiated by the MS or the network, or may be implied during an MS access.

Registration is the means by which an MS informs the immediate service provider of its presence and desire to receive service. The Radio System (RS) informs the Mobile Switching Center (MSC) and other network elements of the request. The MSC and the other network elements perform necessary registration functions. This may result in local service being established along with location records being updated in the Home Location Register (HLR) of the MS.

Registration also takes place as an MS moves across registration areas supported by one or more service providers. A registration area reflects where an MS is geographically registered and thus the location to be searched when locating the MS for call delivery.

Upon receiving the registration request from the MS, the RS constructs the Registration Update Request message and sends it to the network. The Registration Update Request message contains the MS's identification and location information and may contain authentication parameters. The network may respond with a request for authentication (optional) and finally with a Registration Update Response message.

The network sends a Registration Update Response message to the RS when a registration procedure has been successfully completed at the network. This message indicates whether the MS's registration has been accepted or rejected. The message may contain additional parameters to be sent to the MS. Upon receipt of this message, the RS sends the appropriate response to the MS. The three possible results of the registration request are: successful registration, unsuccessful registration, and cancellation of registration. See chapters 8, 9, and 10 for more details.

5.10 TERMINAL AUTHENTICATION

Terminal authentication is a process to verify the authenticity of a MS, permitting denial of service to an invalid MS and thereby deterring fraud. The authentication process involves a common algorithm in the network and MS and a parameter unique to each MS known to the MS and the network. This parameter is called the "private key" and is never broadcast across the air interface. Instead, a publicly available random number (RAND) is sent to the MS. The RAND and private key are used as input to the authentication algorithm at both the MS and the network. The result computed by the MS is sent to the network where it is compared with the network-generated value. If the results match, the authentication process is considered successful; otherwise the authentication is a failure. See chapters 8, 9, and 10 for more details of the authentication process.

5.11 HANDOFF

Handoff is the process used to allow a call in progress to continue as the mobile terminal moves between cells. Handoffs may be based on received signal strength or S/I ratio (measured either at terminal, the RS, or both) or may be based on network resource management needs (for example, a forced handoff to free resources to allow an emergency call to be placed). Handoffs may be performed locally within the RS or may involve several network elements. The handoff process may also involve reregistration and authentication of the terminal. Handoffs can be classified as soft or hard, based on the following definitions.

Soft handoff occurs when the mobile terminal communication is passed to the target radio port without interrupting communications with the current serving radio port. In a soft handoff, the mobile terminal communicates with two radio ports simultaneously, with the signals from the radio ports to the terminal treated as multipath signals that are coherently combined at the mobile.

Hard handoff occurs when the communication to the mobile terminal is passed between disjointed radio systems, different frequency assignments, or different air interface characteristics or technologies. A hard handoff is a "break-before-make" process at the air interface.

Whether a soft or hard handoff can occur is air-interface-technology dependent.

The handoff process maintains the privacy of all calls in progress by a transfer of the session key from the source radio port to the target radio port. The encryption key may remain unchanged at handoff.

The handoff process may be network directed (including network initiated), terminal assisted, or terminal directed. The handoff process may be initiated by the network upon determination that the quality of the radio connection has deteriorated or for management purposes. The quality of the radio connection may be assessed either from network measurements or from measurements provided by the terminal. In addition, the terminal, making its own measurements, may specifically request that a handoff take place. The handoff execution process requires coordination of functionality contained in various network elements. The management occurs at three relative layers.

1. Execution management of handoff occurs when the resources needed for handoff are all within the scope of specific element's ability to control.

2. Higher-level requests occur when the resources needed for the handoff are beyond the scope of the current element's ability, and control should be passed on to an element placed higher in the network to execute the task. The handoff request information is passed on to the higher-level entity, which either performs the execution of the handoff or determines that a still higher-level functionality is required.

3. Peer-level requests occur when an element determines that management of resources is shared within the same level of the network. No special functionality is required of elements placed at a higher level in the network,

but information is transferred between peer elements. The functional capability may be used in conjunction with or in place of higher-level request capability.

The handoff process consists of the following steps:

1. **Initiation:** Either the mobile terminal or network identifies the need for a handoff and alerts the necessary network elements.
2. **Resource reservation:** The appropriate network elements reserve the resources necessary to support the handoff.
3. **Execution:** The actual handoff connection of the network resources takes place.
4. **Completion:** Any unneeded network resources are freed, and access signals are exchanged following a successful handoff.

5.12 SUMMARY

A relationship between the reuse ratio (q) and cell cluster size (N) was developed for the hexagonal cell geometry. Cochannel interference ratios for the omnidirectional and sectorized cell were derived. A numerical example was presented to show that, for a given cluster size, sectorization provides a higher S/I ratio, but reduces the spectral efficiency. However, it is possible to achieve a higher spectrum efficiency by reducing the cluster size in a sectorized system without lowering the S/I ratio below the minimum requirement. The chapter was concluded by discussing the cell splitting, registration, terminal authentication, and handoff procedures used in cellular communications.

5.13 PROBLEMS

1. Sometimes it is possible to delay cell splitting by allowing the blocking probability to increase for a short period of time. This will increase the carried load of the system at the expense of some customer dissatisfaction. If only one or two cells are experiencing high blocking, this may be a viable alternative to cell splitting. For Example 5-1, calculate the increase in carried traffic load when the blocking probability is allowed to increase to 5%, 10%, and 20%. If the wireless system is growing at a rate of 20% per year, how much extra time is gained before cell splitting, for each blocking rate, compared to the normal 2% blocking rate? Is this a viable alternative? Explain your answer.

2. In section 5.5, the S/I was calculated by neglecting the interference from cells other than the nearest ones on the same frequency. Calculate the amount of interference from the next ring of cells. Is it reasonable to neglect this interference?

3. In Example 5-4, the offered and carried load of the analog and digital systems was calculated for 168 digital voice channels per cell site. As the number of channels per cells site is lowered, the trunking efficiency (from the Erlang-B tables) is lower and

the system becomes more expensive per user. If new spectrum was opened up that allowed less than 168 digital channels per cell site but allowed a cost per channel of 25% less than the current system, how many channels would be needed to have the same cost per user as the competing systems with larger channel capacity?

4. A high-speed highway (90 km/h) passes through a central city. If the average call duration is 3 minutes, how many handoffs will a call have as the cell size in the center of the city changes from 24-km (omnidirectional cells) to 1.5-km six-sector cells.

5.14 REFERENCES

1. AT&T Technical Education Center, "Cellular System Design and Performance Engineering I," CC1400, version 1.12, 1993.

2. Chen, G. K., "Effects of Sectorization on the Spectrum Efficiency of Cellular Radio Systems," *IEEE Transactions on Vehicular Technology* 41, no. 3 (August 1992): 217–25.

3. Dersch, U., and W. Braun, "A Physical Mobile Radio Channel Model," Proceedings of IEEE Vehicular Technology Conference, May 1991, 289–94.

4. French, R. C., "The Effects of Fading and Shadowing on Channel Reuse in Mobile Radio," *IEEE Transactions on Vehicular Technology* 28 (August 1979).

5. Lee, W. C. Y., "Elements of Cellular Mobile Radio System," *IEEE Transactions on Vehicular Technology* 35 (May 1986): 48–56.

6. Lee, W. C. Y., *Mobile Cellular Telecommunications System*, New York: McGraw-Hill, 1989.

7. Lee, W. C. Y., "Spectrum Efficiency and Digital Cellular," presented at 38th IEEE Vehicular Technology Conference, Philadelphia, June 1988.

8. Lee, W. C. Y., "Spectrum Efficiency in Cellular," *IEEE Transactions on Vehicular Technology* 38 (May 1989): 69–75.

9. MacDonald, V. H., "The Cellular Concept," *Bell System Technical Journal* 58, no. 1 (January 1979): 15–41.

10. Mehrotra, A., *Cellular Radio Analog & Digital System*, Boston: Artech House, 1994.

11. Whitehead, J. F., "Cellular System Design: An Emerging Engineering Discipline," *IEEE Communications Magazine* 24, no. 2 (February 1986): 8–15.

12. Young, W. R., "Advanced Mobile Phone Service: Introduction, Background, and Objectives," *Bell System Technical Journal* 58, no. 1 (January 1979): 1–14.

Digital Modulation Techniques

6.1 INTRODUCTION

System designers use digital modulation to transmit baseband digital information over a band pass channel. The most common form of digital modulation is binary signaling where the transmitted information is coded into the values 0 or 1 and sent at a rate of 1 bit per T seconds. Two signals $s_0(t)$ and $s_1(t)$ are required to represent the binary digits 0 and 1, respectively. Alternatively, the binary digits can be divided into blocks where each block consists of n bits. Since there are $M = 2^n$ distinct blocks, M different signals are needed to represent the n-bit blocks unambiguously. Each n-bit block is called a symbol with duration $T_s = nT$, where T is the bit interval. This type of transmission is called M-ary signaling.

In this chapter we discuss the transmission of digital information in a band-limited Additive White Gaussian Noise (AWGN) channel with an emphasis on the average probability of bit error. First we study baseband signaling, its bandwidth, and probability of error. We then introduce the concept of modulating the baseband signal onto a radio frequency (RF) carrier and show that the baseband signal can be modulated using the amplitude, frequency, or phase of the carrier. With these concepts in mind, we develop the probability of error for common modulation methods and show their applicability to cellular and PCS systems.

This chapter presents only a brief description of digital modulation techniques. For more details, the reader should refer to references [8] and [9].

6.2 BASEBAND SIGNALING

The baseband outputs of the data transmitters are a series of binary data that cannot be sent directly over a radio link. The communications designer must choose radio signals that represent the binary data and permit the data receiver to decode the data with minimum errors. For the simplest binary signaling system, we choose two signals denoted by $s_0(t)$ and $s_1(t)$ to represent the binary values of 0 and 1, respectively. Since no channel is perfect, the receiver will also have AWGN, $n(t)$. The data receiver (see Figure 6.1) will then process the signal and noise through a filter, $h(t)$ and, at the end of the signaling interval, T, make a determination of whether the transmitter sent 0 or 1.

The energies of $s_0(t)$ and $s_1(t)$ in a T interval are assumed to be finite and denoted as E_0 and E_1, respectively. For simplicity we assume that the noise has a probability density function of amplitude that is Gaussian and that the noise spectral density is flat with frequency (white noise) with the double-sided Power Spectral Density (PSD) of $N_0/2$.

When $s_0(t)$ is present at the filter as an input, its output at $t = T$ is:

$$V = S_0 + N, \text{ with } s_0(t) \text{ present} \tag{6.1}$$

where:
S_0 = the output signal component at $t = T$ for the input $s_0(t)$ and
N = the output noise component.

Similarly, when $s_1(t)$ is present at the filter as an input, its output at $t = T$ is:

$$V = S_1 + N, \text{ with } s_1(t) \text{ present} \tag{6.2}$$

Since $n(t)$ is Gaussian with zero mean (implied by its constant PSD), so also is N. The variance of the noise, σ^2, can be determined as:

$$\sigma^2 = \int_{-\infty}^{\infty} |H(f)|^2 \frac{N_o}{2} df$$

$$\sigma^2 = N_o \int_0^{\infty} |H(f)|^2 df = N_o B_N \tag{6.3}$$

where:
$H(f)$ is the transfer function of the filter and

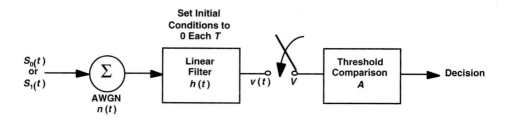

Fig. 6.1 Receiver Structure to Detect Binary Signals in White Gaussian Noise

$B_N = \int_0^\infty |H(f)|^2 df$ is the noise-equivalent bandwidth or simply the noise
bandwidth of the receiver filter function $H(f)$.

Given that $s_0(t)$ is present at the receiver input, the Probability Density
Function (PDF) of V is

$$p(v|s_0) = \frac{1}{\sqrt{2\pi}\sigma}e^{-\left(\frac{(v-S_0)^2}{2\sigma^2}\right)} \tag{6.4}$$

Similarly, the conditional probability function of V, when $s_1(t)$ is present at the
receiver input, is:

$$p(v|s_1) = \frac{1}{\sqrt{2\pi}\sigma}e^{-\left(\frac{(v-S_1)^2}{2\sigma^2}\right)} \tag{6.5}$$

From Figure 6.2, the probability of error, given that $s_0(t)$ is present, is:

$$P(e|s_0) = \int_A^\infty p(v|s_0)\,dv \tag{6.6a}$$

and if $s_1(t)$ is present, it is

$$P(e|s_1) = \int_{-\infty}^A p(v|s_1)\,dv \tag{6.6b}$$

If the a priori probability that $s_0(t)$ was sent is p, and the a priori probability
that $s_1(t)$ was sent is $q = 1 - p$, then the average probability of error is

$$P_e = pP(e|s_0) + qP(e|s_1) \tag{6.7a}$$

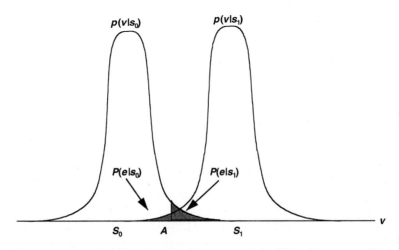

Fig. 6.2 Conditional Probability Density Function of the Filter Output at Time $t = T$

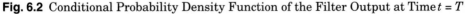

$$P_e = p \int_A^\infty \frac{1}{\sqrt{2\pi}\sigma} e^{-\left(\frac{(v-S_0)^2}{2\sigma^2}\right)} dv + q \int_{-\infty}^A \frac{1}{\sqrt{2\pi}\sigma} e^{-\left(\frac{(v-S_0)^2}{2\sigma^2}\right)} dv \qquad (6.7b)$$

If we simplify Eq. (6.7b), differentiate the result with respect to A, and then set the derivative equal to zero, we can determine the optimum choice for the threshold value, A, to minimize the probability of error, P_e.

$$A = A_{opt} = \frac{\sigma^2}{S_1 - S_0} \ln \frac{p}{q} + \frac{S_o + S_1}{2} \qquad (6.8)$$

In most systems, the values of 0 and 1 are equally likely; if they are not, then the designer usually redesigns the encoding method to ensure that they are equally likely. Thus, $p = q$ and

$$A = A_{opt} = \frac{S_0 + S_1}{2} \qquad (6.9)$$

For the optimum value of A, with $p = q$, the probability of error is:

$$P_e = \frac{1}{2} erfc \left[\frac{S_1 - S_0}{2\sqrt{2}\sigma} \right] = Q \left[\frac{S_1 - S_0}{2\sigma} \right] \qquad (6.10)$$

where $erfc(u)$ is the complementary error function, $= 1 - erf(u) = 2Q(\sqrt{2}u)$. The error function, $erf(u)$ is defined as

$$erf(u) = \frac{2}{\sqrt{\pi}} \int_0^u e^{-t^2} dt$$

and

$$Q(u) = \frac{1}{\sqrt{2\pi}} \int_u^\infty e^{-\frac{x^2}{2}} dx$$

or

$$Q(u) \approx \frac{e^{-\frac{u^2}{2}}}{\sqrt{2\pi}u}, \ u \gg 1 \quad \text{Gaussian Integral}$$

Next, we want to find the filter that provides the minimum probability of error, as expressed by Eq. (6.10). At time t_0, the sample value consists of a signal-related component $g_0(t_0)$ and a noise component $n_0(t_0)$. This filter is known as a matched filter and has the transfer function $H_0(f)$ optimized to provide the maximum SNR ratio at its output at time t_0. M. Schwartz [2] shows that this filter must be the conjugate match of the signal, $s(t)$. Since we have two signals, $s_0(t)$ and $s_1(t)$, we will need two filters in our receiver design.

If we transmit a signal, $s(t)$, then it has Fourier transform, $S(\omega)$, that is a complex function. The optimum filter, the matched filter, must have a frequency response, $H(\omega)$, where

$$H(\omega) = S^*(\omega) e^{-j\omega t_0} \tag{6.11}$$

where:
S^* is the complex conjugate of Fourier transform of the signal.

In general this filter is not realizable, since analysis will show that it must have output before there is input and we have not yet been able to design circuits that predict the future. However, we can design filters that approximate the ideal filter.

The SNR ratio is defined as

$$\xi^2 = \frac{g_o^2(t_o)}{\sigma^2} \tag{6.12}$$

It can also be shown that the maximum value for the SNR ratio ξ is twice the energy of the input signal (E_g) divided by the single-sided input noise spectral density, regardless of the input signal shape.

$$\xi_{max}^2 = \frac{2E_g}{N_0} \tag{6.13}$$

For a binary system, Eq. (6.13) becomes

$$\xi_{max}^2 = \frac{1}{N_0} \int_o^T [s_1(t) - s_0(t)]^2 dt \tag{6.14}$$

Since the signals are zero outside the range $(0, T)$, the probability of error corresponding to the optimum receiver filter becomes

$$P_e = \frac{1}{2} erfc(\sqrt{z}) = Q(\sqrt{2z}) \tag{6.15}$$

where:
$z = \frac{1}{4N_0} \int_o^T [s_1(t) - s_0(t)]^2 dt$.

If the transmitted pulses are allowed to take on any of "M" transmitted levels with equal probability, then the information rate per transmitted pulse is $\log_2 M$ bits. For a constant information rate, the bandwidth of the transmitted system can be reduced by the same factor. With M-ary transmission, we will show that the error rates are higher, but, if we have sufficient SNR, then the higher errors rates will not matter. Thus, we are using excess SNR to code the signal and reduce its bandwidth.

When we add additional levels to a baseband system, we are reducing the distance between detection levels in the receiver output. Thus, the error rate of a multilevel baseband system can be determined by calculating the appropriate reduction in the error distance. If the maximum amplitude is V, the error distance d_e between equally spaced levels at the detector is

$$d_e = \frac{V}{M-1} \tag{6.16}$$

where:
M = number of levels.

Setting the error distance V of a binary system to that defined in Eq. (6.16) provides the error probability of the multilevel system:

$$P_e = \frac{1}{log_2 M} \left[\frac{M-1}{M}\right] erfc \left[\frac{V}{(M-1)\sqrt{2}\sigma}\right] \tag{6.17}$$

where:
the factor $[(M-1)/M]$ reflects that interior signal levels are vulnerable to both positive and negative noise, and
the factor $1/log_2 M$ arises because the multilevel system is assumed to be coded so symbol errors produce single-bit errors ($log_2 M$ is the number of bits per symbol).

The probability of multiple bit errors is assumed to be small and can be neglected.

Eq. (6.17) relates error probability to the peak signal power V^2. To determine the P_e with respect to average power, the average power of an M-level system is determined by averaging the power associated with the various pulse amplitude levels.

$$[V^2]_{avg} = \frac{2}{M} \left[\left(\frac{V}{M-1}\right)^2 + \left(\frac{3V}{M-1}\right)^2 + \dots + V^2\right] \tag{6.18a}$$

$$[V^2]_{avg} = \frac{2V^2}{M(M-1)^2} \sum_{j=1}^{M/2} (2j-1)^2 \tag{6.18b}$$

where:
the levels $\frac{V}{M-1} [\pm1, \pm3, \pm5, \dots, \pm(M-1)]$ are assumed to be equally likely.

If T is the signaling interval for a two-level system, the signaling interval T_M for an M-level system providing the same data rate is determined as

$$T_M = T \, log_2 M \tag{6.19}$$

For a raised cosine filter, the noise bandwidth is $B_N = 1/(2T_M)$
From Eq. (6.3), we get:

$$\sigma^2 = \frac{N_o}{2T_M} \tag{6.20a}$$

$$\sigma = \frac{1}{\sqrt{2}} \left[\frac{N_o}{T_M} \right]^{1/2} \tag{6.20b}$$

Substituting Eq. (6.20b) into Eq. (6.17) we get

$$P_e = \left[\frac{1}{log_2 M} \right] \left[\frac{M-1}{M} \right] erfc \left[\frac{V}{(M-1)\left(\frac{N_o}{T_M}\right)^{1/2}} \right] \tag{6.21}$$

The energy per symbol is given as, $E_s = E_b \, log_2 M = V^2 T_M$, where E_b is the energy per bit.

$$\therefore V^2 = \frac{E_b \, log_2 M}{T_M} \tag{6.22}$$

Substituting for V from Eq. (6.22) into Eq. (6.21), we get

$$P_e = \left[\frac{1}{log_2 M} \right] \left[\frac{M-1}{M} \right] erfc \left[\left(\frac{E_b}{N_o} \right)^{1/2} \frac{(log_2 M)^{1/2}}{M-1} \right] \tag{6.23a}$$

$$SNR = \frac{\text{Signal Power}}{\text{Noise Power}} = \frac{E_b \, (log_2 M) \, (1/T_M)}{N_o \left(\frac{1}{2} T_M \right)}$$

$$\therefore SNR = 2 log_2 M \left(\frac{E_b}{N_o} \right) \tag{6.23b}$$

Another variation of baseband signaling is AntiPodal Baseband Signaling (APBS) where two signals of opposite polarities are sent. If $s_0(t) = -V$ and $s_1(t) = V$ for $0 \le t \le T$, then $s_1(t) - s_0(t) = 2V$.

We then calculate the value of P_e from Eq. (6.15) as

$$z = \frac{1}{4N_o} \int_0^T (2V)^2 dt = \frac{V^2 T}{N_o} = \frac{E_b}{N_o}$$

where E_b is the energy in either $s_0(t)$ or $s_1(t)$, i.e., the bit energy.

$$P_e = \frac{1}{2} erfc \left[\sqrt{\frac{E_b}{N_o}} \right] = Q \left[\sqrt{\frac{2E_b}{N_o}} \right] \tag{6.24}$$

APBS is used to modulate some signals. We will compare the SNR of APBS with other modulation methods.

6.3 MODULATION TECHNIQUES

When we want to send signals over any distance, baseband signaling is not sufficient. We must therefore modulate the signals on to an RF carrier. When we transmit the digital bit stream, we convert the bit stream into the analog signal, $A(t) \cos(\omega t + \theta)$. The characteristic of this signal has amplitude, frequency, and phase; thus, we can change any of the three characteristics to formulate the modulation method. The basic form of the three modulation methods used for transmitting digital signals are:

☞ Amplitude Shift Keying (ASK)

☞ Frequency Shift Keying (FSK)

☞ Phase Shift Keying (PSK)

When ω and θ remain unchanged, we have ASK. When $A(t)$ and θ remain unchanged, we have binary (or M-ary) FSK. When $A(t)$ and ω remain unchanged, we have binary (or M-ary) PSK. Hybrid systems exist where two characteristics are changed with each new symbol transmitted. The most common method is to fix ω and change $A(t)$ and θ. This method is known as Quadrature Amplitude Modulation (QAM). Each of the modulation methods results in a different transmitter and receiver design, different occupied bandwidth, and different error rates. In the remainder of this chapter we will examine most of the common methods used and calculate their error rates. Since all signals have a theoretical bandwidth that is infinite, all modulation methods must be band limited. The band limiting introduces detection errors, and the filter bandwidths must be chosen to optimize trade-offs between bandwidth and error rates. Since this trade-off is often done empirically after much experimental data has been gathered, we will not calculate the bandwidth of each of the modulation methods.

6.3.1 Amplitude Shift Keying

With ASK we modulate the baseband signal into changes in amplitude of the transmitter carrier. When we do our detection, we find that ASK offers no improvement in error rates or bandwidth compared to its corresponding baseband signal. For ASK we transmit one of two signals: $s_0(t) = 0$ for binary "0" and $s_1(t) = A \cos (\omega_0 t)$ for binary "1." The energy in a bit interval T is $E_b = A^2 T / 2$.

Pure ASK sends no signal for binary "0" and a signal for binary "1." Thus, for $0 \leq t \leq T$, $s_0(t) = 0$, and $s_1(t) = V \cos (2\pi f t)$. In Eq. (6.15), we defined the probability of error for a baseband system as $P_e = Q(\sqrt{2z})$. We now proceed to calculate z for ASK.

$$z = \frac{1}{4N_o} \int_0^T \left[V \cos (2\pi f t) \right]^2 dt = \frac{V^2 T}{8N_o}$$

$$\therefore z = \frac{(E_b)_{avg}}{2N_o}$$

$$P_e = Q\left[\sqrt{\frac{(E_b)_{avg}}{N_o}}\right] = \frac{1}{2}erfc\sqrt{\frac{(E_b)_{avg}}{2N_o}} \qquad (6.25)$$

We leave as a homework problem the calculation of the probability of error for an M-ary ASK system. In general, M-ary ASK systems are not used since other modulation methods have better error performance.

6.3.2 Frequency Shift Keying

In FSK, the frequency of the transmitted signal is changed with different baseband inputs. Thus, $s_0(t) = A \cos (\omega + \Delta\omega)t$ for binary "0," $s_1(t) = A \cos (\omega - \Delta\omega)t$ for binary "1." The energy in a bit interval T is $E_b = (A^2T)/2$. For an M-FSK, M different frequencies are required and every n ($M = 2^n$) bits of the binary bit stream are grouped as a signal that is transmitted as $A \cos (\omega + \Delta\omega_j)t, j = 1, M$.

If we choose values for $\Delta\omega$ so that $\Delta\omega$ is properly related to the symbol duration, then we have coherent FSK signaling. For this modulation method, during the interval from $0 \le t \le T$,

$$s_0(t) = V \cos 2\pi ft \text{ and}$$
$$s_1(t) = V \cos 2\pi (f + \Delta f)t$$

where:
$\Delta f = k/2T$ and
k is an integer.

The signals, $s_0(t)$ and $s_1(t)$, are said to be coherently orthogonal since

$$\int_0^T s_0(t) s_1(t) \, dt = 0$$

$$\int_0^T [s_1(t) - s_0(t)]^2 dt = V^2T$$

$$\therefore z = \frac{V^2T}{4N_o} = \frac{E_b}{2N_o}$$

$$P_e = Q\left[\sqrt{\frac{E_b}{N_o}}\right] = \frac{1}{2}erfc\sqrt{\frac{E_b}{2N_o}} \qquad (6.26)$$

FSK has the same performance as ASK in terms of E_b/N_o to give the same P_e. The bandwidth efficiency of coherent and noncoherent M-ary FSK as shown in reference [8] is:

$$\text{Coherent } M\text{-ary FSK: } \frac{R}{B_w} = \frac{2log_2M}{M+3}$$

$$\text{Noncoherent } M\text{-ary FSK: } \frac{R}{B_w} = \frac{2log_2M}{2M}$$

where:
R = data rate and
B_w = bandwidth.

6.3.3 Phase Shift Keying

The amplitude or the frequency of a signal can be changed in only one dimension since they are one-dimensional objects. The phase of the carrier represents a two-dimensional object; therefore we have more opportunities to make bandwidth and error rate trade-offs with PSK. We will therefore see a multiplicity of modulation methods using PSK where we change the phase of the transmitted carrier.

For binary PSK, $s_0(t) = A\cos(\omega t)$ for binary "0" and $s_1(t) = A\cos(\omega t + \pi)$ for binary "1." The energy in a bit interval T is $E_b = (A^2 T)/2$. For M-ary PSK, M different phases are required, and every n ($M = 2^n$) bits of the binary bit stream are coded as a signal that is transmitted as $A\sin(\omega t + \theta_j), j = 1, M$.

The error distance of a PSK system with M phases is $V\sin(\pi/M)$ where V is the signal amplitude at the detector. A detection error occurs if noise of the proper polarity is present at the output of either of the two phase detectors. A detection error, however, is assumed to produce only a single-bit error. The general expression for the error rate for PSK modulation is obtained by modifying Eq. (6.17) as:

$$P_e = \frac{1}{log_2 M} erfc \left[\frac{V\sin\dfrac{\pi}{M}}{\sqrt{2}\sigma} \right] \tag{6.27}$$

The signal amplitude V can be expressed as:

$$V = \left[E_b (log_2 M) \frac{1}{T} \right]^{1/2} \tag{6.28}$$

and the RMS noise σ as:

$$\sigma = \left[N_o \left(\frac{1}{2T} \right) \right]^{1/2} \tag{6.29}$$

for noise in a Nyquist bandwidth.

By substituting Eqs. (6.28) and (6.29) into Eq. (6.27) we get

$$P_e = \left(\frac{1}{log_2 M} \right) erfc \left[\sin\left(\frac{\pi}{M} \right) (log_2 M)^{1/2} \left(\frac{E_b}{N_o} \right)^{1/2} \right] \tag{6.30}$$

The SNR ratio is given as:

$$SNR = log_2 M \left(\frac{E_b}{N_o} \right), \quad \text{For } M > 2 \tag{6.31}$$

Also, as shown in reference [8], the bandwidth efficiency of the M-ary PSK is given as:

$$\frac{R}{B_w} = \frac{log_2 M}{2}$$

where:
R = data rate and
B_w = bandwidth.

We will now examine several variations of PSK.

6.3.3.1 Binary Phase Shift Keying

If we use two signals that have phases 0 and 180 degrees, then we have Binary Phase Shift Keying (BPSK). Thus, for $0 \le t \le T$

$$s_0(t) = -V \cos 2\pi f t$$

and

$$s_1(t) = V \cos 2\pi f t$$

We then calculate the error rate by calculating z for the Q function.

$$z = \frac{1}{4N_o} \int_0^T [2V \cos(2\pi f t)]^2 dt = \frac{V^2 T}{2N_o} = \frac{E_b}{N_o}$$

$$P_e = \frac{1}{2} erfc \sqrt{\frac{E_b}{N_o}} = Q\left[\sqrt{\frac{2E_b}{N_o}}\right] \tag{6.32}$$

BPSK has the same error rate as baseband signaling. This can be explained by examining the signals used in BPSK and realizing that BPSK is ASK with the values of $-V$ and V.

6.3.3.2 Quadrature Phase Shift Keying

If we define four signals, each with a phase shift differing by 90°, then we have Quadrature Phase Shift Keying (QPSK). We have previously calculated the error rates for a general PSK signal with M signal points. For QPSK, $M = 4$, so substituting 4 in Eq. (6.30) we get

$$P_e = \left(\frac{1}{log_2 4}\right) erfc\left[\sin\left(\frac{\pi}{4}\right)(log_2 4)^{1/2}\left(\frac{E_b}{N_o}\right)^{1/2}\right]$$

$$P_e = \frac{1}{2} erfc \sqrt{\frac{E_b}{N_o}} = Q\left[\sqrt{\frac{2E_b}{N_o}}\right] \tag{6.33}$$

The input binary bit stream $\{b_k\}$, $b_k = \pm 1$; $k = 0, 1, 2, ...$, arrives at the modulator input at a rate $1/T$ bits/sec and is separated into two data streams, $a_I(t)$ and $a_Q(t)$, containing even and odd bits, respectively. The modulated QPSK signal $s(t)$ is given as:

$$s(t) = \frac{1}{\sqrt{2}} a_I(t) \cos\left(2\pi f t + \frac{\pi}{4}\right) + \frac{1}{\sqrt{2}} a_Q(t) \sin\left(2\pi f t + \frac{\pi}{4}\right) \tag{6.34a}$$

$$s(t) = A \cos\left[2\pi f t + \frac{\pi}{4} + \theta(t)\right] \tag{6.34b}$$

where:

$$A = \sqrt{(1/2)\,(a_I^2 + a_Q^2)} = 1 \text{ and}$$

$$\theta(t) = -\mathrm{atan}\frac{a_Q(t)}{a_I(t)}.$$

The values of $\theta(t) = 0, -(\pi/2), \pi/2, \pi$ represent the four values of $a_I(t)$ and $a_Q(t)$. On the I/Q plane, QPSK represents four equally spaced points separated by $\pi/2$ (Figure 6.3). Each of the four possible phases of carriers represents two bits of data. Thus, there are two bits per symbol. Since the symbol rate for QPSK is half of the bit rate, twice the information can be carried in the same amount of channel bandwidth as compared to BPSK. This is possible because the two signals I and Q are orthogonal to each other and can be transmitted without interfering with each other.

In QPSK, the carrier phase can change only once every $2T$ seconds. If, from one $2T$ interval to the next one, neither bit stream changes sign, the carrier phase remains the same. If one component $a_I(t)$ or $a_Q(t)$ changes sign, a phase shift of $\pi/2$ occurs. However if both components, I and Q change sign, then a phase shift of π or 180° occurs. When this 180° phase shift is filtered by the transmitter and receiver filters, it generates a change in amplitude of the detected signal and causes additional errors.

If the two bit streams, I and Q, are offset by a 1/2 bit interval, then the amplitude fluctuations are minimized since the phase never changes by 180°.

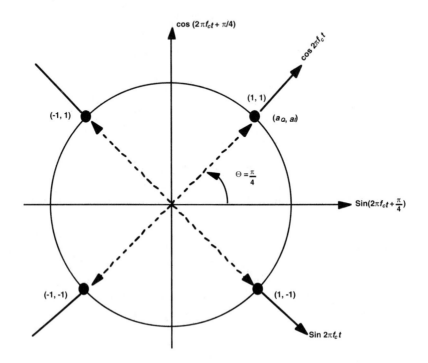

Fig. 6.3 Signal Constellation for 4-PSK (QPSK) and 4-QAM

This modulation scheme, Offset Quadrature Phase Shift Keying (OQPSK), is obtained from the conventional QPSK by delaying the odd bit stream by a half-bit interval with respect to the even bit stream. Thus, the range of phase transition is 0° and 90° and occurs twice as often, but with half the intensity of the QPSK. While amplitude fluctuations still occur in the transmitter and receiver (see Figure 6.4) they have smaller magnitude. The bit error rate and bandwidth efficiency of QPSK and OQPSK are the same as for BPSK.

In theory, quadrature (or offset quadrature) phase shift keying systems can improve the spectral efficiency of mobile communication. They do however require a coherent detector, and, in a multipath fading environment, the use of coherent detection is difficult and often results in poor performance over non-coherently based systems. The coherent detection problem can be overcome by using a differential detector, but then OQPSK is subject to intersymbol interference that results in poor system performance.

6.3.3.3 π/4 Differential Quadrature Phase Shift Keying

We can design a phase shift keying system to be inherently differential and thus solve the detection problems. The π/4 Differential Quadrature Phase Shift Keying (π/4-DQPSK) is a compromise modulation method because the phase is restricted to fluctuate between $\pm(\pi/4)$ and $\pm((3\pi)/4)$ rather than the $\pm(\pi/2)$ phase changes for OQPSK. It has a spectral efficiency of about 20% more than the Gaussian Minimum Shift Keying (see the next section) modulation used for GSM.

π/4-DQPSK is essentially a π/4-shifted QPSK with differential encoding of symbol phases. The differential encoding mitigates loss of data due to phase slips. However, differential encoding results in the loss of a pair of symbols when channel errors occur. This can be translated to approximately a 3-dB loss in E_b/N_o relative to coherent π/4-QPSK.

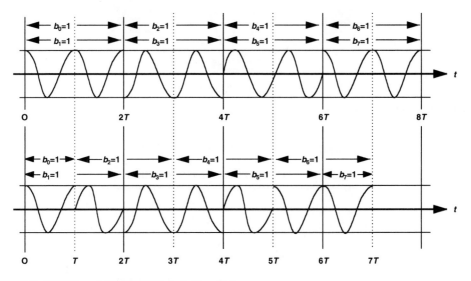

Fig. 6.4 QPSK and OQPSK Modulation Schemes

A π/4-shifted QPSK signal constellation (Figure 6.5) consists of symbols corresponding to eight phases. These eight phase points can be considered to be formed by superimposing two QPSK signal constellations, offset by 45° relative to each other. During each symbol period a phase angle from only one of the two QPSK constellations is transmitted. The two constellations are used alternately to transmit every pair of bits (di-bits). Thus, successive symbols have a relative phase difference that is one of the four phases shown in Table 6.1.

Figure 6.5 shows the π/4-shifted QPSK signal constellation. When the phase angles of the π/4-shifted QPSK symbols are differentially encoded, the resulting modulation is π/4-shifted DQPSK. This can be done either by differential encoding of the source bits and mapping them onto absolute phase angles or, alter-

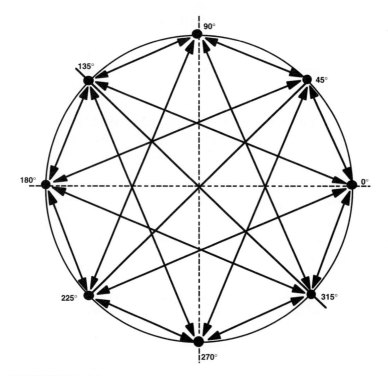

Fig. 6.5 π/4-DQPSK Modulation

Table 6.1 Phase Transitions of π/4-DQPSK

Symbol	π/4-DQPSK Phase Transition
00	45°
01	135°
10	−45°
11	−135°

nately, by directly mapping the pairs of input bits onto relative phase ($\pm(\pi/4)$, $\pm(3\pi)/4$) as shown in Figure 6.6. The binary data stream entering the modulator $b_m(t)$ is converted by a serial-to-parallel converter into two binary streams $b_o(t)$ and $b_e(t)$ before the bits are differentially encoded.

$$I_k = I_{k-1} \cos\Delta\phi_k - Q_{k-1} \sin\Delta\phi_k \tag{6.35a}$$

$$Q_k = I_{k-1} \sin\Delta\phi_k + Q_{k-1} \cos\Delta\phi_k \tag{6.35b}$$

where

I_k and Q_k are the in-phase and quadrature components of the $\pi/4$-shifted DQPSK signal corresponding to the k-th symbol. The amplitudes of I_k and Q_k are ± 1, 0, $\pm(1/\sqrt{2})$. Since the absolute phase of $(k-1)$th symbol is ϕ_{k-1}, the in-phase and quadrature components can be expressed as

$$I_k = \cos\phi_{k-1} \cos\Delta\phi_k - \sin\phi_{k-1} \sin\Delta\phi_k = \cos(\phi_{k-1} + \Delta\phi_k) \tag{6.36a}$$

$$Q_k = \cos\phi_{k-1} \sin\Delta\phi_k + \sin\phi_{k-1} \cos\Delta\phi_k = \sin(\phi_{k-1} + \Delta\phi_k) \tag{6.36b}$$

These component signals (I_k, Q_k) are then passed through baseband filters having a raised cosine frequency response as:

$$|H(f)| = \begin{cases} 1 & 0 \le |f| \le \dfrac{1-\alpha}{2T_s} \\[2mm] \sqrt{\dfrac{1}{2}\left\{1 - \sin\left[\dfrac{\pi T_s}{\alpha}\left(|f| - \dfrac{1}{2T_s}\right)\right]\right\}} & \dfrac{1-\alpha}{2T_s} \le |f| \le \dfrac{1+\alpha}{2T_s} \\[2mm] 0 & |f| \ge \dfrac{1+\alpha}{2T_s} \end{cases} \tag{6.37}$$

where:
α is the roll-off factor and
T_s is the symbol duration.

If $g(t)$ is the response to pulse I_k and Q_k at the filter input, then the resultant transmitted signal is given as:

$$s(t) = \sum_k g(t - kT_s) \cos\phi_k \cos\omega t - \sum_k g(t - kT_s) \sin\phi_k \sin\omega t \tag{6.38a}$$

$$s(t) = \sum_k g(t - kT_s) \cos(\omega t + \phi_k) \tag{6.38b}$$

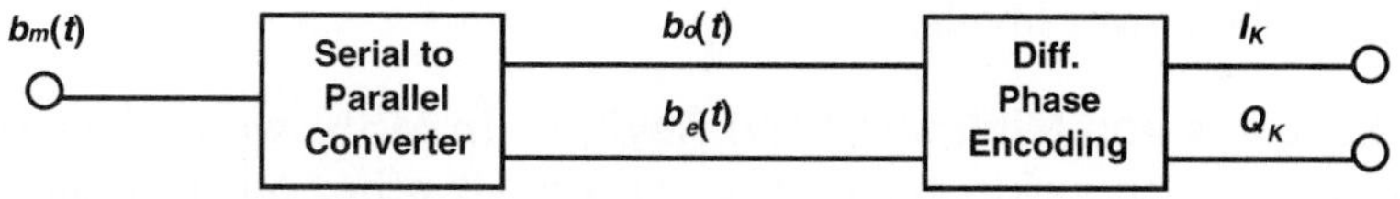

Fig. 6.6 Differential Encoding of $\pi/4$-DQPSK

where $2\pi\omega$ is the carrier frequency of transmission.

The component ϕ_k results from differential encoding (i.e., $\phi_k = \phi_{k-1} + \Delta\phi_k$).

6.3.3.4 Minimum Shift Keying   We previously showed that OQPSK is derived from QPSK by delaying the Q data stream by 1 bit or T seconds with respect to the corresponding I data stream. This delay has no effect on the error rate or bandwidth.

Minimum Shift Keying (MSK) is derived from OQPSK by replacing the rectangular pulse in amplitude with a half-cycle sinusoidal pulse. The MSK signal is defined as:

$$s(t) = a_I(t)\cos\left(\frac{\pi t}{2T}\right)\cos 2\pi ft + a_Q(t)\sin\left(\frac{\pi t}{2T}\right)\sin 2\pi ft \qquad (6.39a)$$

$$s(t) = \cos\left[2\pi ft + b_k(t)\frac{\pi t}{2T} + \phi_k\right] \qquad (6.39b)$$

where:
$b_k = +1$ for $a_I \cdot a_Q = -1$,
$b_k = -1$ for $a_I \cdot a_Q = 1$,
$\phi_k = 0$ for $a_I = 1$, and
$\phi_k = \pi$ for $a_I = -1$.

MSK has the following properties:

1. For a modulation bit rate of R, the high tone, $f_H = f + 0.25\,R$ when $b_k = 1$, and the low tone, $f_L = f - 0.25\,R$ when $b_k = -1$.
2. The difference between the high tone and the low tone is $\Delta f = f_H - f_L = 0.5R$
3. The signal has a constant envelope.

The error probability for an ideal MSK system is:

$$P_e = \frac{1}{2}erfc\sqrt{\frac{E_b}{N_o}} = Q\left[\sqrt{\frac{2E_b}{N_o}}\right] \qquad (6.40)$$

which is the same as for QPSK/OQPSK.

The MSK modulation makes the phase change linear and limited to $\pm(\pi/2)$ over a bit interval T. This enables MSK to provide a significant improvement over QPSK. Because of the effect of the linear phase change, the power spectral density has low side lobes that help to control adjacent-channel interference. However, the main lobe becomes wider than the quadrature shift keying (see Figure 6.7). Thus, it becomes difficult to satisfy the CCIR-recommended value of -60-dB side lobe power levels.

6.3.3.5 Gaussian Minimum Shift Keying   In MSK, we replace the rectangular data pulse with a sinusoidal pulse. Obviously other pulse shapes are possible. A Gaussian-shaped impulse response filter generates a signal with low side lobes and narrower main lobe than the rectangular pulse. Since the filter theoretically has output before input, it can only be approximated by a delayed and

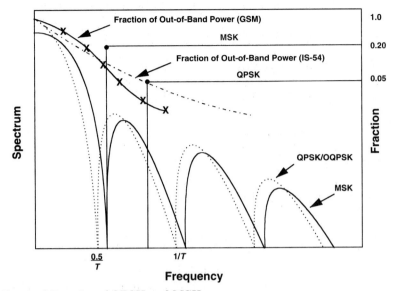

Fig. 6.7 Spectral Density of QPSK and MSK

shaped impulse response that has a Gaussian-like shape. This modulation is called Gaussian Minimum Shift Keying (GMSK).

The relationship between the premodulation filter bandwidth B and the bit period, T defines the bandwidth of the system. If $B > 1/T$, then the waveform is essentially MSK. When $B < 1/T$, the intersymbol interference occurs since the signal cannot reach its next position in the symbol time. However, intersymbol interference can be traded for bandwidth reduction if the system has sufficient SNR. GSM designers used a BT of 0.3 with a channel data rate of 270.8 kbs. DECT designers adopted BT = 0.5 with a data rate of 1.152 Mbs. A choice of BT = 0.3 in GSM is a compromise between a bit error rate and out-of-band interference since the narrow filter increases the intersymbol interference and reduces the signal power.

6.3.3.6 PSK Examples

E X A M P L E   6 – 1

Problem Statement
Find E_b/N_o in dB to have $P_e = 10^{-6}$ for BPSK and coherent FSK.

Solution
For BPSK

$$P_e \approx \frac{e^{-\frac{E_b}{N_o}}}{2\sqrt{\pi\left(\frac{E_b}{N_o}\right)}}$$

$$10^{-6} = \frac{e^{-\xi}}{2\sqrt{\pi\xi}}$$

where:
$\xi = E_b/N_o$

$$\xi = 11.32 = 10.54 \text{ dB}$$

FSK requires 3 dB more in terms of E_b/N_o to give same P_e as BPSK, i.e., 13.54 dB.

E X A M P L E   6 – 2

Problem Statement

If BPSK signaling is used through a channel that adds white noise with single-sided power spectral density $N_o = 10^{-10}$ W/Hz, calculate the amplitude V of the carrier signal to give $P_e = 10^{-6}$ for a data rate of 100 kbs.

Solution

From Example 6-1, $E_b/N_o = 10.54$ dB = 11.32 for $P_e = 10^{-6}$.

$$\frac{E_b}{N_o} = \frac{V^2 T}{2N_o} = \frac{V^2}{2N_o R}$$

$$\therefore V = \left[2N_o\left(\frac{E_b}{N_o}\right)R\right]^{1/2}$$

$$V = [2\times10^{-10} \times 11.32\times10^5]^{1/2}$$

$$V = 1.505 \text{ mV}$$

E X A M P L E   6 – 3

Problem Statement

A 2-Mbs stream is transmitted using QPSK signaling. Even- and odd-indexed bits are associated with $a_I(t)$ and $a_Q(t)$, respectively. Calculate the symbol rate, the required channel bandwidth if only the main lobe of the signal spectrum is to be passed, and the bit error, P_e, if the carrier power is 6 mW and $N_o = 10^{-9}$ W/Hz.

Solution

Since every other bit is assigned to $a_I(t)$ and $a_Q(t)$, it follows that the symbol rate = 1/2 bit rate = 1 mega symbols/second.

The QPSK spectrum has a main lobe bandwidth of $2/T_s = 2R_s$, where T_s = the symbol duration and R_s = the symbol rate.

Therefore the required bandwidth = $2 \times 1 = 2$ MHz.

From Eq. (6.33), the bit error is

$$P_e = Q\left[\left(\frac{2E_b}{N_o}\right)^{1/2}\right] = Q\left[\left(\frac{E_s}{N_o}\right)^{1/2}\right]$$

$$\frac{E_s}{N_o} = \frac{V^2 T_s}{2 N_o} = \frac{P T_s}{N_o} = \frac{(6 \times 10^{-3})\,(1 \times 10^6)^{-1}}{10^{-9}} = 6$$

$$\therefore P_e = Q\,(\sqrt{6}) = 8 \times 10^{-3}$$

E X A M P L E   6 – 4

Problem Statement

If the bit stream is 00100111010011101010100 and the transmitted signal is $V \cos(\omega t + \phi_k)$, calculate ϕ_k for the π/4-DQPSK modulation scheme.

Solution

See Table 6.2.

Table 6.2 Phase Angle for π/4-DQPSK

$(b_o b_e)$	$\Delta\phi_k$ (degree)	ϕ_k (degree)
		0
00	45	45
10	−45	0
01	135	135
11	−135	0
01	135	135
00	45	180
11	−135	45
10	−45	0
10	−45	−45
01	135	90
01	135	225
00	45	270

6.3.4 Quadrature Amplitude Modulation

QAM can be viewed as an extension of multiphase PSK modulation wherein the two baseband signals are generated independently of each other. Thus, two completely independent (quadrature) channels are established, including the baseband coding and detection processes. In the special case where two levels (±1) are used on each channel, the system is identical to 4-PSK.

Higher-level QAM systems, however, are distinctly different from the higher-level PSK systems. Figure 6.8 shows a signal constellation of a 16-QAM system, obtained from four levels on each quadrature channel, and a 16-PSK system. Notice that, in contrast to the PSK signal, the QAM signal does not have a constant envelope.

The spectrum of a QAM system is determined by the spectrum of the baseband signals applied to the quadrature channels. Since these signals have the same basic structure as the baseband PSK signals, QAM spectrum shapes are identical to PSK spectrum shapes with equal numbers of signal points. Specifically, 16-QAM has a spectrum shape that is identical to 16-PSK, and 64-QAM has a spectrum shape identical to 64-PSK. Even though the spectrum shapes are identical, the error performances of the two systems are quite different. With large numbers of signal points, QAM systems outperform PSK systems. The basic reason is that the distance between signal points in a PSK system is smaller than the distance between points in a comparable QAM system.

The distance between adjacent signal points in a unit peak amplitude QAM system with M levels on each axis is:

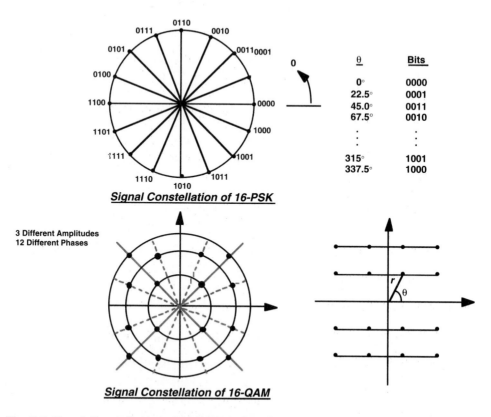

Fig. 6.8 Signal Constellation of 16-PSK and 16-QAM

$$d_e = \frac{\sqrt{2}}{M-1} \tag{6.41}$$

In terms of average power levels, a QAM system has advantages over a PSK system. The peak-to-average power ratio of a QAM system is

$$\frac{\text{Peak Power}}{\text{Average Power}} = \frac{M(M-1)^2}{2\sum\limits_{j=1}^{M/2}(2j-1)^2} \tag{6.42}$$

where the QAM system has M^2 total points.

The error rate for 4-QAM can be calculated by substituting $M = 2$ in Eq. (6.23a):

$$P_e = \frac{1}{log_2 2}\left(\frac{1}{2}\right)erfc\left[\frac{(log_2 2)^{1/2}}{(2-1)}\left(\frac{E_b}{N_o}\right)^{1/2}\right]$$

$$P_e = \frac{1}{2}erfc\sqrt{\frac{E_b}{N_o}} = Q\left[\sqrt{\frac{2E_b}{N_o}}\right] \tag{6.43}$$

When calculating the error rates for simple signal constellations, we have made simplifying assumptions that the noise will result in a detection to the nearest neighbor in the constellation. For low error rates and when the distance to the second nearest neighbor is much greater than the distance to the nearest neighbor, the assumption holds. However, with complex constellations, e.g., 16-QAM, these assumptions are no longer true. While a theoretical calculation is still possible, more complex modulation methods are best analyzed by either computer simulation or experimental data.

6.4 DEMODULATION ISSUES

While we have discussed the various modulation methods, we have ignored issues of synchronization and equalization. All of the systems described require that the data receiver be synchronized to the data transmitter. All of the calculations in this chapter assume that the data receiver is perfectly synchronized with the data transmitter. In practice this is not true, and in the real world error rates will be higher because of the lack of synchronization. Modulation methods that have no signal power at the data clock rate will be harder to synchronize than those that have strong spectral components at the data clock rate. Differential systems are inherently easier to synchronize than absolute systems since the absolute phase of the data clock is not important. The design of the synchronizing circuits is usually an art rather than a science since many real-world problems must be solved. Examples of these problems are: clock oscillator drift with temperature and voltage, signal dropouts during fades, certain data patterns that cause the clock signal to be reduced in amplitude or disappear completely, and certain error patterns that cause the clock phase to slip and propagate the errors.

Some demodulators attempt to improve intersymbol interference by employing equalizers. When the intersymbol interference is caused by limiting the bandwidth of the signal, it is difficult to eliminate it without adding noise to the system. A filter that reduces intersymbol interference would have to have the inverse frequency response to that of the filter in the transmitter. If such a filter were designed, it would have a wider bandwidth than the transmitter filter and would pass noise and interference from adjacent channels. Clearly this approach is not desirable and is therefore not used. When the intersymbol interference is because of the multipath nature of the channel, it is possible to build an equalizing filter that resolves the multipaths. We will discuss the design of the RAKE receiver equalizing filter in chapter 7. Typically it is used only with systems that are inherently wideband such as CDMA.

6.5 SUMMARY

In this chapter we have discussed the different methods used to modulate baseband digital signals. We studied amplitude, frequency, and phase modulation and a hybrid of amplitude and phase modulation. For each of these modulation methods, we discussed the signal constellation and the bit error rate. We will refer to these methods as we discuss the system designs used for cellular and PCS systems in Europe, North America, and Japan.

6.6 PROBLEMS

1. Determine the probability of error for an M-ary ASK system. Use as your signals $s_0(t) = 0$ and $s_n(t) = (V_c/(M-1)) \cos (2\pi f_c t)$.

2. If BPSK signaling is used through a channel that adds white noise with single-sided power spectral density, $N_0 = 10^{-8}$ W/Hz, calculate the signal power required to give $P_e = 10^{-6}$ for a data rate of 1 Mbs.

3. Repeat problem 2 for ASK and FSK modulation scheme.

4. Calculate the channel bandwidth required to transmit data at 1 Mb/s for BPSK, 4-PSK, coherent BFSK, and noncoherent BFSK.

5. Find E_b/N_o for M-ary PSK with $M = 8$ and $P_e = 10^{-6}$.

6. Design an M-ary PSK modulation scheme to provide $P_e = 10^{-6}$ with a data rate of 9.6 kbs and a channel bandwidth 4.8 kHz. Assume $N_0 = 10^{-9}$ W/Hz.

6.7 REFERENCES

1. Proakis, J. G., *Digital Communication*, New York: McGraw-Hill, 1989.

2. Schwartz, M., W. Bennett, S. Stein, *Communications Systems and Techniques*, New York: McGraw Hill, 1966.

3. Sklar, B., *Digital Communications: Fundamental and Applications*, Englewood Cliffs, New Jersey: Prentice-Hall, 1988.

4. "Special Issue on Bandwidth and Power Efficient Coded Modulation," *IEEE Journal of Selected Area in Communications* 7 (August and December 1989) (2 parts).

5. "Special Issue on Bandwidth and Power Efficient Modulation," *IEEE Communications Magazine* 29 (December 1991).

6. Stalling, W., *Data and Computer Communications*, 2nd ed., New York: Macmillan Publishing Company, 1988.

7. Wozencraft, John M., and Irwin M. Jacobs, *Principals of Communications Engineering*, New York: John Wiley and Sons, 1965.

8. Ziemer, R. E., and R. L. Peterson, *Introduction to Digital Communication*, New York: Macmillan Publishing Company, 1992.

9. Ziemer, R. E., and W. H. Tranter, *Principle of Communications*, 3d ed., Boston: Houghton Mifflin, 1990.

CHAPTER 7

Antennas, Diversity, and Link Analysis

7.1 Introduction

Wireless communications systems need antennas at the transmitter and the receiver to operate properly. The design and deployment of antennas can make or break a wireless system, and many poorly performing systems can be traced to improperly installed or placed antennas. In this chapter we will define and discuss the concepts of antenna gain, free space path loss, and path loss over reflecting surfaces (typical of ground-wave propagation). We will then describe the noise that a receiver sees and use that concept to do an analysis of the link between the base station and the mobile telephone and calculate the SNR ratio at a receiver. We will then complete our discourse on antennas by examining the interrelationships between beamwidth, directivity, and antenna gain. After our discussion of antennas is complete, we will explore the concepts of diversity reception where multiple signals are combined to improve the SNR of the system. Time diversity is used for the CDMA systems to improve system performance; therefore we will explore that system in more detail. Finally, we will discuss some practical antennas used in cellular telephones today. Throughout the chapter, we present examples to improve your understanding of the topics.

7.2 Objectives of a Cellular System Antenna

The cellular communications system requires a reliable communication from a fixed base station to a mobile station. Since the cellular radio is a duplex system, the goal of the system designer is to have the same performance in both the transmitting and receiving directions. This is not always possible since the base station typically has a higher output power than the mobile station. Also, the mobile antenna is typically on the street at a height of 5–6 feet compared to the base station antenna height of 150 feet. This results in the mobile receiver having a high noise level because of the interference caused by the ignitions of nearby vehicles, whereas the base station, with its higher antenna, usually sees a quieter radio environment. These factors can combine to favor one direction over the other, and the wireless system designer must carefully consider all factors or the range of the system may be limited by poor performance in one direction.

The types of antennas, their gain and coverage patterns, the available power to drive them, the application of simple or multiple antenna configurations, and polarization are the major factors that are controlled by the system designer. The system designer has no control over the topography between the cell-site and mobile station antennas, the speed and direction of the mobile station, and the location of antenna(s) on the vehicle. Each of these factors significantly affects the system performance. Sometimes the placement of the mobile antenna can severely limit system performance. While the mobile antenna is usually installed by a knowledgeable technician, the owner of the vehicle may force a nonoptimum placement of the antenna. Furthermore, cellular antennas are vertically polarized, but many vehicle antennas are no longer vertical after the vehicle is sent through a car wash.

7.3 ANTENNA GAIN

Gain is the most significant parameter in the design of an antenna system. A high gain is achieved by increasing the aperture area, A, of the antenna. Antennas obey reciprocity; the transmit gain and receive gain are the same, and the antenna can be analyzed by examining it as either a receive or transmit antenna. The amount of the power captured by an antenna is given as:

$$P = pA \qquad\qquad (7.1)$$

where:
p = power density and
A = aperture area.

Antenna gain can be defined either with respect to an isotropic antenna or with respect to a half-wave dipole and is usually analyzed as a transmit antenna. An isotropic antenna is an idealized system that radiates equally in all directions. The gain of an antenna in a given direction is the ratio of the power density produced by it in that direction divided by the power density that would be produced by an isotropic antenna. The term dBi is used to refer to the antenna gain with respect to an isotropic antenna, whereas the term dBd is used to refer to the antenna gain with respect to a half-wave dipole (0 dBd = 2.1 dBi). While most analyses of system performance use a half-wave dipole as the reference, many times antenna gain figures are quoted in dBi to give a falsely inflated gain figure.

The system designer must carefully read data sheets on antennas to use the correct gain figure. As a rule of thumb, if the gain is not quoted in either dBd or dBi, the gain is in dBi, with the dBi left out to inflate the gain figures.

For an antenna mounted high and away from obstructions, the antenna can be analyzed as if it is located at the center of a large sphere (see Figure 7.1). The transmitted power from the antenna is given as the surface integral of the power density. In Eq. (7.2), $p_R(\theta, \phi)$ is the power per unit area (i.e., power density).

$$P_T = \int_s p_R(\theta, \phi)\, ds \tag{7.2}$$

It is related to the normalized field pattern by

$$p_R(\theta, \phi) = \left[\frac{E^2_{\theta n}(\theta, \phi) + E^2_{\phi n}(\theta, \phi)}{Z_o} \right] \tag{7.3}$$

where:

$$E_{\theta n}(\theta, \phi) = \frac{E_\theta(\theta, \phi)}{E_\theta(\theta, \phi)_{max}} \quad \text{and}$$

$$E_{\phi n}(\theta, \phi) = \frac{E_\phi(\theta, \phi)}{E_\phi(\theta, \phi)_{max}}$$

Z_0 is the intrinsic impedance of space = 377.752 Ω, and
$E_\theta(\theta, \phi)_{max}$ and $E_\phi(\theta, \phi)_{max}$ are the maximum value of the field intensity in θ and ϕ directions.

The integral is performed over the surface of the sphere (see Figure 7.2).

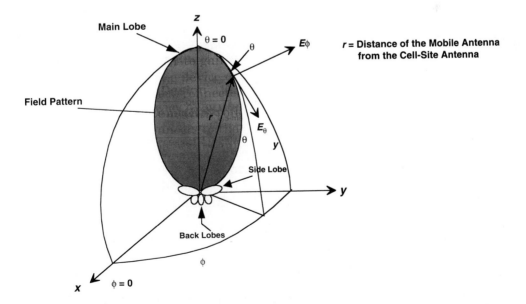

Fig. 7.1 Antenna Field Pattern

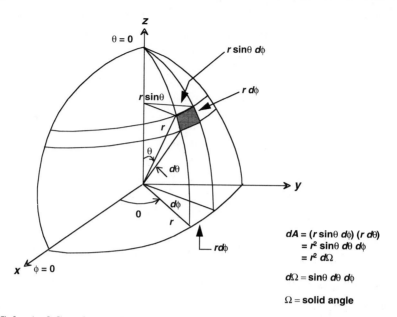

Fig. 7.2 Spherical Coordinate System—Relationship between dA and the Solid Angle $d\Omega$

$$P_T = r^2 \int_0^{2\pi} \int_0^{2\pi} p_R(\theta, \phi) \sin\theta d\theta d\phi \qquad (7.4)$$

For an isotropic antenna, in free space,

$$P_T = p_R r^2 \int_0^{2\pi} \int_0^{2\pi} \sin\theta d\theta d\phi \qquad (7.5)$$

$$p_R = \frac{P_T}{4\pi r^2} \qquad (7.6)$$

An isotropic antenna is an idealized antenna and is used as a reference for other antennas. Real antennas have stronger power densities in some directions and weaker power densities in other directions. The goal of the antenna designer is to shape the pattern of the antenna so the power density is higher in the desired direction and lower in the nondesired direction. We will explore this concept in more detail in section 7.6.

7.4 FREE SPACE PATH LOSS

We consider the system shown in Figure 7.3, where a cell-site transmitter is transmitting at an average power level of P_T. We want to find the received power

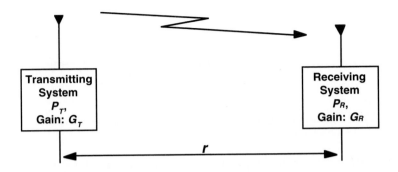

Fig. 7.3 A Simple Model for Path Loss in Free Space

level, P_R, at the receiving antenna (mobile station) located at a distance r from the transmitter.

For an antenna radiating uniformly in all directions (spherical pattern), the power density, p_R, at the receiver is given by Eq. (7.6).

When a directional transmitting antenna with power gain factor, G_T, is used, the power density at the receiver site is G_T times Eq. (7.6).

The amount of power captured by the receiver is p_R times the aperture area, A_R, of the receiving antenna. The aperture area is related to the gain of receiving antenna by (see references [8] and [10] for details).

$$G_R = \frac{4\pi A_R}{\lambda^2} \tag{7.7}$$

where:
$\lambda = c/f$,
f = the transmission frequency in Hz,
$c = 3 \times 10^8$ m/s is the free-space speed of propagation for electromagnetic
        waves, and
A_R is the effective area, which is less than the physical area by an efficiency
        factor ρ_R.

Typical values for ρ_R range from 60% to 80%. The total received power, P_R is:

$$P_R = A_R p_R \tag{7.8}$$

Substituting the value of p_R and A_R from Eq. (7.6) and (7.7) into Eq. (7.8), together with the transmitting antenna gain G_T, we get

$$P_R = \left[\frac{\lambda}{4\pi r}\right]^2 P_T G_T G_R \tag{7.9}$$

Eq. (7.9) includes the power loss only from the spreading of the transmitted wave. If other losses are also present, such as atmospheric absorption or ohmic losses of the waveguides leading to the antennas, Eq. (7.9) is modified as:

$$\frac{P_R}{P_T} = \left[\frac{\lambda}{4\pi r}\right]^2 \frac{G_T G_R}{L_0} \tag{7.10a}$$

$$\frac{P_R}{P_T} = \frac{G_T G_R}{L_p L_0} \tag{7.10b}$$

where the term

$L_p = \left[\frac{4\pi r}{\lambda}\right]^2$ denotes the loss associated with propagation of electromagnetic waves from the transmitter to the receiver.

L_p depends on the carrier frequency and separation distance r. This loss is always present. L_0 = loss factor for additional losses. When we express Eq. (7.10a) in terms of decibels, we get

$$P_R = 20 \log \left[\frac{\lambda}{4\pi r}\right] + P_T + G_T + G_R - L_0 \tag{7.11}$$

The product $P_T G_T$ is called the Equivalent Isotropic Radiated Power (EIRP) and term $20 \log [\lambda/(4\pi r)]$ refers to **free-space loss** (L_p) in dB.

7.5 RECEIVER NOISE

The movement of electrons through a device is not constant but undergoes random fluctuations called thermal noise. The fluctuations are proportional to the noise temperature of the device, the bandwidth being considered, and a constant. Passive devices (e.g., resistors) have noise temperatures equal to their physical temperature. Active devices (transistors, integrated circuits, vacuum tubes, and so on) have noise temperatures that are higher than their physical temperatures. The effects of the thermal noise can be expressed as the average noise power generated internally to the receiver and is

$$P_{N,int} = kT_e B_w \tag{7.12}$$

where:

k is Boltzmann's constant and is 1.38×10^{-23} $J/°K$ (J = Joules of energy, K = temperature in degrees Kelvin),
B_w = bandwidth, and
T_e = effective noise temperature of the device and is given as

$$T_e = T_0 (N_f - 1) \tag{7.13}$$

where:
T_0 = 290°K, and
N_f = noise figure of the receiver.

The noise figure is expressed either in degrees Kelvin or in decibels, with 3 dB corresponding to 290°K.

The noise figure is a property of the receiver and in a properly designed receiver is dominated by the initial stage of the receiver. However, in general, noise from all stages must be taken into account. Lower noise figures are obtained by careful design and selection of the first stages of the receiver. With modern semiconductors, noise figures of less than 1 dB are common.

The noise figure is a measure of the noise generated by a receiver. When extremely good noise performance is needed, lower noise levels are obtained by cooling the receiver to temperatures near absolute zero.

Even if the receiver were perfect and had a 0-dB noise figure, additional noise is caused by the antenna and the feed line from the antenna to the receiver. To include the effect of noise seen by the antenna and the feed line loss between the antenna and the receiver, we add the temperature of the antenna, T_{ant} and the equivalent temperature of the feed line loss, T_{fl} to T_e, to get:

$$P_N = k\,(T_{ant} + T_e + T_{fl})\,B_w = kTB_w \tag{7.14}$$

where:
$T = T_{ant} + T_e + T_{fl}$

The antenna temperature is not a physical temperature; it represents an additional source of noise that depends on where the antenna is pointed and the frequency band of the received signal. For a narrow beamwidth receiving antenna on the ground and pointed at the sky between 10° and 90° with respect to the horizon and signal frequency between 1 and 20 GHz, $T_{ant} \approx 3°$K. This represents the noise of the "big bang" at the creation of universe. If a narrow beamwidth antenna is pointed at the sun or the Milky Way, then the noise of the antenna increases. Moderate rainfall (≤ 10 millimeters/h) changes this value little, but severe rainstorms may increase it by 10° to 50°K because of scattering of the sky background noise into the antenna by raindrops.

When the narrow beamwidth antenna is pointed at the horizon or when a nondirectional antenna (such as on a car or handset) is used, the equivalent temperature of the antenna is the temperature of the ground, 290°K.

By dividing Eq. (7.10a) by Eq. (7.14), we obtain the received SNR at a receiver as:

$$SNR = \frac{P_R}{P_N} = \left[\frac{\lambda}{4\pi r}\right]^2 \frac{P_T G_T G_R}{L_0 kTB_w} \tag{7.15}$$

E X A M P L E   7 – 1

Problem Statement
We consider a base station (BS) transmitting to a mobile station (MS). The following parameters relate to this communication system:

(a) Distance between BS and MS: 8,000 m

(b) BS EIRP ($G_T = 20$ dB; $P_T = 10$ W): 30 dBW

(c) Transmitter frequency: 1.5 GHz($\lambda = 0.2$ m)

(d) MS receiver antenna gain: 3 dB

(e) Total system losses: 6 dB

(f) MS receiver noise figure, $N_f = 5$ dB

(g) MS receiver antenna temperature = 290°K

(h) MS receiver bandwidth, $B_w = 1.25$ MHz

(i) Neglect any feed line loss between the antenna and the receiver

Calculate the received signal power at the MS receiver antenna and the SNR of the received signal.

Solution

$$L_p : \text{Free-space loss} = -20 \log\left[\frac{0.2}{4\pi \times 8000}\right] = 114 \text{ dB}$$

$$P_R = -114 + 30 + 3 - 6 = -87 \text{ dBW} = -57 \text{ dBm} = 2 \times 10^{-6} \text{ mW}$$

$$T_e = 290\,(3.162 - 1) = 627°K$$

$$P_N = 1.38 \times 10^{-23}\,(627 + 290)\,(1.25 \times 10^{-6}) = 1.58 \times 10^{-14} \text{ W} = -138 \text{ dBW}$$

$$SNR = \frac{P_R}{P_N} = -87 - (-138) = 51 \text{ dB}$$

7.6 THE PATH LOSS OVER A REFLECTING SURFACE

In outer space, the path between two antennas has no obstructions and no objects where reflections can occur. Thus the received signal is composed of only one component. When the two antennas are on the earth, then there are multiple paths from the transmitter to the receiver. The effect of the multiple paths is to change the path loss between two points. The simplest case occurs when the antenna heights h_T and h_R are small compared with their separation r and the reflecting earth surface can be assumed to be flat. The received signal can then be represented by a scattered field E_s, that can be approximated by a combination of a direct wave and a reflected wave (refer to Figure 7.4).

$$E_s = \left[1 + a_r e^{j\Delta\theta}\right] E \tag{7.16a}$$

$$E_s = \left[1 + a_r (\cos\Delta\theta + j\sin\Delta\theta)\right] E \tag{7.16b}$$

where:
a_r = coefficient of reflection,
$\Delta\theta$ = phase difference between the direct and reflected path.

$$\Delta\theta = \frac{2\pi}{\lambda}(d_r - d_o) = \left[\frac{2\pi}{\lambda}\right]\Delta d \tag{7.17}$$

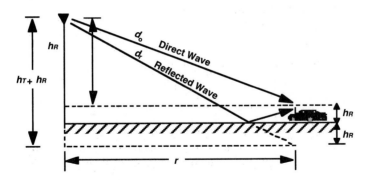

Fig. 7.4 A Simple Model for Path Loss with Reflection

where:
Δd = the difference between direct path (d_o) and reflected path (d_r).

In the mobile radio environment, $a_r = -1$ and $\Delta\theta$ is much less than one radian. Therefore, Eq. (7.16b) can be written as:

$$E_s \approx -Ej\Delta\theta \tag{7.18}$$

Since the received power level P_R is proportional to the square of the field strength, the power level at the antenna output located at a distance, r, from the transmitter, including the path loss, will be:

$$P_R = \left[\frac{\lambda}{4\pi r}\right]^2 \frac{P_T G_T G_R}{L_0}|j\Delta\theta|^2 \tag{7.19}$$

From Figure (7.4):

$$d_o = \sqrt{(h_T - h_R)^2 + r^2}$$

and

$$d_r = \sqrt{(h_T + h_R)^2 + r^2}$$

$$\Delta d = r\left[1 + \left(\frac{1}{2}\right)\left[\frac{h_T + h_R}{r}\right]^2 + \ldots\right] - r\left[1 + \left(\frac{1}{2}\right)\left[\frac{h_T - h_R}{r}\right]^2 + \ldots\right]$$

When h_T and $h_R \ll r$, then

$$\Delta d \approx \frac{2h_T h_R}{r} \tag{7.20}$$

$$\therefore \Delta\theta \approx \left[\frac{4\pi}{\lambda r}\right]h_T h_R \tag{7.21}$$

Substituting for $\Delta\theta$ from Eq. (7.21) into Eq. (7.19), we get

$$P_R = \left[\frac{h_T h_R}{r^2}\right]^2 \frac{P_T G_T G_R}{L_0} \tag{7.22}$$

Expressing Eq. (7.22) in decibels, we get

$$P_R = 20 \log\left[\frac{h_T h_R}{r^2}\right] + P_T + G_T + G_R - L_0 \tag{7.23}$$

It should be observed that Eq. (7.23) is independent of transmitting frequency.

E X A M P L E   7 – 2

Problem Statement

Using the data in Example 7-1 and with antenna height at the BS and MS units to be 30 m and 3 m, respectively, calculate the received signal power at the MS receiver antenna and the SNR of the received signal.

Solution

$$\text{Path Loss} = -20 \log\left[\frac{30 \times 3}{8000^2}\right] = 117 \ \text{dB}$$

$$P_R = -117 + 30 + 3 - 6 = -90 \ \text{dBW} = -60 \ \text{dBm} = 1 \times 10^{-6} \ \text{mW}$$

Note that the path reflection has reduced P_R by 50% (refer to Example 7-1).

$$P_N = -138 \ \text{dBW} \ \text{(from Example 7-1)}$$

$$\therefore SNR = \frac{P_R}{P_N} = -90 - (-138) = 48 \ \text{dB}$$

7.7 THE RELATIONSHIP BETWEEN DIRECTIVITY, GAIN, AND BEAMWIDTH

Real antennas are not isotropic radiators but have a pattern of more and less power in different directions. Antenna engineers consider the pattern of the power radiated in the horizontal and vertical directions. The shape of the pattern describes the directionality of the antenna. The direction for maximum power is called the primary beam, or the major lobe, whereas secondary beams are referred to as the minor lobes (see Figure 7.5). The pattern of the antenna has two desired effects: concentration of the power in a desired direction to improve the signal strength at the receiver and weakening of the power in an undesired direction to reduce interference from or to other receivers. Therefore, the minor lobes provide undesired radiation or reception. Since the major lobe propagates

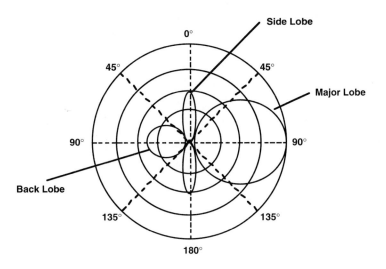

Fig. 7.5 Radiation Pattern—Directional Antenna

and receives the most energy, this lobe is called the "front" lobe. Lobes adjacent to the front lobe are called the "side" lobes and the lobe in a direction exactly opposite to the front lobe is called the "back" lobe. The front-to-back ratio (R_{FB}) of an antenna is defined as:

$$R_{FB} = 10 \log \frac{p_{mM}}{p_{mB}} \qquad (7.24)$$

where:
p_{mM} = maximum power density of the front lobe and
p_{mB} = maximum power density of the back lobe.

The front-to-side ratio (R_{FS}) of an antenna is defined as:

$$R_{FS} = 10 \log \frac{p_{mM}}{p_{mS}} \qquad (7.25)$$

where:
p_{mS} = maximum density of the side lobe.

Since the dipole antenna is the simplest one to build, it is often used as the reference to describe the gain of other antennas (see section 7.3). The average power density for the dipole vertical antenna is given as:

$$p_{avg} = \frac{3P_T}{8\pi r^2} \, sin^2\theta \qquad (7.26)$$

From Eq. (7.26) it can be noticed that the power density at any point depends on the direction θ and the distance r from the dipole. Since there is no dependence on ϕ, the antenna pattern is directional in the x-z and y-z planes, but omnidirectional in the x-y plane.

A dipole mounted in a vertical direction provides an omnidirectional pattern that is useful for the base station antenna of cellular system. While dipoles are often used as simple antennas, real antenna systems use other types of antennas to provide higher gain than a dipole.

7.7.1 The Relationship between Directivity and Gain

The **directivity** of an antenna is defined as the ratio of the maximum power density from the antenna and the power density from an isotropic antenna.

$$D = \frac{|p_{avg}|_{max}}{\left[\dfrac{P_T}{4\pi r^2}\right]} \tag{7.27}$$

The **gain** of an antenna is defined as the ratio of the maximum power density from the antenna and the input power density if the antenna were isotropic.

$$G = \frac{|p_{avg}|_{max}}{\left[\dfrac{P_I}{4\pi r^2}\right]} = \frac{|p_{avg}|_{max}}{\left[\dfrac{P_T}{4\pi r^2 \eta}\right]} = \eta \times D \tag{7.28}$$

where:
η is the efficiency of the antenna and
P_I is the input power.

7.7.2 Relationship between Gain and Beamwidth

The receiver antenna gain G_R can be related to its half-power beamwidth as:

$$G_R = \frac{4\pi}{\theta_{HP}\phi_{HP}} \tag{7.29}$$

where:
θ_{HP} and ϕ_{HP} are the half-power beamwidths in the θ and ϕ planes.

The factor 4π is the solid angle subtended by a sphere in steradians (square radians).

$$4\pi \text{ steradians} = 4\pi \times \left[\frac{180}{\pi}\right]^2 = 41,250 \text{ deg}^2 = \text{solid angle in a sphere}$$

$$G_R = \frac{41,250}{\theta_{HP}\phi_{HP}} \tag{7.30}$$

For an ideal gain antenna, where the power density is uniform inside the 3-dB beamwidth (for both θ and φ) and zero outside the 3-dB beamwidth, the gain G_R can be expressed as:

$$G_R = \frac{41,250}{\theta\phi} \tag{7.31a}$$

where:
θ and φ are in degrees.

For a real antenna, with side lobes, the gain should be calculated using Eq. (7.31b), which includes the effect of side lobes.

$$G_R = \frac{32,400}{\theta\phi} \tag{7.31b}$$

If an antenna is designed with a circular pattern in one direction, i.e., a linear element is used, the approximate gain can be obtained for the vertical 3-dB beamwidth from Eq. (7.31a) as

$$G_R \approx \frac{41,250}{360\theta} = \frac{114.6}{\theta} \tag{7.32a}$$

The corresponding gain equation for a linear element including side lobes is given as:

$$G_R \approx \frac{101.5}{\theta} \tag{7.32b}$$

E X A M P L E   7 – 3

Problem Statement
A Yagi antenna consists of a dipole radiator with passive elements in front and behind the dipole spaced at fractions of a wavelength. The passive elements concentrate the signal and develop a pattern with gain in one direction. Find the gain of a three-element Yagi antenna where: 10 W of power to the antenna result in 2 W of power being dissipated as resistive losses in the antenna and the rest radiated by the antenna; the power at the distant receiver is –44 dBm (40 nW) and the directivity of the antenna is 7 dB. What would be the received power if the Yagi antenna was replaced by an isotropic antenna?

Solution

$$\eta = \frac{10-2}{10} = 0.8$$

$$G = \eta D = 0.8 \times 5 = 4 \ \ (6 \text{ dB})$$

For the isotropic antenna, the received power will be 40 nW/4 = 10 nW (i.e., the received power is reduced from –44 dBm [Yagi] to –50 dBm [isotropic]).

E X A M P L E   7 – 4

Problem Statement
Find the directive gain of a sector antenna with the vertical and horizontal beam widths of 60° and 60°, respectively.

Solution

$$G = \frac{32{,}400}{60 \times 60} = 9.00 = 9.54 \ \text{dBi} = 7.44 \ \text{dBd}$$

E X A M P L E   7 – 5

Problem Statement
Find the 3-dB beamwidth of a linear element antenna with a directive gain of 10 dBd or 12.1 dBi.

Solution

$$\theta = \frac{101.5}{(10)^{1.21}} = 6.25°$$

E X A M P L E   7 – 6

Problem Statement
Calculate the mean signal strength and mean SNR ratio for the downlink and uplink of the antenna system shown in Figure 7.6 using the following parameters:

Downlink ($BS \rightarrow MS$):

(a) P_{TC}: Transmit power of cell site = 25 W (+44 dBm)
(b) L_{TC}: Loss of cell-site feed line = 4.0 dB

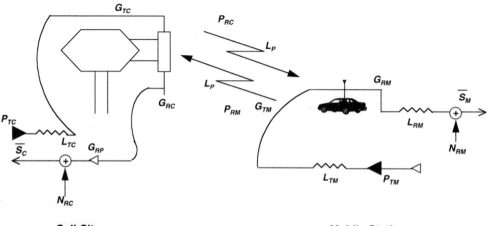

Cell-Site   **Mobile Station**

Fig. 7.6 Uplink and Downlink SNR Ratio Calculation

(c) G_{TC}: Gain of cell-site transmit antenna = 10 dB

(d) P_{RC}: Radiated power of cell site = 100 W EIRP = 50 dBm

(e) L_P: Radio propagation path loss = 150 dB

(f) G_{RM}: Gain of mobile receive antenna = 3 dB

(g) L_{RM}: Loss of mobile antenna cable = 1.5 dB

(h) N_{RM}: Noise floor of mobile receiver = −116 dBm

Uplink ($MS \rightarrow BS$):

(a) P_{TM}: Transmit power of mobile = 2.8 W (+34.5 dBm)

(b) L_{TM}: Loss of mobile antenna cable = 1.5 dB

(c) G_{TM}: Gain of mobile transmit antenna = Gain of mobile receive antenna = 3.0 dB.

(d) P_{RM}: Radiated power of mobile = 4 W EIRP = 36 dBm

(e) L_P: Radio path loss = 150 dB

(f) G_{RC}: Gain of cell-site receive antenna = 10 dB

(g) G_{RP}: Net gain of cell-site receiver preamplifier and feed-line system = 15 dB

(h) N_{RC}: Equivalent noise floor of cell-site receiver (at the end of the feed line) = −110 dBm

Solution
Downlink mean S/N

$$\overline{S_M} = P_{RC} - L_P + G_{RM} - L_{RM} = 50 - 150 + 3 - 1.5 = -98.5 \ \text{dBm}$$

$$\frac{\overline{S}}{N} = \overline{S}_M - \overline{N}_{RM} = 98.5 - (-116) = 17.5 \ \text{dB}$$

Uplink mean S/N

$$\overline{S_C} = P_{RM} - L_P + G_{RC} + G_{RP} = 36 - 150 + 10 + 15 = -89 \ \text{dBm}$$

$$\frac{\overline{S}}{N} = \overline{S}_C - \overline{N}_{RC} = 89 - (-110) = 21 \ \text{dB}$$

Notice that the link performance is different in different directions.

7.8 DIVERSITY RECEPTION

In chapter 4, we described how the various buildings and other obstacles in built-up areas scatter the signal, and, because of the interaction between the several incoming waves, the resultant signal at the antenna is subject to rapid and deep fading. The fading is most severe in heavily built-up areas in an urban environment. In these areas, the signal envelope follows a Rayleigh distribution over short distances and a log-normal distribution over large distances. Diversity reception techniques are used to reduce the effects of fading. In principle, diversity reception techniques can be applied either at the base station or at the mobile, although different problems have to be solved.

The basic idea of a diversity reception is that, if two or more independent samples of a signal are taken, then these samples will fade in an uncorrelated manner. This means that the probability of all the samples being simultaneously below a given level is much less than the probability of any individual sample being below that level. The probability of M samples all being simultaneously below a certain level is p^M, where p is the probability that a single sample is below the level. Thus, it can be seen that a signal composed of a suitable combination of the various samples will have much less severe fading properties than any individual sample alone.

7.9 BASIC COMBINING METHODS

After obtaining the necessary samples, we need to consider the question of processing these samples to obtain the best results. For most communication systems, the process can be broadly classified as the "linear combination of the samples." In the combining process, the various signal inputs are individually weighted and added together as:

$$r(t) = a_1 r_1(t) + a_2 r_2(t) + \ldots + a_M r_M(t) \tag{7.33a}$$

$$r(t) = \sum_{i=1}^{M} a_i r_i(t) \tag{7.33b}$$

where:
$r_i(t)$ is the envelope of the i-th signal and
a_i is the weight factor applied to the i-th signal.

We make the following assumptions in the analysis of a combiner:

1. The noise in each branch is independent of the signal and is additive.
2. The signal amplitudes change because of fading, but the fading rate is much smaller than the lowest modulation frequency present in the signal.
3. The noise components are locally incoherent and have zero mean, with a constant local mean square (i.e., constant noise power).
4. The local mean-square values (powers) of the signals are statistically independent.

Since the goal of the combiner is to improve the noise performance of the system, the analysis of combiners is generally performed in terms of SNR. We will examine several different types of combiners and compare their SNR improvements over no diversity.

7.9.1 Selection Combiner

The selection combiner is the simplest of all the diversity schemes. An ideal selection combiner chooses the signal with the highest instantaneous SNR, so the output SNR is equal to that of the best incoming signal. In practice, the system cannot function on an instantaneous basis; so, to be successful, it is essential that the internal time constants of a selection system are substantially shorter than the reciprocal of the signal fading rate.

We assume that the signal received by each diversity branch is statistically independent of the signals in other branches and is Rayleigh distributed with equal mean signal power P_0. The probability density function of the signal envelope, on branch i, is given by

$$p(r_i) = \frac{r_i}{P_0} e^{-r_i^2/2P_0} \tag{7.34}$$

where:
$2P_0$ = mean-square signal power per branch = $<r_i>$ and
r_i^2 = instantaneous power in the i-th branch.

Let $\xi_i = r_i^2/2N_i$ and $\xi_0 = (2P_0)/(2N_i)$, where N_i is the noise power in the i-th branch.

$$\therefore \frac{\xi_i}{\xi_0} = r_i^2/2P_0 \tag{7.35}$$

The probability density function for ξ_i is given by

$$p(\xi_i) = \frac{1}{\xi_0} e^{(-\xi_i/\xi_0)} \tag{7.36}$$

We assume that the signal in each branch has a constant mean; thus, the probability that the SNR on any one branch is less than or equal to any given value ξ_g is

$$P[\xi_i \le \xi_g] = \int_0^{\xi_g} p(\xi_i)\, d\xi_i = 1 - e^{(-\xi_g/\xi_0)} \tag{7.37}$$

Therefore, the probability that the SNRs in all branches are simultaneously less than or equal to ξ_g is given by

$$P_M(\xi_g) = P[\xi_1, \xi_2, ..., \xi_M \le \xi_g] = \left[1 - e^{(-\xi_g/\xi_0)}\right]^M \tag{7.38}$$

The probability that at least one branch will exceed the threshold SNR value of ξ_g is given by

$$P(\text{at least one branch} \ge \xi_g) = 1 - P_M(\xi_g) \tag{7.39}$$

The percentage of time the instantaneous output SNR ξ_M is below or equal to the threshold value, ξ_g, is equal to $P(\xi_M \le \xi_g)$. We plot results for $M = 1, 2$, and 4 in Figure 7.7. Note that the largest gain occurs for the two-branch diversity combiner. By differentiating Eq. (7.38) we get the probability density function

$$p_M(\xi_g) = (M/\xi_0)\left[1 - e^{(-\xi_g/\xi_0)}\right]^{M-1} e^{(-\xi_g/\xi_0)} \tag{7.40}$$

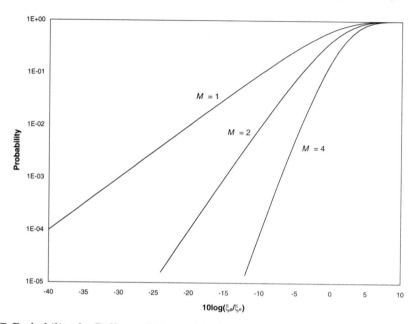

Fig. 7.7 Probability for Different Values of M-Selection Combiner

The mean value of the SNR can be given as

$$\overline{\xi_M} = \int_0^\infty M\left(\frac{\xi_g}{\xi_0}\right)\left[1 - e^{(-\xi_g/\xi_0)}\right]^{M-1} e^{(-\xi_g/\xi_0)} d\xi_g \tag{7.41}$$

Let $x = \xi_g/\xi_0$ and $dx = d\xi_g/\xi_0$

$$\therefore \frac{\overline{\xi_M}}{\xi_0} = M\int_0^\infty x\left[1 - e^{-x}\right]^{M-1} e^{-x} dx \tag{7.42}$$

Substituting $y = 1 - e^{-x}$ or $x = -\ln(1-y)$; then $dy = e^{-x}dx$

$$\therefore \frac{\overline{\xi_M}}{\xi_0} = M\int_0^1 [-\ln(1-y)]\, y^{M-1} dy = M\sum_{K=1}^{\infty} \int_0^1 \frac{1}{K}\, y^{M+K-1} dy \tag{7.43}$$

$$\therefore \frac{\overline{\xi_M}}{\xi_0} = \sum_{K=1}^{M} \frac{1}{K} \tag{7.44}$$

Table 7.1 shows that the mean SNR increases slowly with M.

7.9.2 Maximal-Ratio Combiner

Maximal-ratio combining was first proposed by L. R. Kahn [5]. The M signals are weighted proportional to their signal voltage-to-noise power ratios and then summed.

Table 7.1 Number of Branches vs. Mean SNR (dB)

M	$\dfrac{\overline{\xi_M}}{\xi_0}$	$10 \log \dfrac{\overline{\xi_M}}{\xi_0}$
1	1.000	0.000
2	1.500	1.761
3	1.833	2.632
4	2.083	3.187
5	2.283	3.585
6	2.450	3.892

$$r_M = \sum_{i=1}^{M} a_i r_i(t) \tag{7.45}$$

Since noise in each branch is weighted according to noise power,

$$\overline{n_i^2(t)} = \sum_{j=1}^{M}\sum_{i=1}^{M} a_i a_j \overline{n_i(t)\, n_j(t)} \tag{7.46}$$

The average noise power, $N_T = \displaystyle\sum_{i=1}^{M} a_i^2 \overline{n_i^2(t)} = 2\sum_{i=1}^{M} |a_i|^2 N_i \tag{7.47}$

where:

$$\overline{n_i^2(t)} = 2N_i \tag{7.48}$$

The SNR at the output is given as:

$$\xi_M = \frac{1}{2}\frac{\left| \displaystyle\sum_{i=1}^{M} a_i r_i(t) \right|^2}{\displaystyle\sum_{i=1}^{M} |a_i|^2 N_i} \tag{7.49}$$

We want to maximize ξ_M. This can be done by using the Schwartz inequality.

$$\left| \sum_{i=1}^{M} a_i r_i \right|^2 \le \left[\sum_{i=1}^{M} |r_i^2| \right]\left[\sum_{i=1}^{M} |a_i|^2 \right] \tag{7.50}$$

If $a_i = r_i / \sqrt{N_i}$ then

$$\xi_M = \frac{1}{2} \frac{\sum\limits_{i=1}^{M} r_i^2 \sum\limits_{i=1}^{M} \dfrac{r_i^2}{N_i}}{\sum\limits_{i=1}^{M} r_i^2} \tag{7.51}$$

$$\therefore \xi_M = \frac{1}{2} \sum_{i=1}^{M} \frac{r_i^2}{N_i} = \sum_{i=1}^{M} \xi_i \tag{7.52}$$

Thus, the SNR at the combiner output equals the sum of the SNR of the branches.

$$\overline{\xi_M} = \sum_{i=1}^{M} \xi_i = \sum_{i=1}^{M} \xi_0 = M\xi_0 \tag{7.53}$$

$$\therefore \frac{\overline{\xi_M}}{\xi_0} = M \tag{7.54}$$

The probability density function of the combiner output SNR is given by

$$p(\xi_M) = \frac{\xi_M^{M-1} e^{-\frac{\xi_M}{\xi_0}}}{\xi_0^M (M-1)!} \quad , \quad \xi_M \geq 0 \tag{7.55}$$

The probability that $\xi_M \leq \xi_g$ is given by:

$$P(\xi_M \leq \xi_g) = 1 - e^{-\frac{\xi_g}{\xi_0}} \sum_{K=1}^{M} \frac{\left(\dfrac{\xi_g}{\xi_0}\right)^{K-1}}{(K-1)!} \tag{7.56}$$

$$P(\xi_M > \xi_g) = e^{-\frac{\xi_g}{\xi_0}} \sum_{K=1}^{M} \frac{\left(\dfrac{\xi_g}{\xi_0}\right)^{K-1}}{(K-1)!} \tag{7.57}$$

The plot of P for $M = 1, 2,$ and 4 is shown in Figure 7.8.

7.9.3 Equal-Gain Combining

Equal-gain combining is similar to maximal-ratio combining, but there is no attempt to weight the signal before addition; thus $a_i = 1$. The envelope of the output signal is given by Eq. (7.33b) with all $a_i = 1$.

$$r = \sum_{i=1}^{M} r_i \tag{7.58}$$

and the mean output SNR is given as:

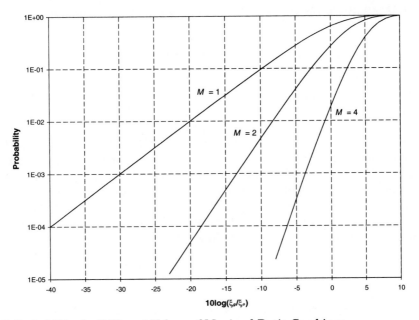

Fig. 7.8 Probability for Different Values of Maximal Ratio Combiner

$$\overline{\xi_M} = \frac{1}{2}\frac{\overline{\left[\sum\limits_{i=1}^{M} r_i\right]^2}}{\sum\limits_{i=1}^{M} \overline{N_i}} \tag{7.59}$$

We assume that mean noise power in each branch is same (i.e., N); then Eq. (7.59) becomes

$$\overline{\xi_M} = \frac{1}{2NM}\overline{\left[\sum\limits_{i=1}^{M} r_i\right]^2} = \frac{1}{2NM}\sum\limits_{j,i=1}^{M} \overline{r_j r_i} \tag{7.60}$$

but $\overline{r_i^2} = 2P_0$; and $\overline{r_i} = \sqrt{\dfrac{\pi P_0}{2}}$.

Since the various branch signals are uncorrelated, $\overline{r_j r_i} = \overline{r_i r_j} = \overline{r_i}\,\overline{r_j}$, for i not equal to j. Therefore Eq. (7.60) will be:

$$\overline{\xi_M} = \frac{1}{2NM}\left[2MP_0 + M(M-1)\frac{\pi P_0}{2}\right] = \xi_0\left[1 + (M-1)\frac{\pi}{4}\right] \tag{7.61}$$

$$\frac{\overline{\xi_M}}{\xi_0} = 1 + (M-1)\frac{\pi}{4} \tag{7.62}$$

For $M = 2$, the probability P can be written in closed form as:

$$P(\xi_M \leq \xi_g) = 1 - e^{-\left(\frac{2\xi_g}{\xi_0}\right)} - \sqrt{\pi\left(\frac{\xi_g}{\xi_0}\right)} \, e^{-\frac{\xi_g}{\xi_0}} \cdot erf\sqrt{\frac{\xi_g}{\xi_0}} \tag{7.63}$$

For $M > 2$, the probability can be obtained by numerical integration techniques. The plot of probability $P(\xi_M \leq \xi_g)$ is given in Figure 7.9.

Table 7.2 shows M vs. SNR at 1% probability for the selection, maximal-ratio, and equal-gain combiner. Table 7.3 shows SNR improvement for $M = 2, 4$, and 6 at 1% probability for the selection, maximal-ratio, and equal-gain combiner. It can be seen that selection diversity scheme has the poorest performance and maximal ratio the best. The performance of equal-gain combining is only marginally inferior to maximum ratio. The implementation complexity for equal-gain combining is significantly less than the maximal-ratio combining because of the requirement of correct weighing factors. The data is compared in Figure 7.10.

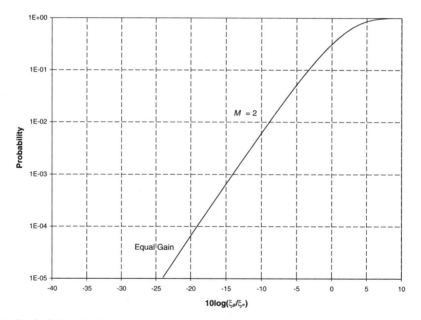

Fig. 7.9 Probability for $M = 2$ Equal-Gain Combiner

Table 7.2 SNR Ratio dB

M	Selection	Maximal Ratio	Equal Gain
1	−20.0	−20.0	−20.0
2	−10.0	−8.5	−9.2
4	−4.0	−1.0	−2.0
6	−2.0	2.0	1.5

Table 7.3 SNR Improvement (dB)

M	Selection	Maximal Ratio	Equal Gain
2	10.0	11.5	10.8
4	16.0	19.0	18.0
6	18.0	22.0	21.5

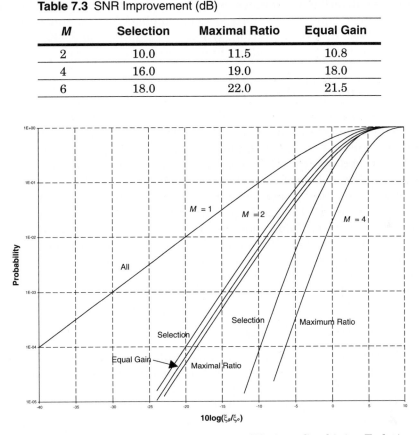

Fig. 7.10 Performance Comparison Improvement of Various Combining Techniques

7.10 TYPES OF DIVERSITY

In this section, we examine the type of diversity that can be used to provide the inputs to a diversity combiner. Typically, the diversity receiver is used in the base station instead of the mobile telephone. The cost of the diversity combiner can be high, especially if multiple receivers are necessary. Also the power output of the mobile telephone is limited by the battery of the vehicle or, as more telephone sets are handheld, by the battery of the handset. Handset transmitters are usually lower power than mobile-mounted transmitters to preserve battery life and reduce radiation into the human body. The base station, however, can increase its power output or antenna height to improve coverage to a mobile telephone.

Also, most diversity systems are implemented in the receiver instead of the transmitter since no extra transmitter power is needed to implement the receiver diversity system. Thus, the discussions in the remainder of this section will focus on base station diversity techniques. Since the path between the mobile and base is assumed to be reciprocal, diversity systems implemented in a mobile will work similarly to those in a base station. When we examine the RAKE receiver for CDMA, both the base station and the handset will use diversity.

7.10.1 Macroscopic Diversity

The local mean signal strength varies because of variations of terrain between the mobile transmitter and the base station receiver. If only one antenna site is used, the traveling mobile unit may not be able to transmit a signal to the base station at certain geographical locations because of terrain variations such as hills or mountains. Therefore, two separated antenna sites can be used to receive two signals and to combine them to reduce long-term fading. The selective combining technique is recommended in the macroscopic diversity scheme since other methods require coherent combining that is difficult to achieve when the receivers are some distance apart. Macroscopic diversity is often used in shortwave radio systems to reduce the effects of fading from the ionosphere. Cellular and PCS system achieve the same effect by handoffs to nearby cell sites when the signal strength becomes weak.

7.10.2 Microscopic Diversity

Microscopic diversity uses two or more antennas that are at the same site (colocated) but designed to exploit differences in arriving signals from the receiver. Once the diversity branches are created, any of the previous combining methods can be used. The following methods are used to derive uncorrelated signals for combining:

1. **Space diversity:** Two antennas separated physically by a short distance d can provide two signals with low correlation between their fades. The separation d in general varies with antenna height h and with frequency. The higher the frequency, the closer the two antennas can be to each other. Typically a separation of a few wavelengths is enough to obtain uncorrelated signals.

2. **Frequency diversity:** Signals received on two frequencies, separated by the coherence bandwidth, B_c, are uncorrelated. To use frequency diversity in an urban or suburban environment for cellular and PCS frequencies, the frequency separation must be 300 kHz or more.

3. **Polarization diversity:** The horizontal and vertical polarization components, E_x and E_y, transmitted by two polarized antennas at the base station and received by two polarized antennas at the mobile unit, can provide two uncorrelated fading signals. Polarization diversity results in a 3-dB power reduction at the transmitting site since the power must be split into two different polarized antennas.

4. **Angle diversity:** When the operating frequency is ≥ 10 GHz, the scattering of the signals from transmitter to receiver generates received signals from

different directions that are uncorrelated with each other. Thus, two or more directional antennas can be pointed in different directions at the receiving site and provide signals for a combiner. This scheme is more effective at the mobile unit than at the base station since the scattering is from local buildings and vegetation and is more pronounced at street level than at the height of base station antennas.

5. **Time diversity:** If the identical signal is transmitted in different time slots, the received signals will be uncorrelated. This system will work for an environment where the fading occurs independent of the movement of the receiver. In a mobile radio environment, the mobile unit may be at a standstill at any location that has a weak local mean or is caught in deep fade. Although fading still occurs even when the mobile is still, the time-delayed signals are correlated and time diversity will not reduce the fades.

7.10.3 RAKE Receiver

In 1958, R. Price and P. E. Green [13] proposed a method of resolving multipaths using wideband pseudorandom sequences modulated onto a transmitter using other modulation methods (AM or FM). The pseudorandom sequence has the property that time-shifted versions of itself are almost uncorrelated. Thus, a signal that propagates from transmitter to receiver over multiple paths (hence multiple different time delays) can be resolved into separately fading signals by cross-correlating the received signal with multiple time shifted versions of the pseudorandom sequence. Figure 7.11 shows a block diagram of a typical system. In the receiver, the outputs are time shifted and therefore must be sent through a delay line before entering the diversity combiner. Various attempts at implementing this system (e.g., see J. E. Wilkes [16]) required that the multipaths be resolved at the expense of significant extra bandwidth of the system and thus low spectrum efficiency. The receiver is called a RAKE receiver since the block diagram looks like a garden rake.

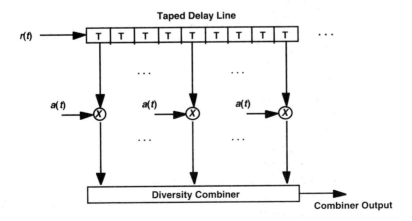

Fig. 7.11 RAKE Receiver

When the CDMA systems were designed for cellular systems, the inherent wide bandwidth signals with their orthogonal Walsh functions (see Appendix B) were a natural for implementing a RAKE receiver. In addition, the RAKE receiver mitigates the effects of fading and is in part responsible for the claimed 10:1 spectral efficiency improvement of CDMA over analog cellular.

In the CDMA system, the bandwidth (1.5 to 15 MHz) is wider than the coherence bandwidth of the cellular or PCS channel. Thus, when the multipaths are resolved in the receiver, the signals from each tap on the delay line are uncorrelated with each other. The receiver can then combine them using any of the combining techniques described in the previous sections. The CDMA system then uses the multipath characteristics of the channel to its advantage to improve the operation of the system.

The performance of the RAKE receiver will be governed by the combining system used. An important factor in the receiver design is obtaining synchronization of the signals in the receiver to that of the transmitted signal. Since adjacent cells are also on the same frequency with different time delays on the Walsh codes, the entire CDMA system must be tightly synchronized. See chapter 8 for a further discussion of the synchronization issues.

7.11 EXAMPLES OF BASE STATION AND MOBILE ANTENNAS

While the simple dipole antenna is the reference example for antenna specifications, most practical antenna designs aim to improve on the gain of the dipole antenna. Practical antenna design must consider the following issues:

☞ **Antenna pattern:** Closely related to the gain of the antenna is the antenna pattern. As gain is increased, the beamwidth is decreased. This can be an advantage or a disadvantage depending on the antenna orientation and the needs of the system design.

☞ **Bandwidth:** The antenna must operate over the full range of frequencies in use for the cellular or PCS system. If the antenna bandwidth is small, channels at the edge of the band may not receive signals as well as those near the band center.

☞ **Gain:** The higher the gain of the antenna, the lower the power that is necessary at the transmitter. Since the antenna is purchased once and the transmitter power is purchased continuously, high-gain antennas are useful for saving money in electricity and help conserve the natural resources used to create the electricity.

☞ **Ground plane:** Some antennas require that they be mounted above a reflecting surface to function correctly. For example, a quarter-wave antenna is one-half of a dipole and requires that the other half of the dipole be developed by a mirror image below a ground plane. This can be used to advantage in designing antennas for vehicles but is a disadvantage when base station antennas (high above the earth) are designed.

☞ **Height:** The higher the antenna, the better the coverage of the system. However, if the coverage of the system is too good, interference from other cells may become troublesome. In an interference-limited system, all levels scale equally so, at the first order, there will not be a problem. However, since radio wave propagation is statistical, there may be locations where good propagation exists from a point far removed from a base station. The higher the base station antenna, the more likely that these anomalous events will occur.

☞ **Input impedance:** Most cables used as feed line from the transmitter/ receiver to the antenna are either 50 ohms or 72/75 ohms. If the input impedance of the antenna is far removed from either of these values, it will be difficult to get the antenna to accept the power delivered to it and its efficiency η will be low.

☞ **Mechanical rigidity:** If the antenna flexes in the wind, it will introduce an additional fading component to the received signal. Ultimately, the continuous flexing will cause metal fatigue and mechanical failure of the antenna.

☞ **Polarization:** For wireless cellular and PCS communications, a vertical antenna is the easiest to mount on a vehicle; therefore vertical polarization has been standardized. In general, horizontal or vertical polarization will work equally well.

With this background, we will examine some simple antennas that are used for base and mobile operation.

7.11.1 Quarter-Wave Vertical

The simplest antenna for a vehicle is the quarter-wave vertical (see Figure 7.12). A length of wire 1/4 wavelength long is mounted on the roof of the vehicle. With metal vehicles (most cars and trucks), the other half of the dipole is developed in the image in the ground plane. Since a vertical dipole antenna has an omnidirectional pattern [6], the quarter-wave vertical has an omnidirectional pattern. The gain of the antenna is the same as that of a dipole 0 dBd or 2.1 dBi.

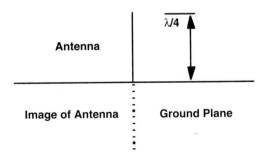

Fig. 7.12 Quarter-Wave Vertical Antenna

The impedance of a quarter-wave vertical is 36.5 ohms and requires a matching transformer for proper feeding of the antenna.

7.11.2 Stacked Dipoles

Since a vertical dipole has an omnidirectional pattern, three or more dipoles can be stacked vertically to produce increased gain and maintain the omni pattern. A typical base station antenna will have four half-wave dipoles spaced by 1 wavelength vertically. The gap can also be zero instead of one-half of a wave length (see Figure 7.13). All the antenna elements are fed with a signal from the transmitter and are fed in phase. The resultant pattern is omnidirectional in the horizontal plane and has 8.6-dBi gain on the horizon (0° elevation) with a vertical beamwidth of ±6.5°. It has an impedance of 63 ohms; thus a matching transformer is necessary, but it is easier to build than the one for the quarter-wave vertical.

A variation of the base station antenna for use on a vehicle uses a half-wave dipole above a quarter-wave vertical (see Figure 7.14). The other half of the antenna is in the ground plane image. The two elements are decoupled from each other by a quarter wavelength long decoupling coil. This is the common cellular antenna seen on most vehicles. It has a gain of 7–10 dBi.

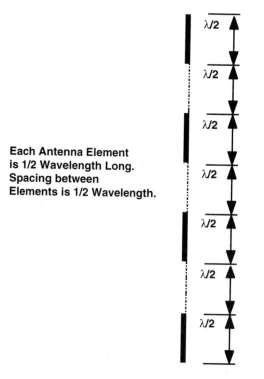

Each Antenna Element is 1/2 Wavelength Long. Spacing between Elements is 1/2 Wavelength.

λ/2

λ/2

λ/2

λ/2

λ/2

λ/2

λ/2

Fig. 7.13 Stacked Vertical Dipoles

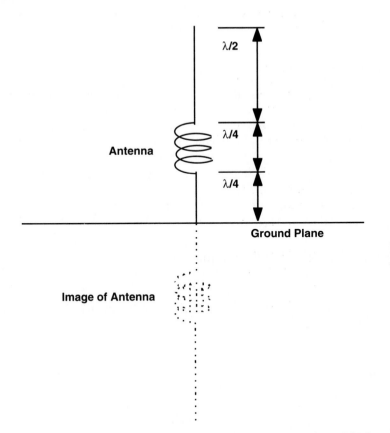

Fig. 7.14 Half-Wave over Quarter-Wave Antenna—When Mounted on Vehicle

7.11.3 Corner Reflectors

The previously discussed antennas have omnidirectional patterns. As described in chapter 3, a directional antenna at a base station can improve the SNR of the system and thus improve the spectral efficiency of the system. In 1939 J. D. Kraus [6] designed the corner reflector antenna consisting of a vertical dipole and two sheets of metal at a 45°, 60°, or 90° angle (see Figure 7.15). The impedance and gain of the antenna depends on the angle of the corner and the spacing from the corner to the dipole antenna. The gain of the antenna varies from 7 to 13 dBd, and the impedance varies from 0 to 150 ohms. A colinear antenna can be used in place of the dipole for additional gain.

7.12 SUMMARY

In this chapter we discussed the role of antennas in the wireless system. We looked at theoretical antennas and practical antennas used in cellular systems.

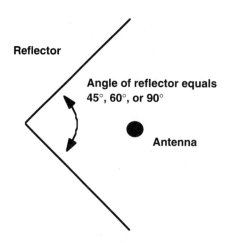

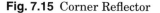

Fig. 7.15 Corner Reflector

We introduced the concepts of antenna gain and beamwidth, receiver noise figure, and propagation loss and used these concepts to do a link analysis of the wireless system and calculate the SNR at the receiver. We also presented the concepts of diversity reception where multiple signals are combined to improve the SNR of the system.

7.13 PROBLEMS

1. In Example 7-1, calculate the SNR ratio when frequencies of 3 GHz and 6 GHz are used for the system. At the higher frequencies use 10 dB for the receiver noise figure.

2. Repeat Example 7-2 for 3 and 6 GHz and use 10 dB for the receiver noise figure. What can you say about system design needs as the frequency is raised?

3. If the 3-dB beamwidth of a linear element antenna is 8°, find its directive gain.

4. A sector antenna has the vertical and horizontal beamwidths of 50° and 60°, respectively. Calculate its directive gain (1) ignoring the side lobes and (2) accounting for the side lobes.

5. Calculate the mean signal strength and mean SNR ratio for the downlink and uplink of the antenna system shown in Figure 7.6, assuming radiated power of cell site = 160 W EIRP (52 dBm) and radiated power of the mobile = 2 W EIRP (33 dBm).

7.14 REFERENCES

1. Brennan, D. J., "Linear Diversity Combining Techniques," Proceedings of the IRE, June 1959, 1075–101.

2. Bryson, W. B., "Antenna System for 800 MHz," IEEE Vehicular Technology Conference, San Diego, May 1982, 287.

3. Halpern, S. W., "The Theory of Operation of an Equal-Gain Predetection Regenerative Diversity Combiner with Rayleigh Fading Channels," *IEEE Transaction on Communications Technology,* COM–22 (8) August 1974, 1099–106.

4. Jakes, W. C., *Microwave Mobile Communications,* New York: John Wiley & Sons, 1974.

5. Kahn, L. R., "Radio Squarer," Proceedings of the IRE, 42, November 1954, 1704.

6. Kraus, J. D., *Antennas,* New York: McGraw-Hill, 1988.

7. Lee, W. C. Y., "Antenna Spacing Requirements for a Mobile Radio Base-Station Diversity," *Bell System Technical Journal* 50, no. 6 (July–August 1971).

8. Lee, W. C. Y., *Mobile Communications Engineering,* New York: McGraw-Hill, 1982.

9. Lee, W. C. Y., "Mobile Radio Performance for a Two-Branch Equal-Gain Combining Receiver with Correlated Signals at the Land Site," *IEEE Transaction* VT-27, November 1978, 239–43.

10. Mahrotra, A., *Cellular Radio Performance Engineering,* Boston: Artech House, 1994.

11. Parsons, J. D., et al., "Diversity Techniques for Mobile Radio Reception," *Radio and Electronic Engineer* 45, no. 7 (July 1975): 357–67.

12. Parsons, J. D., and J. G. Gardiner, *Mobile Communications Systems,* New York: Halsted Press, 1989.

13. Price, R., and P. E. Green, Jr., "A Communications Technique for Multipath Channels," Proceedings of the IRE, March 1958, 555–70.

14. Schwartz, M., et al., *Communication Systems and Techniques,* New York: McGraw-Hill, 1966.

15. Tilston, W. V., "On Evaluating the Performance of Communication Antennas," *IEEE Communications Society Magazine,* (September 1981).

16. Wilkes, J. E., "An Antimultipath Communications System Applied to Analog FM," Ph.D. diss., Polytechnic Institute of Brooklyn (now Polytechnic University), June 1971.

CHAPTER 8

North American Cellular and PCS Systems

8.1 INTRODUCTION

The North American cellular systems have evolved to two digital standards—IS-136 (enhanced IS-54) using TDMA and IS-95 using CDMA. The cellular versions of the standards support both analog AMPS and the digital protocols. Thus, there are three types of phones in general use for cellular in North America: an analog-only AMPS phone, a dual-mode AMPS and TDMA phone, and a dual-mode AMPS and CDMA phone.

When work was started on PCS, each of the two groups that supported TDMA and CDMA made modifications to their standards to remove the analog capabilities and support the new PCS frequencies. The standards also support a dual frequency phone that can function on cellular or PCS frequencies.

Other groups did not want to use the existing CDMA and TDMA protocols and set about to define a new set of protocols and services for PCS. Oki, a Japanese company with an extensive market share throughout the world, has set about defining a wideband CDMA protocol that will be used in the United States and Asia.

Motorola, Bellcore, and others have worked extensively in defining a protocol for use by wireline companies that want to provide switching functions while letting radio companies provide the radio portion of the call. This protocol is called PACS and uses TDMA. Its primary use is for low-mobility applications in residential wireless.

Omnipoint has defined a combined CDMA/TDMA protocol, PCS2000, for use in some systems.

While PCS was initially thought of as a new service, clearly different companies have different interests in what PCS is and how they will provide services. Existing cellular companies are looking at PCS as a means to provide wide coverage areas and fill in gaps in their current service offerings. For these companies, PCS is an extension of cellular to the 1900-MHz band using identical standards for both bands. Other companies see PCS as the opportunity to offer new services that will compete with cellular by offering lower costs, additional services, and better quality of existing services. These companies do not want the existing protocols but want new ones or extensions of the old ones to support the new services.

This chapter discusses the original analog cellular air interface and five digital air interfaces and compares the characteristics of them. All of these air interfaces share a common heritage from two sources. The original analog cellular system was defined in 1979 by the EIA. All of the digital air interfaces inherit their characteristics from the analog protocol. In addition, all of the air interfaces discussed in this chapter have services based on the IS-41 intersystem communications protocol. This is the protocol supported in North America for communications between wireless systems. IS-41 and IS-104 (for supplementary services) define the functionality for a Mobile Application Part (MAP) for U.S.-based PCS. Since the GSM system (discussed in chapter 9) supports a different MAP, chapter 12 presents the differences between the two systems and discusses issues and solutions for interworking between the two systems.

First this chapter examines the TR-46 and T1P1 reference models and compares them. We discuss both the elements of the models and the interfaces. Then we examine the basic services from the T1P1 service description and the supplementary services from IS-104. Call flows for basic call processing are discussed with operations traced from the personal station to the radio system to the PCS switch and other network elements. Since each of the six air interfaces accomplish the same purpose with different protocols, we examine each interface and describe the similarities and differences of the protocols.

Work on PCS has progressed in five different standards groups:

☞ **T1P1:** This committee is under the ATIS and is responsible for PCS services requirements.

☞ **TR-46**: This committee is under the TIA and is responsible for PCS services and protocols.

☞ **T1S1:** This is an ATIS body responsible for the signaling protocols and is undertaking the role of the upper-layer signaling protocols between various elements of the PCS system.

☞ **T1M1:** This is an ATIS body responsible for operations, administration, maintenance, and provisioning (OAM&P) and is undertaking the role of the OAM&P services and protocols for PCS.

☞ **The Joint Technical Committee of T1P1 and TR-46:** This committee is result of a cooperative effort between T1P1 and TR-46 and has the responsibilities of developing the air interface requirements for PCS.

Clearly with five standards bodies examining the problem, there is overlap between some of the bodies.

8.2 PCS REFERENCE MODELS

Key to the North American systems is the use of a common reference model. Both TR-46 and T1P1 have a reference model, but each model can be converted into the other one. All of the North American systems follow the reference model. The names of each of the network elements are similar, and some of the functionality is partitioned differently between the models. The main difference between the two reference models is how mobility is managed. Mobility is the capability for users to place and receive calls in systems other than their home system. In the T1P1 reference model, the user data and the terminal data are separate; thus users can communicate with the network via different radio personal terminals. In the TR-46 reference model, only terminal mobility is supported. A user can place or receive calls at only one terminal (the one the network has identified as owned by the user). The T1P1 functionality is migrating toward independent terminal and user mobility, but all aspects of it are not currently supported. Although these models are for PCS, they apply equally well to cellular systems.

8.2.1 TR-46 Reference Model

The main elements of the TR-46 reference model (see Figure 8.1) are:

☞ **The Personal Station (PS):** terminates the radio path on the user side and enables the user to gain access to services from the network. The PS can be a stand-alone device or can have other devices (e.g., personal computers, fax machines) connected to it.

☞ **The Radio System (RS):** often called the base station, terminates the radio path and connects to the personal communications switching center. The RS is often segmented into the base transceiver system, and the base station controller:

✗ **The Base Transceiver System (BTS):** consists of one or more transceivers placed at a single location and terminates the radio path on the network side. The BTS may be colocated with a base station controller or may be independently located.

✗ **The Base Station Controller (BSC):** the control and management system for one or more BTSs. The BSC exchanges messages with both the BTS and the personal communications switching center. Some signaling messages may pass through the BSC transparently.

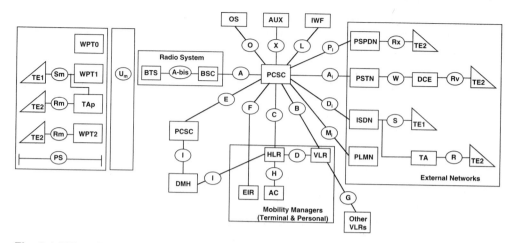

Fig. 8.1 TR-46 Reference Model

☞ **The Personal Communications Switching Center (PCSC):** an automatic system that interfaces the user traffic from the wireless network to the wireline network or other wireless networks.

☞ **The Home Location Register (HLR):** the functional unit used for management of mobile subscribers by maintaining all subscriber information (e.g., Electronic Serial Number (ESN), Directory Number (DN), International Mobile Subscriber Identity (IMSI), user profiles, current location). The HLR may be colocated with a PCSC, an integral part of the PCSC, independent of the PCSC. One HLR can serve multiple PCSCs, or an HLR may be distributed over multiple locations.

☞ **Data Message Handler (DMH):** used for billing and is described in chapter 11.

☞ **The Visited Location Register (VLR):** is linked to one or more PCSCs. The VLR is the functional unit that dynamically stores subscriber information (e.g., ESN, DN, user profile information) obtained from the user's HLR, when the subscriber is located in the area covered by the VLR. When a roaming mobile station enters a new service area covered by a PCSC, the PCSC informs the associated VLR about the PS by querying the HLR after the PS goes through a registration procedure.

☞ **The Authentication Center (AC):** manages the authentication or encryption information associated with an individual subscriber. As of this writing, the details of the operation of the AC have not been defined. The AC may be located within an HLR or PCSC or may be located independently of both.

☞ **The Equipment Identity Register (EIR):** provides information about the PS for record purposes. As of this writing, the details of the operation of the EIR have not been defined. The EIR may be located within a PCSC or may be located independently of it.

☞ **Operations System (OS):** is responsible for overall management of the wireless network. See chapter 11 for a full description of the OAM&P functions of a OS.

☞ **Interworking Function (IWF):** enables the PCSC to communicate with other networks. See chapter 12 for details on interworking.

The external networks are other communications networks—the Public Switched Telephone Network (PSTN), the Integrated Services Digital Network (ISDN), the Public Land Mobile Network (PLMN), and the Public Switched Packet Data Network (PSPDN).

The following interfaces are defined between the various elements of the system:

☞ RS to PCSC (A-Interface): The interface between the RS and the PCSC supports signaling and traffic (both voice and data). A-Interface protocols have been defined using SS7, ISDN BRI/PRI, and frame relay.

☞ The BTS to BSC Interface (A-bis): If the RS is segmented into a BTS and BSC, this internal interface is defined.

☞ The PCSC to PSTN Interface (Ai): This interface is defined as an analog interface using either Dual Tone Multifrequency (DTMF) signaling or Multifrequency signaling.

☞ PCSC to VLR (B-Interface): This interface is defined in the TIA IS-41 protocol specification.

☞ PCSC to HLR (C-Interface): This interface is defined in the TIA IS-41 protocol specification.

☞ HLR to VLR (D-Interface): This interface is the signaling interface between an HLR and a VLR and is based on SS7. It is currently defined in the TIA IS-41 protocol specification.

☞ PCSC to ISDN (D_i-Interface): This is the digital interface to the public telephone network and is a T1 interface (24 channels of 64 kbs) and uses Q.931 signaling.

☞ PCSC to PCSC (E-Interface): This interface is the traffic and signaling interface between wireless networks. It is currently defined in the TIA IS-41 protocol specification.

☞ PCSC to EIR (F-Interface): Since the EIR is not yet defined, the protocol for this interface is not defined.

☞ VLR to VLR (G-Interface): When communication is needed between VLRs, this interface is used. It is defined by TIA IS-41.

☞ HLR to AC (H-Interface): The protocol for this interface is not defined.

☞ DMH to PCSC (I-Interface): This interface is described in chapter 11.

☞ PCSC to the IWF (L-Interface): This interface is defined by the IWF. See chapter 12 for examples of data interworking interfaces.

☞ PCSC to PLMN (M_i-Interface): This interface is to another wireless network.

☞ PCSC to OS (O-Interface): This is the interface to the OS. It is currently being defined in ATSI standard body T1M1.

☞ PCSC to PSPDN (P_i-Interface): This interface is defined by the packet network that is connected to the PCSC.

☞ Terminal Adapter (TA) to Terminal Equipment (TE) (R-Interface): These interfaces will be specific for each type of terminal that will be connected to a PS. See chapter 12 for a discussion of this interface.

☞ ISDN to TE (S-Interface): This interface is outside the scope of PCS and is defined within the ISDN system.

☞ RS to PS (U_m-Interface): This is the air interface. Later sections in this chapter will discuss this interface in detail.

☞ PSTN to Data Communication Equipment (DCE) (W-Interface): This interface is outside the scope of PCS and is defined within the PSTN system.

☞ PCSC to Auxiliary (AUX). (X-Interface): This interface depends on the auxiliary equipment connected to the PCSC.

8.2.2 T1P1 PCS Reference Architecture

The T1P1 architecture (Figure 8.2) is similar to the TR-46 model but has some differences. The following elements are defined in the model:

☞ Radio Personal Terminal (RPT): is identical to the PS of the TR-46 model.

☞ Radio Port (RP): is identical to the BTS of the TR-46 model.

☞ Radio Port Intermediary (RPI): provides an interface between one or more RPs and the radioport controller. The RPI allocates radio channels and may control handoffs. It is dependent on the air interface.

☞ Radio Port Controller (RPC): identical to the BSC of the TR-46 model.

☞ Radio Access System Controller (RASC): performs the radio-specific switching functions of call delivery and origination, handoff control, registration and authentication, and radio access management (control of signaling channels).

☞ PCS Switching Center (PSC): similar to the PCSC of the TR-46 reference model. Some of the functions of the PCSC in the TR-46 model are distributed into other elements.

☞ Terminal Mobility Controller (TMC): provides the control logic for terminal authentication, location management, alerting, and routing to RPTs. This function is supported by the VLR and the PCSC in the TR-46 reference model.

☞ Terminal Mobility Data Store (TMD): maintains the associated terminal data and is similar to the VLR in the TR-46 reference model.

☞ Personal Mobility Controller (PMC): provides the control logic for user authentication, location management, alerting, and routing to users. This function is supported by the HLR and the PCSC in the TR-46 reference model.

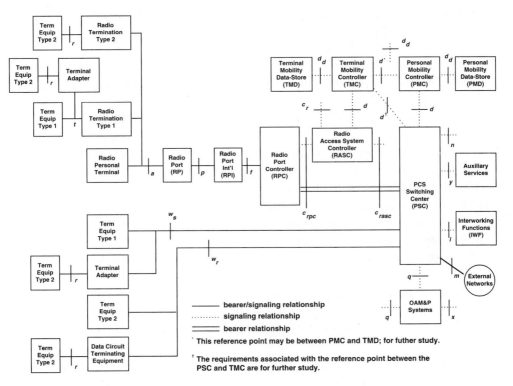

Fig. 8.2 T1P1 Reference Model

☞ Personal Mobility Data Store (PMD): maintains the associated user data and is similar to the HLR in the TR-46 reference model.

☞ OAM&P Systems: identical to the OS in the TR-46 reference model.

☞ Auxiliary Services: a variety of services such as voice mail and paging that may be provided by the PSC.

☞ Interworking Function (IWF): identical to the IWF of the TR-46 reference model.

☞ External Networks: those networks (i.e., wired and wireless) that are not part of the described wireless network.

The following interfaces are described between the elements in the system:

☞ RP to RPT (a-Interface): This interface is identical to the U_m-Interface of the TR-46 model.

☞ PSC to RPC (c-Interface): This is similar to the A-Interface of the TR-46 model. If an RASC is used, then the c_{rpc}-, c_{rasc}-, and c_r-interfaces are defined.

☞ TMC to other elements (d-Interface): The d-Interface is between the TMC and the RASC and between the PMC and the PSC. The d_d-Interface is between the TMC and the TMD and between the TMC and the PMC.

☞ RPI to RPC (f-Interface): This interface is between the RP and the RPI; it may or may not be internal to a radio system.

☞ PSC to interworking functions (l-Interface): This is the same as the L-Interface in the TR-46 reference model. As in the TR-46 model, this interface is defined by the IWF. See chapter 12 for examples of data IWFs.

☞ PSC to external networks (m-Interface): This is the same as the TR-46 reference model to external networks except that TR-46 segments this interface into type of network (A_i, D_i, M_i, and P_i).

☞ PSC to other PSC (n-Interface): This interface is to other PSCs.

☞ RPI to RP (p-Interface): This interface carries baseband bearer and control information, contained in the air interface, between the RPI and the RP.

☞ OAM&P systems to PSC (q-Interface): This is the same as the O-Interface in the TR-46 reference model.

☞ RT to TE (t-Interface): This interface depends on the type of equipment and is the same as the R-Interface of the TR-46 reference model.

☞ PSC to TE (w-Interface): This interface depends on the type of equipment and allows terminal equipment to be connected directly to the PSC. There is no equivalent in the TR-46 reference model.

☞ OAM&P to craft terminal (x-Interface): This interface provides capabilities to access the OS to display, edit, add/delete information.

☞ PSC to auxiliary services (y-Interface): This interface, which is the same as the X-Interface of the TR-46 reference model, depends on the auxiliary equipment connected to the PCSC.

8.3 SERVICES

With the reference models described above, there are enough capabilities to support a wide range of telecommunications services over PCS. Many of these services are similar to those of the wireline network; some are specific to the untethered approach the PCS provides. The services defined here are based on an MAP that is supported by the IS-41 Intersystem Communications Protocol. Most of the IS-41-based services are conceptually the same as those offered by GSM-based systems (see chapter 9). Since these services were originally designed for the North American market, they may not work the same as GSM, which was designed for the European market. See chapter 12 for a more detailed discussion of the differences between the two systems.

8.3.1 Basic Services

The standards body T1P1 is in the process of defining basic call functions and supplementary services for PCS. The T1P1 Stage 2 Service description [2] defines 15 basic services (information flows) that can be grouped as follows:

☞ Registration and deregistration functions to support the process where a PS informs a PCS system of its desire to receive service and its approximate location:

✗ Automatic registration

✗ Terminal authentication and privacy (using private key cryptography)

✗ Terminal authentication and privacy (using public key cryptography)

✗ User authentication and validation

✗ Automatic personal registration

✗ Automatic personal deregistration

✗ Personal registration

✗ Personal deregistration

☞ The Registration and deregistration process requires that a PS identify itself to the PCS network and requires that the PCS network communicate with the home PCS network to obtain security and service profile information. The operation of the registration process will be discussed in detail in chapter 10 on Security.

☞ Roaming: This is the process where a PS registers and receives service in a PCS system other than its home system. This is discussed in more detail in section 8.4.2.5 of this chapter.

☞ Call establishment, call continuation, and call clearing procedures:

✗ Call origination

✗ Call delivery (call termination)

✗ Call clearing

✗ Emergency (E911) calls

✗ Handoff

Each of these procedures will be discussed in later sections of this chapter and again in chapter 12 as the issues of interworking of services are discussed.

8.3.2 Supplementary Services

Supplementary services are defined in the IS-104 Personal Communications Service Descriptions for 1800 MHz (PN-3168). The IS-41 C specification defines those services that can be made available to users as they roam. Obviously, these services would also be available to users in their home system. Additional services may be available in a specific home PCS or cellular system, but users would not necessarily have them available in other systems since no common set of procedures and protocols have been defined to support other services.

These services are:

☞ **Automatic Recall** allows a wireless subscriber calling a busy number to be notified when the called party is free and have the PCS system re-call the number.

☞ **Automatic Reverse Charging** (ARC) allows a wireless subscriber to be charged for calls the special ARC number. This service is similar to wireline 800 service in North America.

☞ **Call Hold and Retrieve** allows a wireless subscriber to interrupt a call and return to the call.

☞ **Call Forwarding-Default** represents the ability to redirect a call to a PS handset in three situations: "unconditional," "Busy," and "No Answer." The PS call forwarding (PS-CF) features build upon the PS call terminating capability. Under all of these features, calls may be forwarded by the network to another PS or to a DN associated with a wireline interface. There are no additional information flows for PS-CF beyond the information flow for PS call terminating.

☞ **Call Forwarding-Busy** permits a called PCS subscriber to have the system send incoming calls addressed to the called personal communications subscriber's personal number to another personal, terminal, or directory number when the PCS subscriber is engaged in a call.

   With "personal call forwarding-busy" activated, a call incoming to the PCS subscriber will be automatically forwarded to the forward-to number whenever the PCS subscriber is already engaged in a prior call.

☞ **Call Forwarding-No Answer** permits a called PCS subscriber to have the system send all incoming calls addressed to the called PCS subscriber's personal number to another personal, terminal, or directory number when the PCS subscriber fails to answer or doesn't respond to paging.

   With "personal call forwarding-no answer" activated, a call incoming to the PCS subscriber will be automatically forwarded to the designated forward-to number whenever the PCS subscriber does not respond to the page or if the PCS subscriber does not answer within a specified period after transmission of the alert indication.

☞ **Call Forwarding-Unconditional** permits a PCS user to send incoming calls addressed to the PCS subscriber's personal number to another personal, terminal, or directory number (forward-to number). The ability of the served PCS subscriber to originate calls is unaffected. If this service is activated, calls are forwarded independent of the state of the PS (busy, idle, and so on).

☞ **Call Transfer** permits a PCS user to transfer a call to another number on or off the PCS switch. When a call is transferred, the PCS personal terminal is then available for other calls.

☞ **Call Waiting** provides notification to a PCS subscriber of an incoming call while the user's personal station is in the busy state. Subsequently, the user can either answer or ignore the incoming call.

   With call waiting activated, the PCS user who is already engaged in conversation on a prior call will receive a notification signal of a new incoming call attempt. This may be repeated a short time later if the PCS user takes no action. The calling party will hear an audible ringing signal either until they abort the call attempt or until the PCS user acknowledges the waiting call. The PCS user may indicate acceptance of the waiting call by: 1) placing the existing call on hold or 2) releasing the existing call.

☞ **Calling Number Identification Presentation (CNIP)** is a supplementary service offered to a called party. It provides to the called party the number identification of the calling party. If the calling party has subscribed to Calling Number Identification Restriction (CNIR), the calling number will not be presented.

☞ **Calling Number Identification Restriction (CNIR)** is a supplementary service offered to a calling party that restricts presentation of that party's calling number identification to the called party. When the CNIR service is applicable and activated, the originating network provides the destination network with a notification that the calling number identification is not allowed to be presented to the called party.

   CNIR may be offered with several options. Subscription options applied are: not subscribed (inactive for all calls); permanently restricted (active for all calls); temporary restricted (specified by user per call), default—restricted; temporary allowed (specified by user per call), default—allowed.

☞ **Conference Calling** is similar to three-way calling except more than three parties are involved in the call.

☞ **Do Not Disturb** allows a wireless subscriber to direct that all incoming calls stop at the PCS switch and not page the PS.

☞ **Flexible Alerting** allows a call to a directory number to be branched into multiple attempts to alert several subscribers. The subscribers may have wireless or wireline terminations.

☞ **Message Waiting Notification** is the service where a message is sent to the PS to inform the user that there are messages stored in the network that the user can access.

☞ **Mobile Access Hunting** is the service where call delivery is presented to a series of terminating numbers. If the first number is not available, the system will try the second and continue down a list. The terminating numbers can be mobile or nonmobile numbers anywhere in the world.

☞ **Multilevel Precedence and Preemption (MLPP)** permits a group of wireless subscribers to have access to wireless service where higher-priority calls will be processed ahead of lower-priority calls and may preempt (i.e., force the termination of) lower-level calls. Only calls within the same group will override each other.

☞ **Password Call Acceptance** is the service where calls to the wireless subscriber are interrupted and the calling party is asked to correctly enter a password before the PS is paged.

☞ **Preferred Language** is the capability for a user to hear all network announcements in their preferred language.

☞ **Priority Access and Channel Assignment** allows the PCS service provider to provide capabilities to a subscriber that allows priority access to radio resources. This service permits emergency services personnel (e.g., police, fire, and rescue squads) priority access to the system. Multiple levels of access may be defined.

☞ **Remote Feature Call** permits a wireless subscriber to call a special directory number (from a wireless or wireline phone) and, after correctly entering account code information and a Personal Identification Number (PIN), change the operation of one or more features of the service. For example, the selective call list can be modified by this capability.

☞ **Selective Call Acceptance** is the service where a wireless subscriber can form a list of those directory numbers that will result in the PS being paged. All other directory numbers will be blocked.

☞ **Subscriber PIN Access** is the ability to block access to the PS until the correct PIN is entered into the PS.

☞ **Subscriber PIN Intercept** is the ability of a wireless subscriber to bar outgoing calls unless the correct PIN is entered. This feature can be implemented in the network or in the PS.

☞ **Three-Way Calling** permits a PCS user authorized for three-way calling to add a third party to an established two-way call regardless of which party originated the established call. To add a third party, the PCS user sends a request for three-way calling service to the service provider, which puts the first party on hold. The PCS user then proceeds to establish a call to the third party. A request by the controlling user for disconnection of the third party (i.e., the last added party) will release that party and will cause the three-way connection to be disconnected and return the call to its original two-way state.

　　If either of the noncontrolling parties to an established three-way call disconnects, the remaining two parties are connected as a normal two-way call. If the controlling PCS user disconnects, all connections are released.

☞ **Voice Message Retrieval** is service where the user can retrieve voice message stored in the network. These messages are typically left by parties calling the user while the user was busy, did not answer, or was not registered with a system.

☞ **Voice Privacy** is the service where the user's voice traffic over the radio link is encrypted to prevent casual eavesdropping. In PCS in the United States, this is a required feature and is not optional.

☞ **Short Message Service** permits alpha and alphanumeric short messages to be sent to or from a PS.

8.4 OPERATION OF A PCS SYSTEM

This section describes the operation of a PCS system. We trace call flows from a PS to an RS to the PCS switch to other network elements. The flows are based on the TR-46 reference model and an A-Interface based on the ISDN. The ISDN model assumes that there are ISDN terminals associated with the switch, one for each directory number on the switch. Since PCS allows PSs to be associated with any RS, there is not a one-for-one correspondence between PSs and ISDN terminals. Thus, with the ISDN model, each PS registered at an RS is assigned a temporary directory number (also called virtual terminal number or interface directory number) that the RS and PCS switch use to refer to the PS in the ISDN signaling messages, while that PS is registered at the RS. The ISDN A-Interface defines a PCS Application Protocol (PCSAP) that uses ISDN signaling. With basic ISDN and PCSAP, the PCSC can support terminal mobility. The RS and switch can interact in either of two methods. In method one, the RS is equivalent to a Private Branch Exchange (PBX) and the switch is an ISDN switch that does minimum call control. In method two, the RS is a virtual ISDN terminal with all call control in the ISDN switch. We have shown call flows for method two and leave the construction of call flows for method one as homework assignments. Either method will work, though some systems designers will support one over the other.

8.4.1 General Operation of a Personal Station

All of the air interfaces share a common approach to the operation of the mobile telephone, called a Personal Station (PS) for the remainder of this chapter. This section will discuss the common characteristics of the operation and later sections will discuss how each particular air interface meets those characteristics.

Each of the air interfaces defines control channels that are used for data communications between the PS and the PCS system and traffic channels that are used for user-to-user communications (voice or data). When a PS is first powered up, it will find and decode data on a control channel before any further processing can be done.

The operation of the PS can be segmented into the following tasks (see Figure 8.3):

☞ **Power-up:** Whenever power is applied to the PS, it will initialize itself and then enter the Scan Control Channels task.

☞ **Scan Control Channels and Decode Data:** Under certain conditions (i.e., the PS first powers up, loses communications with the PCS system, is directed to by an order or moves into a new system) the PS will analyze the transmissions from the RS. It does this by scanning the control channels in the system, choosing the best one (usually based on signal strength), and determining the system characteristics.

☞ **Idle:** When the PS is in the Idle task, the PS is waiting for further action to occur. It needs to process data on the control channel and monitor for its own ID to be paged or for other global orders to be received.

☞ **Registration/DeRegistration:** The registration process informs the PCS system of the location of the PS. Thus, the PS can be paged in the correct system or region of the system without having to resort to paging in multiple areas. This reduces the paging load in a large system. Registration also results in the visited PCS system obtaining security and services data from the home PCS system's database and populating the data into the visited system's database. These databases are known as the Home Location Regis-

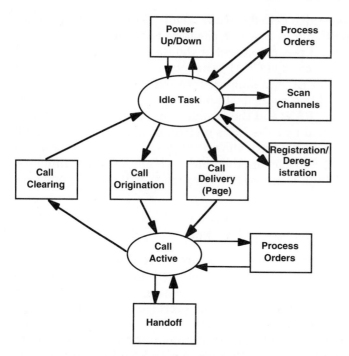

Fig. 8.3 General Operation of a PS

ter (HLR) and Visited Location Register (VLR), respectively. See chapter 10 on security for the information communicated between the databases.

When the PS first powers up, when a registration timer expires, when it enters a new system, or when directed to by a specific or general order on the control channel, the PS will register with the system. The registration consists of sending information to the PCS system to properly identify and authenticate itself.

When the PS is powered down (user pushes the on/off button to "off"), some PSs (but not all, depending on the air interface protocols) will send a deregistration message to the PCS system. The PS powers down only after the request is confirmed (or a time-out occurs). Those PSs not supporting this feature will power down without deregistering.

☞ **Call Origination:** When the user wishes to place a call, dialed digits are collected by the PS and sent to the PCS network in an origination message. The PS then waits for an order to go to a traffic channel and enters the call active task.

☞ **Call Delivery (Page):** Whenever the user receives a call, the network will send a page message to the PS. The PS will respond to the page message and wait for an order to go to a traffic channel. On the traffic channel, the PS will alert the user. When the user answers, the PS enters the call active task.

☞ **Call Active:** When a call is active, the user can communicate with the network over the PS and the voice or data information is sent over a traffic channel. In the call active state, the network or the PS can process orders, the PS can be handed off to another traffic channel, or the call can be terminated.

☞ **Handoff:** When the PS is out of range of one RS, either the PS or the network can initiate a set of procedures to establish communications on another traffic channel. This process, called a handoff, is discussed in more detail in a later section of this chapter.

☞ **Process Orders:** During either the idle task or the call active task, the PS or the network can initiate or respond to orders. Typical orders are three-way calling, call hold, conference, short message service, and caller identification.

☞ **Call Clearing:** When either party in the call wants to end the call, the call is cleared. Either the PS initiates the action or the network initiates the action. The PS then enters the call clearing task and ends the call.

Each of the five digital air interfaces and the analog cellular air interface supports the basic functions of the PS in different ways but accomplishes the same goal of supplying service to the user.

8.4.2 Basic Services

Before a PS can originate or receive a call, it will register with the PCS system. An exception is made for emergency (911) calls. During the registration process, the PS is given a Temporary Mobile Subscriber Identity (TMSI)[*] that is used for all subsequent call processing. The registration process is covered in detail in chapter 10. For the call flows in the remainder of this chapter (except for Emergency Calls) the Personal Station is assumed to be registered on the PCS system.

8.4.2.1 Call Origination Call origination is the service where the PS user calls another telephone on the worldwide telephone network. It is a cooperative effort among the PCS switch, the VLR, and the RS.

The detailed call flow steps are (see Figure 8.4 for the call flow diagram):

1. The PS processes an Origination Request from the user and sends it to the RS.
2. The RS sends a PCSAP Qualification Request to the VLR.
3. The VLR returns a Qualification Request Response to the RS.
4. The RS then processes an ISDN Setup message and sends it to the PCSC.
5. The PCSC sends an SS7 (ISDN User Part) Initial Address Message (IAM) to terminating switch (wireline or wireless).
6. At the same time the PCSC returns an SS7 Call Proceeding message to the RS.
7. The RS assigns a traffic channel to the PS.
8. The PS tunes to the traffic channel and confirms the traffic channel assignment.

---

[*]Most cellular phones and some early PCS phones do not support TMSI and therefore will use a Mobile Identification Number (MIN) to identify themselves to the network.

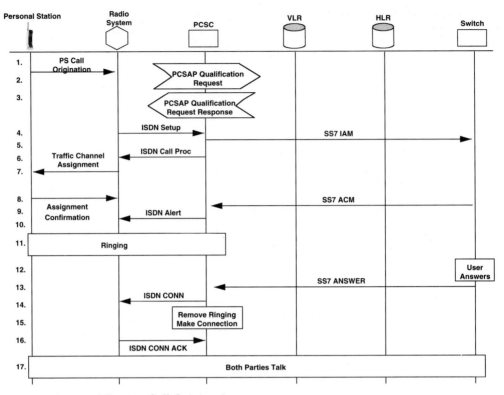

Fig. 8.4 Personal Station Call Origination

9. The terminating switch checks the status of the called telephone and returns an SS7 Address Complete Message (ACM) to the PCSC.

10. The PCSC returns an ISDN alert message to the RS.

11. The PCSC provides audible ringing to the user.

12. The terminating user answers.

13. The terminating switch sends an SS7 ANSWER Message to the PCSC.

14. The PCSC sends an ISDN CONNect message to the RS.

15. The PCSC removes audible ringing and makes the network connection.

16. The RS returns an ISDN CONNect ACKnowlege message.

17. The two parties establish their communications.

8.4.2.2 Call Termination Call termination is the service where a PS user receives call from other telephones in the worldwide telephone network. The following discussion is for calls terminating to a PS registered at its home PCSC. Calls terminating to roaming PSs will be discussed in section 8.4.2.5.

Call termination is a cooperative effort among the PCS switch, the VLR, and the RS.

The detailed call flow steps are (see Figure 8.5 for the call flow diagram):

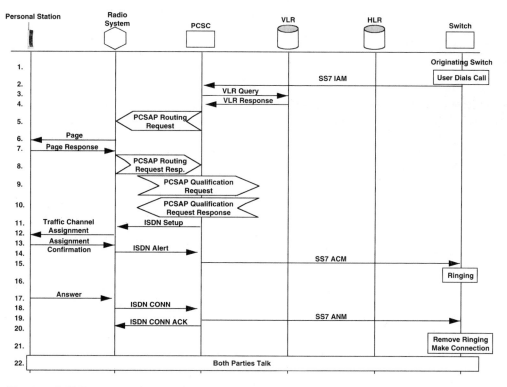

Fig. 8.5 Call Termination to a PS

1. A user in the worldwide phone network (wired or wireless) dials the directory number of the PS.
2. The originating switch sends an SS7 IAM to the PCSC.
3. The PCSC queries the VLR for the list of RSs (one or more) where the PS will be paged, and for the TMSI of the PS.
4. The VLR returns with the TMSI and list of RSs.
5. The PCSC sends an PCSAP Routing Request message to all RSs on the list.
6. Each RS broadcasts a page message on appropriate control channels.
7. The PS responds to the page with a page response message at one RS.
8. The RS sends a PCSAP routing request response to the PCSC.
9. The RS sends a PCSAP Qualification Directive message to the VLR.
10. The VLR responds with a PCSAP qualification request response.
11. The PCSC sends an ISDN setup message to the RS.
12. The RS sends a traffic channel assignment to the PS.
13. The PS tunes to the traffic channel and sends a traffic channel assignment confirmation message.
14. The RS sends an ISDN alert message to the PCSC.

15. The PCSC sends an SS7 Address Complete Message (ACM) to the originating switch.

16. The originating switch applies audible ringing to the network.

17. The user answers and the PS sends an Answer message to the RS.

18. The RS sends an ISDN CONNect message to the PCSC.

19. The PCSC sends an SS7 ANswer Message (ANM) to the originating switch.

20. The PCSC sends and ISDN CONNect ACKnowlege message to the RS.

21. Audible ringing is removed.

22. The two parties establish their communications.

8.4.2.3 Call Clearing   When either party in a conversation wishes to end a call, then the call clearing function is invoked. The exact call flows depend on which side ends the call first.

Call clearing is a cooperative effort among the PCS switch, the VLR, and the RS.

The detailed call flow steps for a PS-initiated call clearing are (see Figure 8.6 for the call flow diagram):

1. The PS user hangs up.

2. The PS sends a release message to the RS.

3. The RS sends an ISDN DISConnect message to the PCSC.

4. The PCSC sends an SS7 RELease message to the other switch.

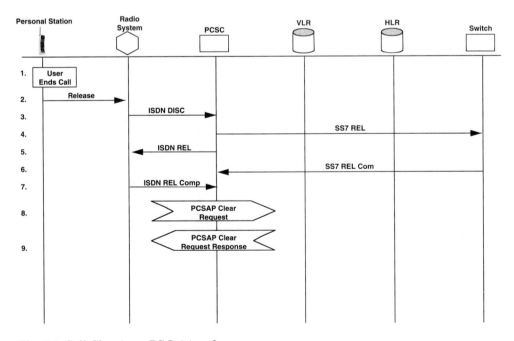

Fig. 8.6 Call Clearing—PS Initiated

5. The PCSC sends an ISDN RELease message to the RS.
6. The other switch sends an SS7 RELease Complete message to the PCSC.
7. The RS sends an ISDN RELease Complete Message to the PCSC.
8. The RS sends a PCSAP Clear Request message to the VLR.
9. The VLR closes the call records and sends a PCSAP Clear Request response message to the RS.

The detailed call flow steps for a far-end-initiated call clearing are (see Figure 8.7 for the call flow diagram):

1. The far-end user hangs up.
2. The other switch sends an SS7 RELease message to the PCSC.
3. The PCSC sends an ISDN DISConnect message to the RS.
4. The RS sends a Release message to the PS.
5. The PS confirms the message and disconnects from the traffic channel.
6. The RS sends an ISDN RELease message to the PCSC.
7. The PCSC sends an ISDN RELease Complete message to the RS.
8. The PCSC sends an SS7 RELease Complete message to the other switch.
9. The RS sends a PCSAP Clear Request message to the VLR.
10. The VLR closes the call records and sends a PCSAP Clear Request Response message to the RS.

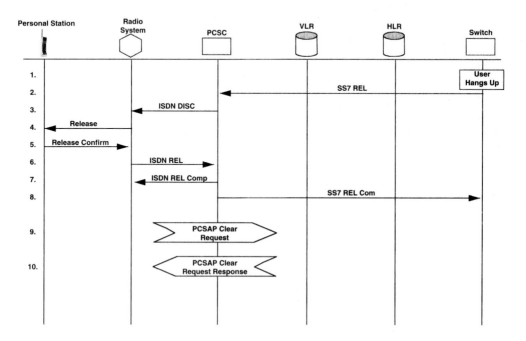

Fig. 8.7 Call Clearing—Far-End Initiated

8.4.2.4 Emergency Calls   Emergency calling is a service that enables a user of a PS to reach an emergency service operator through a simple procedure of dialing 911 or pushing an emergency button on the PS. Emergency calling is offered to unregistered and/or unsubscribed PSs (at the service providers option). The goal is to process the call independent of any failures that may occur. Thus, authentication failures are ignored during call processing for emergency calls. For emergency calling to operate with the following procedures, a emergency call indication will be set in the origination message from the PS. If that bit is not set, then the call will be processed with normal handling.

The following are the steps for processing an emergency call (see Figure 8.8 for the call flow diagram):

1. The user dials 911-SEND or pushes an emergency button on the PS.
2. The PS recognizes the unique number 911 or the emergency button depression and forms an origination message with the emergency calling indication set.
3. The RS sends a PCSAP Emergency Request Message to the VLR.
4. The VLR sends a PCSAP Emergency Request Response Message to the RS.
5. The RS forms an ISDN Setup Message to PCSC.
6. The PCSC sends an SS7 IAM to the Personal Safety Access Point (PSAP).
7. The PCSC sends an ISDN Call Proceeding message to the RS.

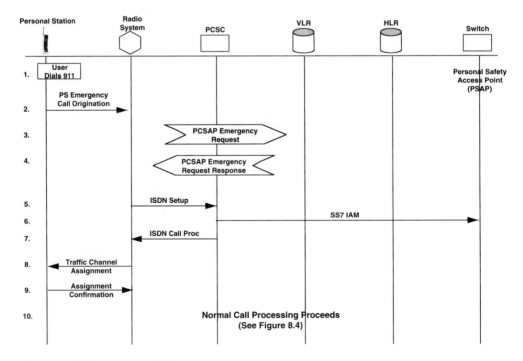

Fig. 8.8 PS Emergency Calling

8. The RS sends a traffic channel assignment to the PS. If no channel is available, the call will go to the top of the queue, if one is maintained. Alternately, the RS may force a handoff to free a traffic channel. Calls in progress are not dropped.

9. The PS tunes to the traffic channel and confirms the traffic channel assignment.

10. Call processing proceeds at step 9 of call origination (see section 8.4.2.1).

8.4.2.5 Roaming Roaming is the ability to deliver services to PSs outside of their home area. When a PS is roaming, registration, call origination and call delivery will take extra steps. Whenever data will be retrieved from the VLR, and the data is not available, then the VLR will send a message to the appropriate HLR to retrieve the data. The data consists of IMSI to MIN conversion, service profiles, Shared Secret Data (SSD) for authentication, and other data needed to process calls. The most logical time to retrieve this data is when the PS registers with the system (see chapter 10 for a description of the registration process).

Once the data on a roaming PS is stored in the VLR, then call processing for any originating services (basic or supplementary) is identical to that of home PSs. However, there may be times when the PS originates a call before registration has been accomplished or when the VLR data is not available. At those times, an extra step will be added for the VLR to retrieve the data from the HLR. Thus any originating service has two optional steps where the VLR sends a message (using IS-41 signaling over SS7) to the HLR requesting data on the roaming PS. The HLR will return a message with the proper call information.

Call delivery is not possible to an unregistered PS since the network does not know where the PS is located. Once the PS is registered with a system, then call delivery to the roaming PS is possible. This section will discuss call delivery to roaming PSs in detail.

There are two cases of call delivery to roaming PS: the PS has a geographic-based directory number (indistinguishable from a wireline number); the PS has a nongeographic number. We will describe the call flows for both operations.

When the PS has a geographic number, then the PCSC is assigned a block of numbers that are within the local numbering plan for the area of the world where the PCSC is located. Call routing to the PS is then done according to the procedures for that of a wireline telephone.* If a PS associated with a PCSC is not in its home area, the PCSC will query the HLR for the location of the PS. The PCSC then invokes call forwarding to the PCSC where the PS is located, and the connection is made to the second PCSC where call terminating services are delivered according to the procedures in section 8.2.4.2. This procedure is inefficient

---

*For example, in New Jersey, 908–610-XXXX is used by the local cellular provider for cellular phones in Monmouth County. The wireline network routes calls to those numbers in a normal fashion, and calls terminate on the cellular switch.

because it results in two sets of network connections—originating switch to home PCSC, and home PCSC to visited PCSC.

Call delivery to a roaming PS is a cooperative effort between the home and visited PCSC, the VLR and HLR, and the RS.

The detailed call flow steps for call delivery to a roaming PS with a geographic directory number are (see Figure 8.9 for the call flow diagram):

1. A user in the worldwide phone network (wired or wireless) dials the directory number of the PS.
2. The originating switch sends an SS7 IAM to the home PCSC.
3. The home PCSC queries the HLR for the location of the PS.
4. The HLR returns the location of the visited system.
5. The PCSC invokes call forwarding to the PCSC in the visited system and the forwarding (home) PCSC switch sends an SS7 IAM to the visited PCSC.
6. Call processing proceeds at step 3 of the terminating call flow (see section 8.4.2.2).

When the PS has a nongeographic number, then calls can be directed from an originating switch directly to the visited switch. Call delivery to a nongeographic number requires that the originating switch recognize the number as a nongeographic number and do special call processing for routing. This special processing is known as Advanced Intelligent Network (AIN) processing. If the originating switch does not support AIN, then it will route the call to a switch that supports AIN. With AIN support, the originating switch will recognize the nongeographic number and send an SS7 message to the HLR with a request for the location of the PS. The HLR will return a temporary directory number (on

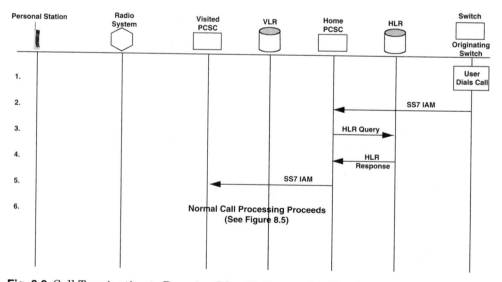

Fig. 8.9 Call Termination to Roaming PS with Geographic Number

the visited PCSC) that can be used to route to the PS in the visited system. Calls then proceed according to normal terminating call flows.

Call delivery to a roaming PS with a nongeographic number is, therefore, a cooperative effort between the visited PCSC, the VLR and HLR, and the RS.

The detailed call flow steps for call delivery to a roaming PS with a nongeographic directory number are (see Figure 8.10 for the call flow diagram):

1. A user in the worldwide phone network (wired or wireless) dials the directory number of the PS.
2. The originating switch recognizes the number as a nongeographic number and sends an SS7 query message to the HLR at the home PCSC.
3. The HLR returns the location of the visited system with a directory number to use for further call processing.
4. The originating switch sends an SS7 IAM to the visited PCSC.
5. Call processing proceeds at step 3 of the terminating call flow (see section 8.4.2.2).

8.4.3 Supplementary Services

IS-41 supports several supplementary services (see section 8.3.2). However, only the call flows for a few of the more common ones are described herein. Others will be assigned as homework problems.

8.4.3.1 Call Waiting   Call waiting provides notification to a PCS subscriber of an incoming call while the user's PS is in the busy state. Subsequently, the user can either answer or ignore the incoming call. Once the call is answered, the

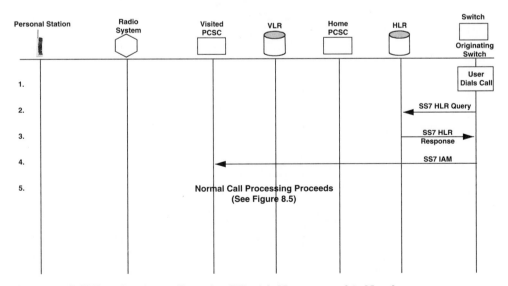

Fig. 8.10 Call Termination to Roaming PS with Nongeographic Number

user can switch between the calls until one or more parties hang up. When either distant party hangs up, then the call reverts to a normal (non-call-waiting) call. If the PS user hangs up, then both calls are cleared according to normal call-clearing functions.

The detailed call flow steps for call waiting delivery are (see Figure 8.11 for the call flow diagram):

1. User dials a call.
2. The originating switch sends an SS7 IAM to the PCSC.
3. The PCSC queries the VLR.
4. The VLR returns with a location of the PS that is within the serving system. If it is not, then the call will be forwarded to the serving PCSC.
5. The PCSC determines that the PS is busy and subscribes to call waiting and thus applies a call waiting tone.
6. The user presses the "flash" button (may be "send" on some PSs) to answer the call-waiting indication, and the PS sends a flash message to the RS.
7. The RS sends an ISDN Hold message to the PCSC.
8. The PCSC puts the first call on hold and connects the second call.
9. The PCSC sends a Hold Acknowledge to the RS.

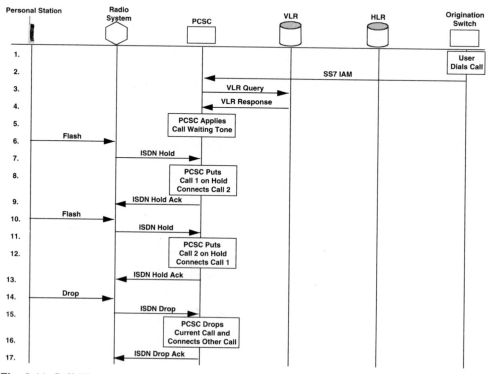

Fig. 8.11 Call Waiting

10. The user presses the flash button (may be SEND on some PSs) to talk to caller 1, and the PS sends a flash message to the RS.

11. The RS sends an ISDN Hold message to the PCSC.

12. The PCSC puts the second call on hold and connects the first call.

13. The PCSC sends a Hold Acknowledge to the RS.

14. The user wants to drop the current call (either 1 or 2) and pushes the "drop" (or "END") key, and the PS sends a Drop message to the RS.

15. The RS sends an ISDN Drop message to the PCSC.

16. The PCSC drops the current call and connects the other call (the one currently on hold).

17. The PCSC sends an ISDN Drop Acknowledge message to the RS.

8.4.3.2 Call Forwarding Call forwarding represents the ability to redirect a call to a PS handset in three situations: Unconditional, Busy, and No Answer. The call forwarding features build upon the PS call terminating capability. Call forwarding is separate from call delivery to a roaming PS; thus, if a roaming PS has call forwarding invoked, two stages of call forwarding may occur.

The detailed call flow steps for call forwarding unconditional are (see Figure 8.12 for the call flow diagram):

1. A user in the worldwide phone network (wired or wireless) dials the directory number of the PS.

2. The originating switch sends an SS7 IAM to the PCSC.

3. The PCSC determines that the PS has invoked call forwarding for all calls.

4. The PCSC sends an SS7 IAM message to the destination switch.

5. Call processing proceeds according to normal procedure for a wireline or wireless user.

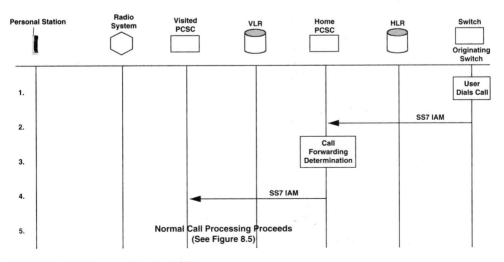

Fig. 8.12 Call Forwarding to a PS

When the user has invoked call forwarding busy or no answer, additional steps between 2 and 3 above are needed to page the PS or to determine that it is busy.

8.4.3.3 Three-Way Calling   Three-way calling is the service where a PS user can connect to two parties simultaneously and all three parties are part of the call. If the PS subscribes to three-way calling, then the PS can add a third party independent of who originated the call. When a PS user wants to place a three-way call, the user puts the first party on hold, calls the second party, and then requests a three-way connection. When either of the two other parties disconnects, the call reverts to a normal two-way call. If the PS user disconnects, then all connections are released.

The detailed call flow steps for three-way calling establishment are (see Figure 8.13 for the call flow diagram):

1. The PS user signals a flash indication to the RS.
2. The RS signals to the PCSC a ISDN Information message with a feature activator of three-way calling.
3. The PCSC allocates resources for the three-way call.

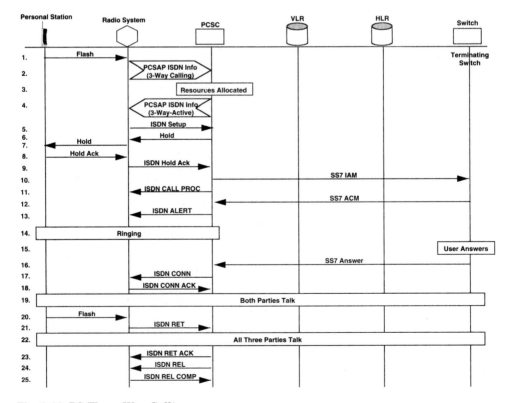

Fig. 8.13 PS Three-Way Calling

4. The PCSC sends an ISDN Information message to the RS with a feature identifier of three-way calling and active status.

5. The RS sends an ISDN Setup message to the PCSC with appropriate call information (e.g., calling party ID, call reference, B channel used).

6. The PCSC sends an ISDN Hold message to the RS.

7. The RS sends a Hold message to the PS.

8. The PS sends a hold acknowledge to the RS.

9. The RS sends an ISDN Hold Acknowledge message to the PCSC.

10. The PCSC begins call processing for the new call by sending an SS7 Initial Address Message (IAM) to the terminating switch.

11. The PCSC sends a ISDN CALL PROCeeding message to the RS.

12. The terminating switch returns an SS7 Address Complete Message (ACM).

13. The PCSC sends a ISDN ALERT message to the RS.

14. The network provides audible ringing.

15. The terminating party answers.

16. The terminating switch sends an SS7 Answer Message to the PCSC.

17. The PCSC sends an ISDN CONNect message to the RS.

18. The RS sends an ISDN CONNect ACKnowlege message to the PCSC.

19. The two users begin talking.

20. The PS user sends a flash message to the RS and recognizes the flash as a request to join the two calls.

21. The RS sends a ISDN RETrieve message to the PCSC with the call record for the held call.

22. The PCSC forms a three-way call from the two call records.

23. The PCSC sends an ISDN Retrieve ACKnowlege message to the RS.

24. The PCSC sends an ISDN RELease message to the RS to release the unneeded second call reference.

25. The RS sends an ISDN RELease COMPlete message to the PCSC.

When the user desires to release the connection to the last added party, the user signals a flash request to the network, which then releases the last added party. If the user disconnects, the network releases both other parties.

8.5 AIR INTERFACE UNIQUE CAPABILITIES

This section will discuss how the analog cellular and each of five digital air interfaces (the U_m-Interface in the reference model) support the functions of basic encoding of the digital data being sent (i.e., the framing of the channel), speech encoding method, the use of the frequency allocation for PCS, the output power of the PS and how it is controlled, the modulation method, and support for handoffs.

8.5.1 Framing of Digital Signal

Analog cellular uses a digital control channel with analog voice channels. During a voice call, digital data is sent via blanking the audio and sending the data (blank and burst) during the blanked audio period. Data transmission is handled via either voice band modems or Cellular Digital Packet Data (CDPD; see chapter 14).

For the digital air interfaces, the digital format of each interface will multiplex voice and data traffic, control information, and other data as needed. Each of the interfaces uses a different approach to this multiplexing. After the signal is multiplexed, it will be encoded and then modulated before transmission. The receiver will undo each step. This section discusses how each of the air interfaces support the sending and receiving of the data.

8.5.1.1 Framing of Analog Cellular

The analog cellular system uses a digital format at 10 kbs sent on a separate control channel. The basic frame is 463 bits long.[*] Figure 8.14 shows the framing on the forward control channel (RS to PS). Synchronization of the data receiver is established via a 10-bit 1010101010 (dotting) pattern and 11-bit 11100010010 framing pattern. Messages are then

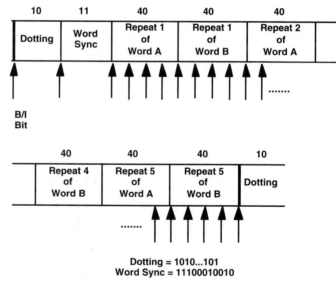

Dotting = 1010...101
Word Sync = 11100010010

Note:
1) **A given mobile reads only one of the two interleaved messages**
2) **Busy-idle bits are inserted at each arrow**

Fig. 8.14 Forward Control Channel Framing

---

[*]The frame rate was established in the early 1970s before the widespread use of microprocessors to do data transmission. Obviously, 463 is a prime number and not the best choice for use with a 4-, 8-, 16-, etc., bit microprocessor.

sent on an A or B frame and repeated five times. Each message uses a $(40, 28, 5)^*$ BCH code. The combination of the BCH code and the five repetitions of the message provide the error detection and correction in the data receiver. After the data is encoded, it undergoes a further encoding, called Manchester Encoding, using two bits per baud.

Every 10 bits in the frame, a busy/idle bit is sent to inform PSs of the status (busy or idle) of the reverse control channel. A PS can transmit on the reverse control channel, only when the channel is idle, as indicated by the busy/idle bits sending an idle status.

Figures 8.15, 8.16, and 8.17 show the frame information of the reverse control channel and the forward and reverse blank and burst channel. Different channels use a different number of repeats of the message and a different length of the dotting pattern. On the reverse channel a $(48, 36, 5)$ BCH code is used.

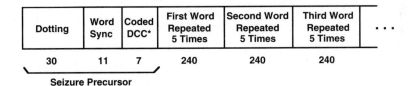

Dotting = 1010...101
Word Sync = 11100010010

* Digital Color Code

Fig. 8.15 Reverse Control Channel Framing

Dotting	W.S.	Repeat 1 of Word	Dot.	W.S.	Repeat 2 of Word
101	11	40	37	11	40

Dot.	W.S.	Repeat 9 of Word	Dot.	W.S.	Repeat 10 of Word	Dot.	W.S.	Repeat 11 of Word
37	11	40	37	11	40	37	11	40

Dotting = 1010...101
Word Sync = 11100010010

Fig. 8.16 Forward Voice Channel Framing

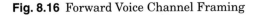

*28 data bits, 12 parity check bits for a total of 40 bits and capability of detection of five errors.

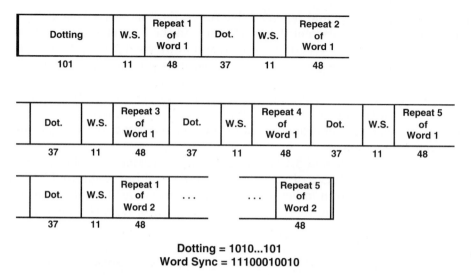

Dotting = 1010...101
Word Sync = 11100010010

Fig. 8.17 Reverse Voice Channel Framing

The analog control channel was designed in the 1970s when most data receivers were implemented in hardware, and therefore it uses inefficient coding systems. The overall throughput of the various channels (forward and reverse) is 300 to 600 bits per second even though the signaling is done at 20 kilobaud per second.

The digital cellular and PCS protocols have taken advantage of improvements in microprocessor technology and coding algorithms to implement more efficient coders and decoders.

The analog cellular systems are based on a frequency reuse factor (N) of 7. Nearby cells on the same frequency are coded with a different Digital Color Code (DCC) on the control channels and a different Supervisory Audio Tone (SAT) on the voice channels.

8.5.1.2 Framing of CDMA   In CDMA, the entire 1.25-MHz transmission bandwidth is occupied by every station (see section 8.6.3.2 for the frequency bands used by a radio system). Different channels are chosen by the PS or the RS by selecting a particular Walsh function to decode. The Walsh functions are chosen so that the set of functions are all orthogonal to each other. All radio systems in the system are on the same frequency and use time-shifted Walsh functions. Walsh functions cross correlate to low values when a time shifted version of the Walsh code is considered (see appendix B for more information on Walsh functions). Thus, CDMA uses a frequency reuse factor (N) of 1. A code channel that is modulated by Walsh function n will be assigned to code channel number n ($n = 0$ to 63). Walsh function time alignment is defined so that the first Walsh chip, designated by 0 in the column headings of Table B.1, begins at an even second time mark referenced to radio system transmission time. The Walsh function spreading sequence will repeat with a period of 52.083 μsec (= 64/1.2288 million chips per second [Mcps]), which is equal to the duration of one forward traffic channel modulation symbol.

The forward CDMA channel has the overall structure shown in Figures 8.18 and 8.19. One channel code is assigned to the pilot channel; one is assigned to the sync channel; a maximum of seven are assigned to paging channels; the remainder are assignable to the forward traffic channels. The information (voice or data) on each channel is modulated by the appropriate Walsh function and then is modulated by a quadrature pair of pseudonoise (PN) sequences at a fixed chip rate of 1.2288 Mcps. The pilot channel is always assigned to code channel number zero. When the sync channel is present, it will be assigned code channel number 32. If paging channels are present, they will be assigned to code channel numbers one through seven (inclusive) in sequence. The remaining code channels are available for assignment to the forward traffic channels. When a radio system supports multiple forward CDMA channels, then frequency division multiplex is used.

Figure 8.20 shows a sample assignment of the code channels transmitted by a radio system. The example includes: the pilot channel (that is always required), 1 sync channel, 7 paging channels (the maximum number allowed), and 55 traffic channels. Alternately, where a radio system does not need paging

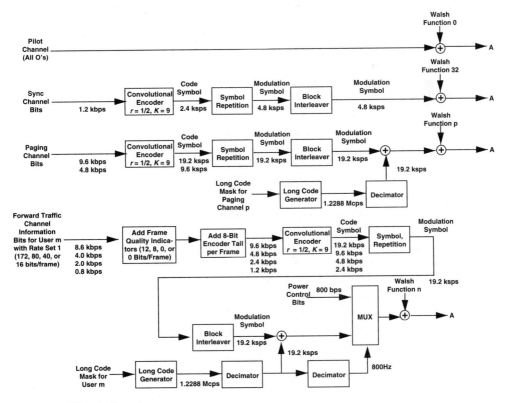

Fig. 8.18 CDMA Framing

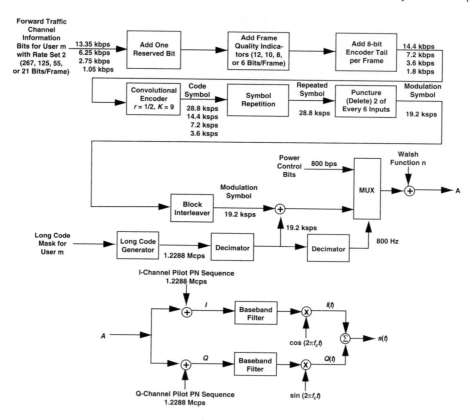

Fig. 8.19 CDMA Framing (Part 2)

and sync channels, the assignments could be one pilot channel, no paging channels, no sync channels, and 63 traffic channels.

The sync channel always operates at a fixed rate of 1,200 bps. The paging channel operates at a fixed data rate of 9,600 or 4,800 bps.

The forward traffic channels' data rates are grouped into sets called "rate sets." Rate Set 1 contains four elements—9,600, 4,800, 2,400, and 1,200 bps. Rate Set 2 contains four elements—14,400, 7,200, 3,600, and 1,800 bps.

All radio systems support Rate Set 1 on the forward traffic channel. Rate Set 2 is optionally supported on the forward traffic channel. When a radio system supports a rate set, all four rates of the rate set are supported.

CDMA uses frequency division multiplex for the reverse path from a PS to an RS (see section 8.5.3.2). Signals on the reverse CDMA channel (Figure 8.21) are either access signals or reverse traffic signals. All personal stations accessing an RS over the reverse channel share the same CDMA frequency assignment. Each PS transmits a different Walsh code; therefore the RS can correctly decode the transmissions from an individual PS.

Figure 8.22 shows an example of all of the signals received by an RS. If more than 1.25 MHz is assigned to CDMA at an RS, the additional channels are assigned in 1.25-MHz increments using frequency division multiplex.

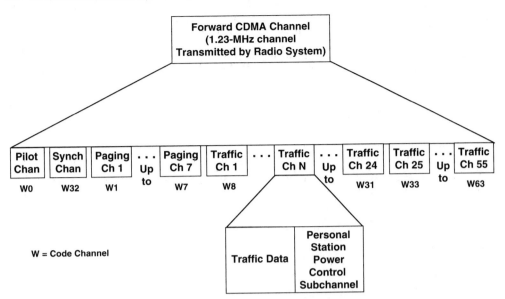

Fig. 8.20 Example of a Forward CDMA Channel Transmitted by a RS

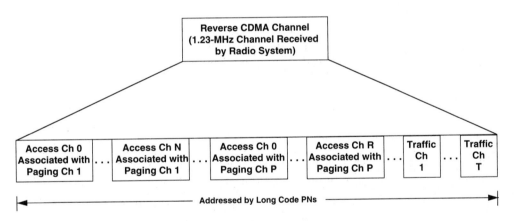

Fig. 8.21 Example of Logical Reverse CDMA Channel Received at an RS

The reverse CDMA channel has the overall structure shown in Figures 8.23 and 8.24. Data transmitted on the reverse CDMA channel is grouped into 20-ms frames. All data transmitted on the reverse CDMA channel is convolutionally encoded, block interleaved, modulated by the 64-ary orthogonal modulation, and direct-sequence spread prior to transmission. This process will be examined in more detail in section 8.5.5.2.

8.5.1.3 Framing of PACS The personal access communications system, or PACS, is a low-power system designed primarily as a low-mobility (less than 30 mph) solution to PCS for use in residential wireless applications and other areas where the handset moves slowly.

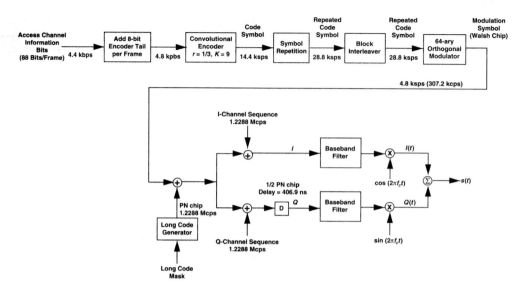

Fig. 8.22 Reverse CDMA Channel Structure for the Access Channel

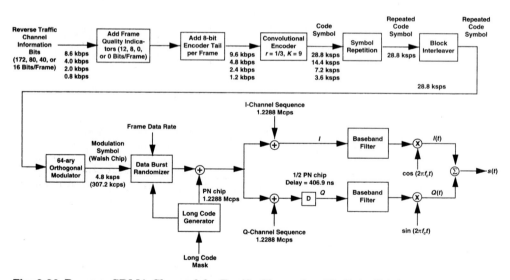

Fig. 8.23 Reverse CDMA Channel for Traffic Channels with Rate Set 1

PACS uses Time Division Multiple Access (TDMA) with Frequency Division Duplex (FDD) on 300-kHz channels. Although the band plan (see section 8.5.3.3) shows channels on 100-kHz spacing, only every third channel can be used in a given region. The extra channels are identified to solve guard-channel problems at band edges. Once the channel at the band edge is established, all other channels increment in 300-kHz spacing.

PACS uses a basic frame structure (see Figures. 8.25 and 8.26) of 960 bits

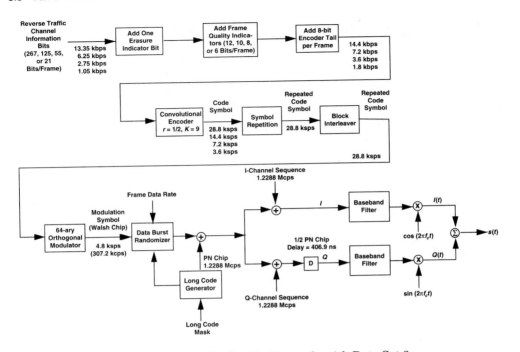

Fig. 8.24 Reverse CDMA Channel for Traffic Channels with Rate Set 2

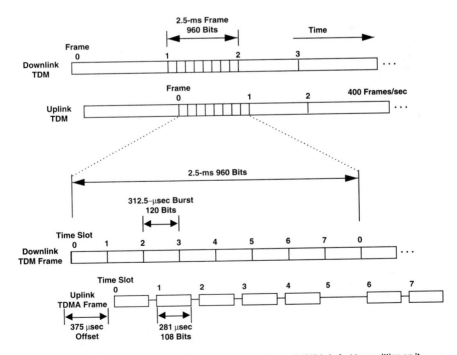

Note: Time slots will be "filled" only on the uplink if a subscriber unit (SU) is in fact transmitting on it.

Fig. 8.25 PACS TDM/TDMA Frame Structure

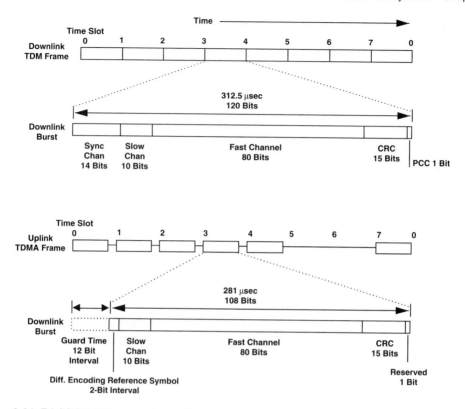

Fig. 8.26 PACS TDM/TDMA Burst Structure

(2.5 ms) to minimize speech transmission delay. Within each frame are 8 TDMA bursts of 312.5 μsec each. A super frame consists of a group of 8 frames and supports subrate multiplexing. The overall bit rate is 384 kbs that is transmitted at 2 bits per symbol (see Section 8.5.5.3).

The PS transmits a 281.2-microsecond burst with a 31.3-microsecond guard time. To account for propagation delays, the PS delays its burst by 375 ± 5.2 microseconds with respect to the received burst from the RS (see Figure 8.27). The delay also allows the PS to operate without a duplexer.

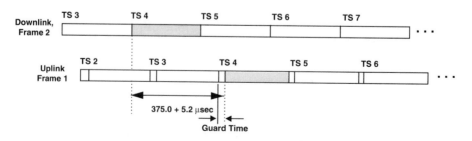

Fig. 8.27 Frame and Burst Offset as measured at the PS's Antenna

The RS transmits bursts (Figure 8.28) of 120 bits as follows:

☞ **Synchronization channel**—14 bits, used as:

    ✗ Unique code (7 bits) that carries an initial frame synchronization that may vary from frame to frame.

    ✗ Reserved bit (1 bit)

    ✗ Frame number (3 bits) numbered sequentially from 000 to 111.

    ✗ Time slot number (3 bits) numbered sequentially from 000 to 111.

☞ **Slow Channel (SC)**—10 bits, used for signaling and control. The nominal rate is 4.0 kbs. Bit 1 of this channel is the Word Error Indicator (WEI) that is set when the RS detects an error on the uplink.

☞ **Fast Channel (FC)**—80 bits, used for bearer information (speech or data) or signaling and control. The nominal data rate is 32 kbs.

☞ **Cyclic Redundancy Code (CRC)**—15 bits, used to check the SC and FC on each burst using a (105, 90) code.

☞ **Power Control Channel (PCC)**—1 bit, that controls the PS power on the uplink.

The PS transmits on its assigned time slot on the uplink and sends the following information in each burst (Figure 8.29):

☞ **Guard time**—equivalent to 12 bits, to avoid overlapping with adjacent transmission from other PSs. No signal is sent during the guard time.

☞ **Differential decoder start bit**—2 bits, with value undefined, that serve to derive a phase reference to decode subsequent bits.

☞ **Slow channel**—10 bits, used for signaling and control. The nominal rate is 4.0 kbs. Bit 1 of this channel is the WEI that is set when the PS detects an error on the downlink.

☞ **Fast channel (FC)**—80 bits, used for bearer information (speech or data) or signaling and control. The nominal data rate is 32 kbs.

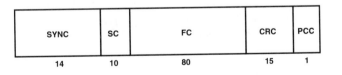

Fig. 8.28 PACS RS TDMA Burst

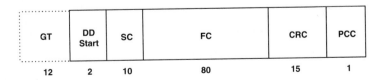

Fig. 8.29 PACS RS TDMA Burst

☞ **Cyclic redundancy code**—15 bits, used to check the SC and FC on each burst using a (105, 90) code.

☞ **Reserved bit**—1 bit, will be set to "0."

Each time slot of 2.5 ms has 80 bits of data; thus, there are 400 frames per second and the nominal data rate of the system is 32 kbs. This is the normal data rate used for the ADPCM speech (see section 8.5.2). The PACS system can transmit data at rates of 4 kbs, 8 kbs, and 16 kbs by using subrate multiplexing. With subrate multiplexing, a PS transmits in fewer than 8 frames per superframe. This lower data rate can be used for data traffic or for alternative speech encoding methods. At initial deployment, PACS will support only ADPCM.

The rate of a channel and its time slot assignment are identified by:

1. The letters "TS" followed by
2. The number of the time slot (0 to 7) and
3. The letters "a" to "p."

Table 8.1 summarizes the use of a traffic channel at different data rates. See Figure 8.30 for examples of channel usage at each data rate.

Table 8.1 Traffic Channel Designation for PACS

Traffic Channel Data Rate	PS or RS Uses Frame Numbers	Time Slot Designation	Example, if RS or PS Uses Time Slot 4
32 kbs	0, 1, 2, 3, 4, 5, 6, 7	TS #-a	TS 4-a
16 kbs	0, 2, 4, 6	TS #-b	TS 4-b
16 kbs	1, 3, 5, 7	TS #-c	TS 4-c
8 kbs	0, 4	TS #-d	TS 4-d
8 kbs	1, 5	TS #-e	TS 4-e
8 kbs	2, 6	TS #-f	TS 4-f
8 kbs	3, 7	TS #-g	TS 4-g
4 kbs	0	TS #-h	TS 4-h
4 kbs	1	TS #-i	TS 4-i
4 kbs	2	TS #-j	TS 4-j
4 kbs	3	TS #-k	TS 4-k
4 kbs	4	TS #-l	TS 4-l
4 kbs	5	TS #-m	TS 4-m
4 kbs	6	TS #-n	TS 4-n
4 kbs	7	TS #-o	TS 4-o

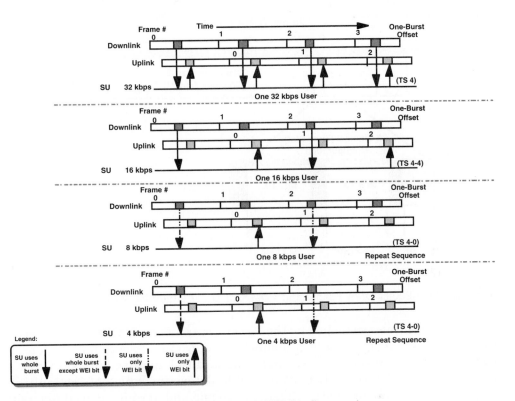

Fig. 8.30 PACS Subrate Traffic Channel and WEI Bit Sequencing

8.5.1.4 Framing of PCS2000 The Omnipoint, composite CDMA/TDMA, PCS2000 system employs a combination of TDMA, FDMA, and CDMA for multiple user access to the PCS network. Within a PCS cell, TDMA is used to separate users. FDMA is used to provide a greater area of coverage, or to provide a greater capacity for densely populated regions. In those areas, multiple cells or sectorized cells use FDMA. The Direct Sequence Spread Spectrum (DSSS) method of CDMA is also used for each RF link to reduce cochannel interference between cells reusing the same RF carrier frequency.

In PCS2000, the TDMA frame and time slot (channel) structure is based on a 20-ms polling loop for user access to the RF link (see Figure 8.31). Both the PS and the RS transmit on the same frequency using Time Division Duplex (TDD). The 20-ms frame supports 32 full duplex channels within the frame. Each time slot (channel) supports an 8-kbs full duplex user.

The RS transmits in the first half of the TDMA/TDD time slot, and the PS transmits during the second half of the slot. Thus, the PS receives during the first half of the time slot and transmits during the last half. A small portion of each time slot, designated the "guard time," is allocated to the end of a transmission to account for propagation delays in the system.

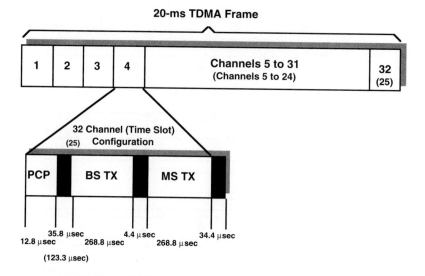

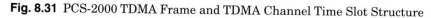

Fig. 8.31 PCS-2000 TDMA Frame and TDMA Channel Time Slot Structure

Power control is managed by the Power Control Pulse (PCP) signal received from the PS. The PCP serves as a channel sounding pulse to determine link propagation loss and to serve as a measurement of link quality for the PCS2000 power control subsystem. It is also used to determine which of the multiple antennas to use for the spatial diversity scheme and permits spatial diversity control to be updated during each TDMA time slot period.

In a large cell system, where propagation delays are long, an RS may optionally support 25 channels (time slots). At the boundary between large and smaller cell diameters (32- and 25-channel cells), the frequencies are assigned in nonoverlapping groups to the 32-channel cells and the adjacent 25-channel cells. After sufficient separation between the 32-channel/25-channel boundary, the channels can then be reused with no restrictions.

Each of the 32 full duplex channels per frame are provided with an 8-kbs data capability (for voice or data) for use during the 20-ms frame period. This requires a TDMA burst data rate of 512 kbs in TDD mode. Guard timing and additional overhead bits required for link protocols and control functions make the total TDMA burst rate of 781.25 kbs.

An individual PS may negotiate for multiple or submultiple slots in the polling loop to have data rates higher or lower than 8 kbs. The negotiation may take place at any time via signaling traffic. The slots, if available, are assigned by the RS to a PS. Those PSs using higher rates than 8 kbs use additional channels (time slots) per TDMA frame (see Figure 8.32). For example, a PS using two channels (time slots) operates at a 16-kbs data rate, versus 8 kbs for one channel (time slot). The maximum data rate supported per user is 256 kbs full duplex or 512 kbs half duplex. If a PS wants a data rate of less than 8 kbs, time slots may be granted in frames separated by an integral number of intermediate frames

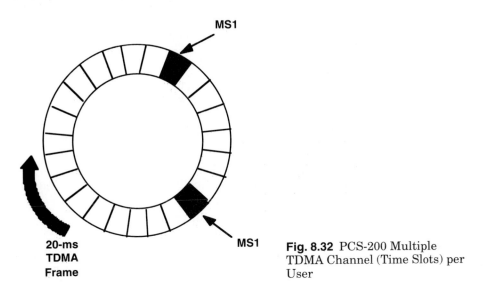

Fig. 8.32 PCS-200 Multiple TDMA Channel (Time Slots) per User

(see Figure 8.33). The maximum limit on the separation of slots allocated to a single PS is 0.5 seconds for a data rate of 320 bits per second.

To maximize system capacity, the TDMA frame times for all RSs within the same geographical area will be synchronized. The most common method for synchronization is to utilize a Global Positioning System (GPS) receiver at the RS controller to generate the primary-reference timing marker for the TDMA frame timing.

The synchronization of all RSs within a given area allows an RS controller to temporarily turn off any TDMA time slot of a given cell that may be interfering with a neighboring cell. It also facilitates switching a PS to a different time slot if a current time slot has interference from an adjacent cell using the same time slot.

The PS can synchronize to a new RS within one channel (time slot) and is capable of synchronizing with multiple RSs when those RSs are synchronized to a common digital network.

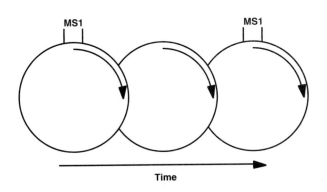

Fig. 8.33 PCS-2000 Submultiple TDMA Channels (Time Slots) per User

Both the RS and the PS can transmit either of two packet types: base poll packets and base traffic packets. Poll packets (Figure 8.34) are 200 bits in length and consist of 16 bits of a header, 168 bits of data and 16 bits of a frame check word. For traffic packets (see Figures 8.35, 8.36, and 8.37), the system can operate in any of three modes: the RS sends long packets, and PS sends short packets, RS sends short packets and PS sends long packets, or RS and PS both send symmetric packets (symmetric bandwidth signaling). In all three cases, one combined RS/PS time slot is used and the D channel is used for signaling and the B channel for bearer traffic. The system can optionally remove error checking and use the additional 16 bits for B-channel data.

The frequency reuse pattern for the PCS-2000 system is three (see Figure 8.38). Adjacent cells differ either by frequency or code word; nearby cells on the same frequency use a different code to avoid interference.

8.5.1.5 Framing of TDMA   In the TDMA system, as supported at both 900 MHz for cellular and at 1,800 MHz for PCS, the overall 30-kHz RF channel structure from analog cellular is maintained. Each RF channel has six time slots and supports three PSs. In the future with half-rate channels, an RF channel will be able to support up to six PSs. As in analog cellular, the channels are designated as control channels or traffic channels.

Fig. 8.34 PCS-2000 Poll Packets

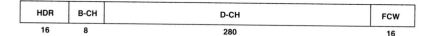

Fig. 8.35 PCS-2000 Traffic Packets (High Bandwidth Signaling)

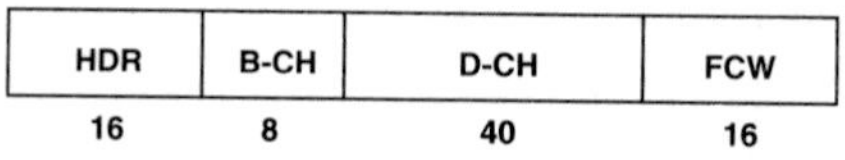

Fig. 8.36 PCS-2000 Traffic Packets (Low Bandwidth Signaling)

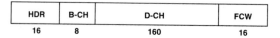

Fig. 8.37 PCS-2000 Traffic Packets (Symmetric Bandwidth Signaling)

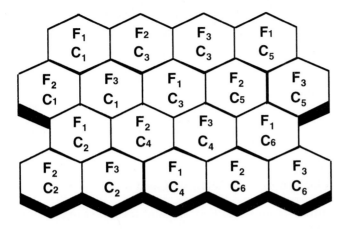

Fig. 8.38 Frequency and Code Reuse for PCS-2000— $N = 3$

Figure 8.39 shows the frame structure of the digital TDMA RF channel frame. The frame length is 40 ms (i.e., 25 frames per second), with each frame being 1,944 bits (972 symbols) long. The frame is segmented into six equally sized time slots (1–6) with 162 symbols (or 324 bits). Each full-rate traffic channel utilizes two equally spaced time slots of the frame (i.e., 1 and 4, 2 and 5, or 3 and 6), whereas each half-rate channel uses only one time slot of the frame.

Both control and traffic channels use a common frame format with a slot format that is slightly different depending on whether the transmission is on a control channel or a traffic channel.

The slot format on the traffic channel is defined in Figures. 8.40 and 8.41. The fields in the RS to PS slot are defined as:

☞ **SYNC:** 28 bits of synchronization

☞ **SACCH:** 12 bits for the Slow Associated Control CHannel

☞ **DATA:** two 130-bit fields of data; the data can be digitized voice or data.

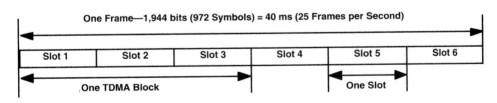

Fig. 8.39 TDMA Frame Structure

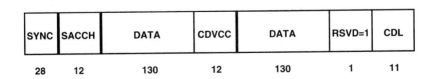

Fig. 8.40 Time Slot Format: RS to PS on Traffic Channel

G	R	DATA	SYNC	DATA	SACCH	CDVCC	DATA
6	6	16	28	122	12	12	122

Fig. 8.41 Time Slot Format: PS to RS on Traffic Channel

☞ **CDVCC:** 12 bits of a Coded Digital Verification Color Code (similar to the SAT tone for analog cellular)

☞ **RSVD:** 1 bit of a reserved field

☞ **CDL:** 11 bits for a Coded Digital Control Channel Locator

On the reverse traffic channel (from PS to RS), the slot format (Figure 8.41) has the following field definitions:

☞ **G:** equivalent of 6 bits guard time (the PS does not transmit during this time)

☞ **R:** equivalent of 6 bits of ramp time (the PS is powering up)

☞ **SYNC:** 28 bits of synchronization

☞ **DATA:** one field of 16 bits plus two fields of 122 bits; the data can be digitized voice or data

☞ **SACCH:** 12 bits for the slow associated control channel

☞ **CDVCC:** 12 bits of a coded digital verification color code (similar to the SAT tone for analog cellular)

On the forward digital control channel (RS to PS), the fields in each slot are (see Figure 8.42):

☞ **SYNC:** 28 bits of synchronization

☞ **SCF:** two fields of 12 bits of Shared Channel Feedback (i.e., status of reverse channel)

☞ **DATA:** two 130-bit fields of data

☞ **CFSP:** 12 bits of Coded Super Frame Phase

☞ **RSVD:** 2 bits of a reserved field

On the reverse digital control channel (from PS to RS), the PS can transmit with a normal format (Figure 8.43) or an abbreviated format (Figure 8.44). For these slots, the field definitions are:

SYNC	SCF	DATA	CFSP	DATA	SCF	RSVD
28	12	130	12	130	10	2

Fig. 8.42 Time Slot Format: RS to PS on Digital Control Channel

G	R	PREAM	SYNC	DATA	SYCH+	DATA
6	6	16	28	122	24	122

Fig. 8.43 Time Slot Format: PS to RS on Digital Control Channel

G	R	PREAM	SYNC	DATA	SYCH+	DATA	R	AG
6	6	16	28	122	24	78	6	38

Fig. 8.44 Abbreviated Time Slot Format: PS to RS on Digital Control Channel

☞ **G:** equivalent of 6 bits guard time (the PS does not transmit during this time)

☞ **R:** equivalent of 6 bits of ramp time (the PS is powering up)

☞ **PREAM:** 16 bits of a preamble

☞ **SYNC:** 28 bits of synchronization

☞ **DATA:** two fields of 122 bits of data for a normal slot or one field of 122 bits and one field of 78 bits for an abbreviated slot; the data can be traffic data (e.g., voice) or control data

☞ **SYNC+:** additional synchronization

☞ **R:** equivalent of 6 bits of ramp time (the PS is powering down)

☞ **AG:** guard time for abbreviated burst

The PS will adjust its timing to account for delays in the system. The offset is equal to one time slot plus 45 symbols (i.e., 207 symbol periods or 414 bit periods). Time slot 1 of frame N of the PS occurs 207 symbol periods (or 414 bit periods) before time slot 1 of frame N of the RS.

The delay can be further refined under command of the RS. When the PS is transmitting on a reverse control channel (access channel), guard times are needed to prevent interference. The PS may be commanded to send abbreviated bursts when handing off to other channels where the timing delays are not known.

8.5.1.6 Framing of Wideband CDMA   The wideband CDMA (W-CDMA) system is similar to the CDMA system of IS-95. W-CDMA can support bandwidths of 5 MHz, 10 MHz, and 15 MHz. The basic operation of the system is the same independent of the bandwidth chosen. The overall block diagrams of the system are similar to the CDMA system (refer to Figures 8.8 to 8.13 for the operation of the system). The frequency reuse factor for W-CDMA is 1 as it is for CDMA. Tables 8.2 and 8.3 summarize the characteristics of the W-CDMA system.

Table 8.2 System Characteristics of Wideband CDMA at Various Bandwidths

Bandwidth (MHz)	Number of Channels	PN Rate (Mcps)	Code Rate (Mcps)	Data Rates Supported (kbs)	Channel Selection Method
5.0	256	4.096	4.096	2.0–64.0	Walsh $64 \times 4^*$
10.0	512	8.192	8.192	2.0–64.0	Walsh $64 \times 8^\dagger$
15.0	768	12.288	12.288	2.0–64.0	Hadamard 48×4 and Hadamard $96 \times 8^\ddagger$

*The 64 Walsh codes are repeated four times in normal and inverted patterns to construct a total of 256 different codes.

†The 64 Walsh codes are repeated eight times in normal and inverted patterns to construct a total of 512 different codes.

‡The codes are based on Hadamard 48 codes to generate a Hadamard 96 code which is then repeated in normal and inverted patterns to form up to 768 different codes.

Table 8.3 Channel Usage of W-CDMA at Various Bandwidths

Bandwidth (MHz)	Number of Channels	Pilot Channel Number	Sync Channel Number	Paging Channel Numbers
1.25	64	0	32	1–7 (no skipping)
5.0	256	0 & 64	32 &/or 96	1–7 (no skipping)
10.0	512	0 & 128	64 &/or 192	1–7 (no skipping)
15.0	768	0	384	1–7 (no skipping)

5-MHz Bandwidth System   In the 5-MHz bandwidth system, a fixed chip rate of 4.096 Mcps is used with 128 Walsh codes. A code channel that is modulated by Walsh function n will be assigned to code channel number n (n = 0 to 64 for 64 kbs, n = 0 to 127 for 32 kbs, and n = 0 to 255 for 16 kbs). Tables B.2 to B.4 in the appendix define the codes used for 64 kbs, 32 kbs, and 16 kbs, respectively. Walsh codes of length greater than 64 are generated by combinations of the basic Walsh 64 and inverted versions of the Walsh 64 code (see Tables B.1 to B.3). The Walsh code spreading sequence will repeat with a period of the duration of one forward traffic channel code symbol.

In a manner similar to the 1.25-MHz CDMA system, the pilot channel is always assigned to code channel number 0 and/or 64. When the sync channel is present, it will be assigned code channel number 32 and/or 96. When paging channels are present, they will be assigned to code channel numbers 1 through 7 (inclusive) in sequence. The remaining code channels are available for assignment to the forward traffic channels. When an RS supports multiple forward CDMA channels, then frequency division multiplex is used.

The maximum number of channels for 64 kbs is 64, for 32 kbs is 128, and for 16 kbs is 256. The same Walsh code cannot be used simultaneously for two different data rates.

10-MHz Bandwidth System   In the 10-MHz system, a fixed chip rate of 8.192 Mcps is used with 256 Walsh codes. A code channel that is modulated by Walsh function n will be assigned to code channel number n (n = 0 to 127 for 64 kbs, n = 0 to 255 for 32 kbs, and n = 0 to 511 for 16 kbs). One of the time-orthogonal Walsh codes as defined in Tables B.3, B.4, and B.5 will be used for 64 kbs, 32 kbs, and 16 kbs, respectively.

In a manner similar to the other CDMA bandwidths, the pilot channel is always assigned to code channel number 0 and/or 128, the sync channel is assigned to 32 and/or 96. and the paging channels are assigned to code channel numbers 1 through 7 (inclusive) in sequence, with the remaining code channels available for assignment to the forward traffic channels.

The maximum number of channels for 64 kbs is 128, for 32 kbs is 256, and for 16 kbs is 512. The same Walsh code cannot be used simultaneously for two different data rates.

15-MHz Bandwidth System   In the 15-MHz system, 768 Hadamard codes are used with a fixed chip rate of 12.288 Mcps. A code channel that is spread using Hadamard code n will be assigned to code index number n (n = 0 to 191 for 64 kbs, n = 0 to 383 for 32 kbs, and n = 0 to 767 for 16 kbs). One of the time-orthogonal Hadamard codes as defined in Tables B.7, B.8, and B.9 will be used for 64 kbs, 32 kbs, and 16 kbs, respectively. The longer Hadamard codes are constructed like the longer Walsh codes.

Similar to the other CDMA bandwidths, code index 0 will always be assigned to the pilot channel. When the sync channel is present, it will be assigned code index 384. When paging channels are present, they will be assigned to code indexes 1 through 7 (inclusive) in sequence. The remaining code indexes are available for assignment to the forward traffic channels.

There maximum number of channels for 64 kbs is 192 channels, for 32 kbs is 384, and for 16 kbs is 768. As with the other bandwidths, the same Hadamard code cannot be used simultaneously for two different data rates.

8.5.2 Speech Coding

The wireline network is based on sending voice using digital Pulse Code Modulation (PCM) at 64 kbs and sending data at rates of 64 kbs or higher multiples of 64 kbs. Many older analog facilities still exist, especially in residential areas, and use voice band modems at rates up to 28.8 kbs for data and analog electrical signals for voice. At the central office the analog voice and analog data is converted to digital signals using PCM or, optionally, using modem pools for data.

It would be nice if the identical systems could be used for PCS. Unfortunately the error rates on the PCS channels are many orders of magnitude higher than those of the copper or fiber-optic cables. In addition, PCM is inefficient for use over scarce and expensive radio channels. Therefore, all of the digital PCS systems use some efficient method of voice coding and extensive error recovery techniques to overcome the harsh nature of the radio channel. This section will cover the various means used for each of the North American protocols to send signals over the radio channel.

The simplest form of waveform-coding scheme is the linear PCM, in which the speech signal is sampled, quantized, and encoded. This approach is widely used for analog-to-digital conversion of a signal. The speech signal is band limited to the frequency range of 200 to 3,300 Hz. To achieve telephone quality speech, 12 bits per sample are required at a sampling rate of 8,000 samples per second. By using logarithmic PCM, 8 bits per sample are sufficient. Each sample is quantized into one of 256 levels. Two widely used variations of PCM for telephone quality speech (μ-law and A-law PCM) are based on a nonuniform quantization of the signal amplitude according to a logarithmic scale rather than a linear scale. Such coders utilize the static characteristics of amplitude nonstationary in speech to achieve good quality at a bit rate of 64 kbs. This is the basis of PCM.

High bit rates are not attractive for wireless systems. Better results are obtained with differential coders, in which a dynamic range of compression can be applied, such as Adaptive Predictive Coding (APC) and Adaptive Differential Pulse Code Modulation (ADPCM). The reason for these coders is to achieve a better signal-to-quantization noise performance over PCM.

Differential coders generate error signals as the difference between the input speech samples and the corresponding prediction estimates. The error signals are quantized and transmitted. ADPCM and APC differential coders are often used for intermediate bit rates—16 to 32 kbs.

ADPCM employs a short-term predictor that models the speech spectral envelope. ADPCM achieves network-quality speech (Mean Opinion Score [MOS] of 4.1 or better) at 32 kbs. This is a low-complexity coder of reasonable robustness with channel bit error rates in the range of 10^{-3} to 10^{-2}. The ADPCM coder is well suited for wireless access applications.

APC employs both short- and long-term prediction in a differential coding structure. APC outperforms ADPCM at 16 kbs and offers communication-quality speech (MOS 3.5 to 4.0) at a bit rate as low as 10 kbs. Introduction of noise shaping and post filtering in ADPCM and APC reduces the subjective loudness of quantization noise.

PCM, ADPCM, and APC operate in the time domain. No attempt is made to understand or analyze the information that is being sent. Redundancy removal techniques have been used successfully in the frequency domain. Frequency domain waveform-coding algorithms decompose the input speech signal into sinusoidal components with varying amplitudes and frequencies. Thus, the speech is modeled as a time-varying line spectrum. Frequency domain coders are systems of moderate complexity and operate well at a medium bit rate (16 kbs). When designed to operate in the range of 4.8 to 9.6 kbs, the complexity of the approach used to model the speech spectrum increases considerably.

The other class of speech-coding techniques consists of algorithms called vocoders which attempt to describe the speech production mechanism in terms of a few independent parameters serving as the information-bearing signals. These parameters attempt to model the creation of the voice by the vocal tract, decompose the information, and send it to the receiver. The receiver attempts to model an electronic vocal tract to produce the speech output.

The model operates this way: Vocoders consider that speech is produced from a source-filter arrangement. Voice speech is the result of exciting the filter with a periodic pulse train similar to the pulses generated by the vocal tract. Vocoders operate on the input signal, using an analysis process based on a particular speech production model, and extract a set of source-filter parameters which are encoded and transmitted. At the receiver, they are decoded and used to control a speech synthesizer that corresponds to the model used in the analysis process. Provided that all the perceptually significant parameters are extracted, the synthesized signal, as perceived by the human ear, resembles the original speech signal. Nonspeech signals are often not modeled well, so this method works poorly for analog modems.

Vocoders are medium-complexity systems and operate at low bit rates, typically 2.4 kbs, with synthetic-quality speech. Their poor-quality speech is due to the oversimplified source model used to drive the filter and the assumption that the source and filter are linearly independent.

In the bit rates from about 5 kbs to 16 kbs, the best speech quality is obtained by using hybrid coders that use suitable combinations of waveform-coding techniques and vocoder techniques. A simple hybrid coding scheme for telephone-quality speech with a few integrated digital signal processors is the Residual Excited Linear Prediction (RELP) coding. This belongs to a class of coders known as analysis-synthesis coder based on Linear Predictive Coding (LPC).

RELP systems employ short-term (and in certain cases, long-term) linear prediction to formulate a difference signal (residual) in a feed-forward manner. RELP systems are capable of producing communications-quality speech at 8 kbs. These utilize either pitch-aligned high-frequency regeneration procedures or full-band pitch prediction in time domain to remove the pitch information from the residual signal prior to band limitation/decimation. At bit rates < 9.6 kbs, the quality of the recovered speech signal can be improved significantly by an Analysis by Synthesis (AbS) optimization procedure to define the excitation signal. In these systems both filter and the excitation are defined on a short-term basis using a closed-loop optimization process that minimizes a perceptually weighted error measure formed between the input and decoded speech signals.

Table 8.4 summarizes the speech-coding method used by each of the North American protocols.

8.5.3 Frequency Allocation

The frequency allocation for cellular is 824–849 and 869–894 MHz. In this band, there are two carriers: a wireline carrier (B-band) and a nonwireline carrier (A-band). The frequencies used for each are defined in Table 8.5.

The frequency allocation for PCS is 1850–1910 MHz and 1930–1990 MHz with a constant 80-MHz spacing between transmit and receive frequencies. The spectrum is segmented into six blocks. Table 8.6 summarizes the frequency bands of each block. Each of the air interfaces use the spectrum in different ways. TDMA segments the spectrum into channels, whereas CDMA uses a wider bandwidth than TDMA and uses different codes to select a channel.

Table 8.4 Speech Coding Used in PCS

System	Coding Type	Coding Rate	Frame Size	Frame/sec.
Analog	none	analog	-	-
CDMA	QCELP (QUALCOMM CELP)	14.4 kbs	-	-
PACS	ADPCM	32 kbs	80 bits	400
TDMA	VSELP	8 kbs	320 bits	25
PCS-2000	ADPCM	32 kbs	640 bits	50 (4 slots used)
	PCM	64 kbs	640 bits	50 (2 slots used)
	PCS HCA	8 kbs	640 bits	50 (1 slot used)
W-CDMA	ADPCM	32 kbs	-	-

Table 8.5 AMPS Frequency Spectrum (after Expansion)

Frequency Band	MHz	PS Transmitter Band (MHz)	RS Transmitter Band (MHz)
A"	1.0	824.000–825.000	869.000–870.000
A	10.0	825.000–835.000	870.000–880.000
A'	1.5	845.000–846.500	890.000–891.500
B	10.0	835.000–-845.000	880.000–890.000
B'	2.5	846.500–849.000	891.500–894.000

Table 8.6 Definition of Bands for PCS

Block Designator	PS Transmit Band (MHz)	RS Transmit Band (MHz)	Bandwidth (MHz)
A	1850–1865	1930–1945	15
D	1865–1870	1945–1950	5
B	1870–1885	1950–1965	15
E	1885–1890	1965–1970	5
F	1890–1895	1970–1975	5
C	1895–1910	1975–1990	15

8.5.3.1 Channel Spacing and Frequency Tolerance for Analog Cellular

Analog cellular uses digital control channels and analog voice channels with the channels defined in Table 8.7.

Table 8.7 AMPS Channels with Expanded Spectrum

Frequency Band	Width (MHz)	Number of Channels	Channel Number	PS Transmitter Center Frequency (MHz)	RS Transmitter Center Frequency (MHz)
Not Used		1	(990)	(824.01)	(869.01)
A"	1.0	voice: 991–1023 (33)	991–1023	824.04–825.00	869.04–870.00
A	10.0	voice: 001–312 (312) control: 313–333 (21)	1–333	825.03–834.99	870.03–879.99
A'	1.5	voice: 667–716 (50)	667–716	845.01–846.48	890.01–891.48
B	10.0	control: 334–354 (21) voice: 355–666 (312)	334–666	835.02–844.98	880.02–889.98
B'	2.5	voice: 717–799 (83)	717–799	846.51–848.97	891.51–893.97

The frequency tolerance for an analog cellular PS is ±2.5 parts per million (±2,000 Hz at 800 MHz). The frequency tolerance for an analog cellular RS is ±1.5 parts per million (±1,200 Hz at 800 MHz).

8.5.3.2 Channel Spacing and Frequency Tolerance for CDMA

CDMA uses a bandwidth of 1.25 MHz and defines a set of channels on 50-kHz spacing. However, the recommended channels are on 1.25-MHz channel spacing. The basic frame signal (see section 8.5.2.2) is encoded with a Walsh function. The Walsh function has the property that, within the set of Walsh functions, each function is orthogonal to all others. Thus different signals are selected by choosing a different Walsh function to decode. After Walsh function encoding, the CDMA signal is modulated by a 1.23-MHz spreading signal called the long code.

The channels for CDMA are described in Tables 8.8, 8.9, and 8.10 for PCS frequencies and Table 8.11 and 8.12 for cellular frequencies.

The frequency tolerance stated as the PS transmit carrier frequency will be below the RS transmit frequency, as measured at the PS receiver, by 80 MHz ±150 Hz for PCS frequencies and 45 MHz ±300 Hz for cellular frequencies. The RS transmitter will maintain its frequency to within ±5 parts per 100 million (±100 Hz at 2000 MHz).

8.5.3.3 Channel Spacing and Frequency Tolerance for PACS

PACS uses TDMA with FDD on 300-kHz channels. Although the band plan (Figure 8.14) shows channels on 100-kHz spacing, only every third channel can be used in a given region. The extra channels are identified to solve guard channel problems

Table 8.8 Definition of Channel Numbers and Frequencies for CDMA at 1900 MHz

Band	Valid CDMA Frequency Assignments	CDMA Channel Number	PS Transmit Center Frequency (MHz)	RS Transmit Center Frequency (MHz)
A (15 MHz)	Not Valid	0–24	1850.000–1851.200	1930.000–1931.200
	Valid	25–275	1851.250–1863.750	1931.250–1933.750
	Cond. Valid	276–299	1863.800–1864.950	1933.800–1934.950
D (5 MHz)	Cond. Valid	300–324	1865.000–1866.200	1945.000–1946.200
	Valid	325–375	1866.250–1883.750	1946.250–1943.750
	Cond. Valid	376–399	1868.800–1869.950	1948.800–1949.950
B (15 MHz)	Cond. Valid	400–424	1870.000–1871.200	1950.000–1951.200
	Valid	425–675	1871.250–1883.750	1951.250–1963.750
	Cond. Valid	676–699	1883.800–1884.950	1963.800–1964.950
E (5 MHz)	Cond. Valid	700–724	1885.000–1886.200	1965.000–1966.200
	Valid	725–775	1886.250–1883.750	1966.250–1963.750
	Cond. Valid	776–799	1888.800–1889.950	1968.800–1969.950
F (5 MHz)	Cond. Valid	800–824	1890.000–1891.200	1970.000–1971.200
	Valid	825–875	1891.250–1893.750	1971.250–1973.750
	Cond. Valid	876–899	1893.800–1894.950	1973.800–1974.950
C (15 MHz)	Cond. Valid	900–924	1895.000–1896.200	1975.000–1976.200
	Valid	925–1175	1896.250–1908.750	1976.250–1988.750
	Not Valid	1176–1199	1908.800–1909.950	1988.800–1989.950

Table 8.9 CDMA Preferred Set of Frequency Assignments

Block Designator	Preferred Set Channel Numbers
A	25, 50, 75, 100, 125, 150, 175, 200, 225, 250, 275
D	325, 350, 375
B	425, 450, 475, 500, 525, 550, 575, 600, 625, 650, 675
E	725, 750, 775
F	825, 850, 875
C	925, 950, 975, 1000, 1025, 1050, 1075, 1100, 1125, 1150, 1175

at band edges. Once the channel at the band edge is established, all other channels increment in 300-kHz spacing.

The frequency stability of the PS and the RS will be within ±1 ppm (±2,000 Hz at 2000 MHz) of their nominal values over the operating temperature ranges and voltage ranges. Short-term stability over a TDMA frame will not exceed ±0.1 ppm.

Table 8.10 Channel Number Designations for CDMA PCS at 1900

Transmitter	CDMA Channel Number	Center Frequency of CDMA Channel (MHz)
PS	$1 < N < 1199$	$0.050\,N + 1850.000$
RS	$1 < N < 1199$	$0.050\,N + 1930.000$

Table 8.11 Definition of Valid Channel Numbers for CDMA at 800 MHz

Frequency Band	MHz	Valid Regions	Channel Number
A"	1	not valid	991–1012
		valid	1013–1023
A	10	valid	1–311
		not valid	312–333
A'	1.5	not valid	667–688
		valid	689–694
		not valid	695–716
B	10	not valid	334–355
		valid	356–644
		not valid	645–666
B'	2.5	not valid	717–738
		valid	739–777
		not valid	778–799

Table 8.12 Definition of Preferred Channel Numbers and Frequencies for CDMA at 800 MHz

Band	Preferred CDMA Channel Number	PS Transmit Center Frequency (MHz)	RS Transmit Center Frequency (MHz)
A	283	833.490	878.490
A'	691	845.730	890.730
A"	—	—	—
B	384	836.520	881.520
B'	777	848.310	893.310

Channels for PACS are described in Table 8.13.

8.5.3.4 Channel Spacing and Frequency Tolerance for PCS2000 The PCS2000 system uses 5-MHz channels separated by 5 MHz. Table 8.14 defines the potential channels (on a 2.5-MHz spacing). Obviously, in any given region, only sets that are 5 MHz apart can be used.

Table 8.13 Definition of Channel Numbers and Frequencies for PACS at 1900 MHz

Band	Bandwidth (MHz)	Number of Channels	Channel Number	PS Transmit Center Frequency (MHz)	RS Transmit Center Frequency (MHz)
			1	1850.1	1930.1
			2	1850.2	1930.2
			3	1850.3	1930.3
A	15	149	...	...	...
			147	1864.7	1944.7
			148	1864.8	1944.8
			149	1864.9	1944.9
			150	1865.0	1945.1
			151	1865.1	1945.2
			152	1865.2	1945.3
D	5	50	...	...	...
			197	1869.7	1949.7
			198	1869.8	1949.8
			199	1869.9	1949.9
			200	1870.1	1950.1
			201	1870.2	1950.2
			202	1870.3	1950.3
B	15	150	...	...	...
			347	1884.7	1964.7
			348	1884.8	1964.8
			349	1884.9	1964.9
			350	1885.1	1965.1
			351	1885.2	1965.2
			352	1885.3	1965.3
E	5	50	...	...	...
			397	1889.7	1969.7
			398	1889.8	1969.8
			399	1889.9	1969.9
			400	1890.1	1970.1
			401	1890.2	1970.2
			402	1890.3	1970.3
F	5	50	...	...	...
			447	1894.7	1974.7
			448	1894.8	1974.8
			449	1894.9	1974.9
			450	1895.1	1975.1
			451	1895.2	1975.2
			452	1895.3	1975.3
C	15	50	...	...	...
			597	1909.7	1989.7
			598	1909.9	1989.8
			599	1909.9	1989.9

Table 8.14 PCS-2000 Channel Plan

Band	Channel Number (Hex)	Frequency (MHz)
A	0	1852.50
	1	1855.00
	2	1857.50
	3	1860.00
	4	1862.50
	5	1932.50
	6	1935.00
	7	1937.50
	8	1940.00
	9	1942.50
B	A	1872.50
	B	1875.00
	C	1877.50
	D	1880.00
	E	1882.50
	F	1952.50
	10	1955.00
	11	1957.50
	12	1960.00
	13	1962.50
C	14	1897.50
	15	1900.00
	16	1902.50
	17	1905.00
	18	1907.50
	19	1977.50
	1A	1980.00
	1B	1982.50
	1C	1985.00
	1D	1987.50
D	1E	1867.50
	1F	1897.50
E	20	1887.50
	21	1967.50
F	22	1897.50
	23	1972.50
Unlicensed	28	1920.625
	29	1921.875
	2A	1923.125
	2B	1924.375
	2C	1925.625
	2D	1926.875
	2E	1928.125
	2F	1929.375

At an RS, the radio frequency signals and the data clock will be generated from a common frequency reference. The RS frequency will be kept within ±1 ppm (±2,000 Hz at 2000 MHz) of the nominal channel frequency when not synchronized to the data network and ±10 ppm (±20,000 Hz at 2000 MHz) of the nominal channel frequency when synchronized to the data network.

8.5.3.5 Channel Spacing and Frequency Tolerance for TDMA For TDMA, the channel spacing is 30 kHz. There are 1,999 TDMA channels in the PCS band with the channels designated as 1 to 1999. Tables 8.15 and 8.16 define the channel numbers and frequencies for TDMA for the 1900-MHz PCS band. When TDMA uses cellular frequencies, the channels are the same as the analog cellular channels.

The PS transmit carrier frequency will be 80 MHz ±200 Hz below the RS transmit frequency. The RS transmitter will maintain its frequency to within ±0.25 ppm (±500 Hz at 2000 MHz).

Table 8.15 Definition of Channel Numbers and Frequencies for TDMA at 1900 MHz

Band	Bandwidth (MHz)	Number of Channels	Boundary Channel Number	PS Transmit Center Frequency (MHz)	RS Transmit Center Frequency (MHz)
A	15	499	1	1850.040	1930.020
			499	1864.980	1944.960
Not used		1	500	1865.010	1944.990
D	5	165	501	1865.040	1945.020
			665	1869.960	1949.940
Not used		1	666	1869.990	1949.970
Not used		1	667	1870.020	1950.000
B	15	498	668	1870.050	1950.030
			1165	1884.960	1964.940
Not used		1	1166	1884.990	1964.970
Not used		1	1167	1885.020	1965.000
E	5	165	1168	1885.050	1965.030
			1332	1899.970	1969.950
Not used		1	1334	1890.000	1970.980
Not used		1	1334	1890.030	1970.010
F	5	165	1335	1890.060	1970.040
			1499	1894.980	1974.960
Not used		1	1500	1895.010	1975.990
C	15	499	1501	1895.040	1975.020
			1999	1909.980	1989.960

Table 8.16 Channel Number Designations for TDMA PCS at 1900

Transmitter	Channel Number	Center Frequency (MHz)
PS	$1 < N < 1999$	$0.030\,N + 1850.010$
RS	$1 < N < 1999$	$0.030\,N + 1929.990$

8.5.3.6 Channel Spacing and Frequency Tolerance for Wideband CDMA

The wideband CDMA system is similar to the CDMA system except than any one of three wider bandwidths can be supported—5 MHz, 10 MHz, or 15 MHz. With different allocations of bandwidths (5 or 15 MHz) in the United States, then the bands will be populated as described in Tables 8.17 and 8.18.

For a W-band CDMA transmitter, the PS transmit carrier frequency will track within ± 200 Hz of a frequency value 80 MHz lower than the frequency of the corresponding RS transmit signal, as measured at the PS receiver. The RS transmitter will maintain its frequency to within ±5 parts per 100 million (±100 Hz at 2000 MHz).

8.5.4 Power Output Characteristics of the PS and the RS

All of the air interfaces use power control to minimize interference in the cellular system. TDMA uses discrete levels that are changed in response to local propagation conditions and distance from the PS to the RS. CDMA systems require instantaneous power control to maintain minimum interference levels at the RS. Each air interface defines various classes of PS transmitters that are permitted different power outputs. Handheld units typically have lower power outputs than units mounted in vehicles. The power output of handheld units is kept lower to increase battery life and to reduce RF radiation into the body of the user. The FCC has established standards for handheld transmitters to avoid radiation hazards. Two classes of users have been established—one for users who are aware they are using transmitters and may have received training (e.g., police,

Table 8.17 Use of 5-MHz Spectrum with Wideband CDMA

0 MHz	
	5 MHz Wideband CDMA
5 MHz	

Table 8.18 Sharing of 1.25-MHz CDMA with Wideband CDMA

0 MHz	
	1.25 MHz Narrowband CDMA
	1.25 MHz Narrowband CDMA
	1.25 MHz Narrowband CDMA
	1.25 MHz Narrowband CDMA
5 MHz	

fire, rescue) and one for those who may not realize that they are using a transmitter (e.g., a cellular or PCS user). The radiation levels for the second class of user are lower. All handheld PCS PSs will meet the second class of requirements.

RS power is typically higher than PS power and depends on the cell size. It too is limited by FCC rules.

8.5.4.1 Power Output Characteristics of Analog Cellular

Four classes of personal stations are defined for analog cellular with the fourth class being for a dual-mode TDMA/analog cellular phone (see Table 8.19). Power control is done under control from the RS based on propagation conditions between the PS and the RS and based on cell size.

All PSs are required to adjust power level on command from an RS specifying power level 0 to 7. PSs in classes IV–VIII are required to change power to levels 0 to 10 by a physical layer control message from the RS.

The power levels 0 to 7 are maintained within the range of + 2 dB/–4 dB of its nominal level over the ambient temperature range of –30°C to + 60°C and over the supply voltage range of ± 10% from the nominal value.

For power levels 8 through 10, RF power emission is maintained within the range +2 dB/–6 dB of the initial power level unless a physical layer control message is received, over the same temperature and supply voltage conditions stated earlier.

Table 8.19 Personal Station Nominal Power Levels

PS Power Level (dBW)	Mobile Attenuation	PS Power Class			
(PL)	Code (MAC)	I	II	III	IV*
0	000	6	2	–2	–2
1	001	2	2	–2	–2
2	010	–2	–2	–2	2
3	011	–6	–6	–6	–6
4	100	–10	–10	–10	–10
5	101	–14	–14	–14	–14
6	110	–18	–18	–18	–18
7	111	–22	–22	–22	–22
Dual mode only					
8					–26 ±3 dB
9					–30 ±6 dB
10					–34 ±9 dB

*Class IV is available only in dual-mode PSs.

The RS EIRP depends on Height Above Average Terrain (HAAT) and power output.

8.5.4.2 Power Output Characteristics of CDMA

For a CDMA PS, each class of transmitter has an absolute maximum effective isotropic radiated power (EIRP). For any class of personal station transmitter, it is 3 dBW (2.0 W). Table 8.20 defines the maximum output power and the minimum power output for each class of PS when it is commanded to its maximum power. The PS attempts to control the power output based on received signal strength (open-loop control), and the RS sends power control messages to the PS and controls the power about once every millisecond (closed-loop control). The net effect is to control the power received at the RS to within 1 dB for all PSs being received at that RS. The fine level of power control is necessary for proper operation of the CDMA system.

The RS will not transmit more than 1,640 W of EIRP in any direction in a 1.25-MHz band for antenna HAAT less than 300 meters. The RS antenna height may exceed 300 meters with a reduction in EIRP according to current FCC rules.

The transmitter output power of the RS in any 1.25-MHz band of the RS's transmit band between 1930 and 1990 MHz and in any direction will not exceed 100 W.

8.5.4.3 Power Output Characteristics of PACS

In PACS, the RS transmits continuously on a channel with a maximum power output of 800 mW (+29 dBm).

The PS* transmits in TDMA bursts with an average power determined by the power control process. The maximum average output power is 200 mW (+23 dBm). There are no classes of PSs defined for PACS. The RS sends power control messages to the PS that control the power in 1-dB steps (±0.5 dB) over a range of at least 30 dB. If a failure of RS control of power output occurs, the PS will transmit at its maximum power output.

Table 8.20 EIRP at Maximum Output Power for a CDMA PS

PS Class	EIRP at Maximum Output Will Exceed	EIRP at Maximum Output Will Not Exceed
I	−2 dBW (0.63 W)	3 dBW (2.0 W)
II	−7 dBW (0.20 W)	0 dBW (1.0 W)
III	−12 dBW (63 mW)	−3 dBW (0.5 W)
IV	−17 dBW (20 mW)	−6 dBW (0.25 W)
V	−22 dBW (6.3 mW)	−9 dBW (0.13 W)

---

*Called a subscriber unit (SU) for PACS.

8.5.4.4 Power Output Characteristics of PCS2000

The PCS2000 system uses a maximum of 1 W EIRP for the PS and a maximum of 1,640 W EIRP for the RS in the licensed PCS band. Actual power output of the RS transmitter is 2 W and an antenna with a gain of 29.15 dB (plus feed-line losses) is permitted. Typical power output for the PS is 10 mW in each 8-kbs time slot.

The peak power of the RS is controllable over a 33-dB range in steps of 3 dB. The peak power of the PS is controllable, by the RS, over a range of 33 dB in steps of 3 dB. The combined PCS2000 system does not define power classes of PSs.

8.5.4.5 Power Output Characteristics of TDMA

The maximum EIRP with respect to a half-wave dipole for any class of PS transmitter is 2 dBW (1.58 W). The nominal EIRP for each class of transmitter is shown in Table 8.21. Power is controlled by the RS (via an order message) and accounts for the changes in propagation between the RS and the PS. No instantaneous power control is used.

The RS power is not specified in the TDMA specification but will be coordinated with other TDMA systems in an area and depends on HAAT and cell size.

8.5.4.6 Power Output Characteristics of Wideband CDMA

The W-CDMA system uses power control similar to that of the CDMA system except that lower power is used and fewer classes of transmitters are used. Table 8.22 summarizes the PS power levels for all three bandwidths for the W-CDMA system.

Table 8.21 PS Power Levels for Wideband CDMA

PS Power Level	Mobile Attenuation Code	Nominal EIRP (dBW) for PS Power Class						
(PL)	(DMAC)	II	III	IV	V[*]	VI[*]	VII[*]	VIII[*]
0	0000	2	−2	−2	—	—	—	—
1	0001	2	−2	−2	—	—	—	—
2	0010	−2	−2	2	—	—	—	—
3	0011	−6	−6	−6	—	—	—	—
4	0100	−10	−10	−10	—	—	—	—
5	0101	−14	−14	−14	—	—	—	—
6	0110	−18	−18	−18	—	—	—	—
7	0111	−22	−22	−22	—	—	—	—
8	1000	−22	−22	−26 ±3 dB	—	—	—	—
9	1001	−22	−22	−30 ±6 dB	—	—	—	—
10	1010	−22	−22	−34 ±9 dB	—	—	—	—

[*]Classes V, VI, VII, and VIII are reserved for future use.

Table 8.22 EIRP at Maximum Output Power

PS Class	EIRP at Maximum Output Will Exceed
I	23 dBm (200 mW)
II	13 dBm (20 mW)
III	3 dBm (2 mW)

A RS for W-CDMA will have no more than 1,000 W of EIRP.

8.5.5 Modulation Methods

Table 8.23 summarizes the modulation methods used by each of the North American protocols. We describe the information in more detail in the following sections.

Table 8.23 Summary of Modulation Methods for Cellular and PCS

Protocol	Analog	CDMA	PACS	PCS2000	TDMA	W-CDMA
Type of modulation	MFM-FSK	QPSK	$\pi/4$ DQPSK	QAM	$\pi/4$ DQPSK	QPSK
Number of points in signal constellation	2	4	8	32	8	4
Frame data rate	10 kbs	1.288	384 kbs	5 Mcps for CDMA, 781.25 kbs (for TDMA frame)	48.6 kbs	4.096 Mcps, 8.192 Mcps, 12.288 Mcps
Bits/sec/baud	1/2	—	2	—	2	—
Bandwidth	30 kHz	1.25 MHz	288 kHz	5 MHz	30 kHz	5 MHz, 10 MHz, 15 MHz
Bits/sec/Hz	0.333	—	0.75	—	0.62	—

8.5.5.1 Analog Cellular Modulation Analog cellular uses FM for voice band and analog data transmission with several voice processing stages:

☞ **Speech compressor:** A 2:1 compressor is used. For every 2-dB change in input, the change in output level is a nominal 1 dB. The compressor will have a nominal attack time of 3 ms and a nominal recovery time of 13.5 ms as defined by the CCITT [6]. A calibration frequency of 1,000 Hz is used, and the level is set by the typical level from the PS microphone. This level will produce a nominal ±2.9-kHz peak frequency deviation of the PS transmitted carrier.

☞ **Pre-emphasis:** Since baseband noise in an FM RS increases with increasing (baseband) frequency, a pre-emphasis network is used. The pre-emphasis characteristic will have a nominal +6 dB/octave response between 300 and 3,000 Hz.

☞ **Deviation limiter:** To avoid overmodulating and interfering with adjacent channels the maximum modulation will be limited to ±12 kHz. The SAT can add to this deviation.

☞ **Post-deviation limiter filter:** The deviation limiter will be followed by a low-pass filter whose attenuation characteristics will exceed the characteristics in Table 8.24.

The receiver will have de-emphasis and an expander to match the transmitter.

When wideband data is sent for signaling, either on the control channel or the voice channel, the wideband data streams will be further encoded (Figure 8.45) so that each non-return-to-zero binary one is transformed to a zero-to-one transition, and each non-return-to-zero binary zero is transformed to a one-to-zero transition.[*] The filtered wideband data stream will then be used to modulate the transmitter carrier using direct binary FSK. A one (i.e., high state) into

Table 8.24 Post-Deviation Limiter Characteristics

Frequency Band	Attenuation Relative to 1,000 Hz for PS	Attenuation Relative to 1,000 Hz for RS
3,000–5,900 Hz	40 log (f/3,000) dB	40 log (f/3,000) dB
5,900–6,100 Hz	35 dB	40 log (f/3,000) dB
6,100–15,000 Hz	40 log (f/3,000) dB	40 log (f/3,000) dB
above 15,000 Hz	28 dB	28 dB

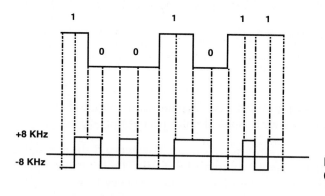

Fig. 8.45 Non-Return to Zero (NRZ) to Manchester Encoding

---

[*]The modulation scheme is known as Manchester encoding and develops a spectrum that has no DC component, independent of the number of zeros or ones sent. This is important for a modulation like FM that has no ability to transmit zero frequency (DC) baseband signals.

the modulator will correspond to a nominal peak frequency deviation 8 kHz above the carrier frequency, and a zero into the modulator will correspond to a nominal peak frequency deviation 8 kHz below the carrier frequency.

When the SAT is sent, it will have a deviation of 1/3 radian (±2 kHz).

The signaling tone used to signal a disconnect or a flash will have a tone of 10 kHz ±1 Hz and a deviation of ±8 kHz.

8.5.5.2 CDMA Modulation The CDMA signal transmitted by a RS consists of up to 64 channels, each on a different Walsh function. Channel 0 is the pilot channel and uses Walsh function 0. Channel 32, the sync channel, uses Walsh function 32 and is used to synchronize the receiver. Channels 1–7 (Walsh 1–7) are the paging channels. A minimum of one and a maximum of seven paging channels are used in order. The remainder of the channels are used as traffic channels.

Traffic channel data enters the modulation stages as 0.8–8.6 kbs (Rate Set 1) or 1.05–13.35 kbs (Rate Set 2). It then goes through the following stages (see Figures. 8.18 and 8.19).

☞ **Add one reserved bit (Rate Set 2 only):** This step is needed only for Rate Set 2.

☞ **Add frame quality indicator:** The number of bits available depend on the data rate. No bits are added at the highest data rates.

☞ **Add 8-bit encoder tail per frame:** These bits simplify the decoding of the digital voice.

☞ **Rate one-half convolutional encoding:** This step is used for error detection in the receiver.

☞ **Symbol repetition:** If the data rate is not the maximum rate, repeat each symbol until a uniform data rate is maintained (19.2 kbs for Rate Set 1 and 28.8 kbs for Rate Set 2).

☞ **Puncture (delete) 2 of 6 bits (Rate Set 2 only):** This step lowers the date rate to 19.2 kbs.

☞ **Block interleave the bits:** This step mixes up the transmission of the bits (see appendix on coding) for additional error protection capabilities. The data rate remains the same.

☞ **Exclusive OR with decimated long code:** The long code is different for each user and provides privacy of communications.

☞ **Multiplex with power control bits and further decimated long code:** This step adds the power control to the signal.

☞ **Exclusive OR with Walsh code for the channel:** The Walsh function selects the channel used for the traffic.

☞ **Add all channels together:** This step adds all 64 Walsh channels.

☞ **Form in-phase and quadrature signals from pilot PN signal** (at 1.2288 Mcps): This step modulates the signal up to the data rate of the channel.

☞ **Baseband filter the signals:** This steps prevents modulation components from appearing outside of the channel for the CDMA signal.

☞ **Modulate with the in-phase and quadrature carriers:** This step modulates the signals to the CDMA channel.

The pilot and paging channels enter the rate 1/2 encoder directly.

The receiver undoes each of these steps. A PS transmits on one Walsh function at a time and uses a rate 1/3 convolutional encoder. Other minor differences are shown in Figures. 8.22 to 8.24.

The net signal from the CDMA modulator is a four-phase quadrature signal. We have previously described this modulation in chapter 6.

8.5.5.3 PACS Modulation

PACS uses $\pi/4$ differentially encoded quadrature shift keying ($\pi/4$-DQPSK) with 2 bits per symbol. The transmitter has eight points in the signal constellation. We studied this modulation method in chapter 6. The data rate is 384 kbs or 192 k-symbols/sec and the bandwidth is 288 kHz. The data signals are differentially encoded and result in a change in phase of transmitted signal.

The baseband signals are shaped with a linear phase Nyquist 50% bandwidth expansion square root raised cosine spectral shaping. The baseband filter has the frequency response described in chapter 6, with a roll-off factor equal to 0.5 for PACS.

8.5.5.4 PCS2000 Modulation

The TDMA bursts of PCS 2000 are modulated with a DSSS signal that uses 32 QAM (Figure 8.46) at a rate of 5 Mcps.

8.5.5.5 TDMA Modulation

The TDMA protocol uses $\pi/4$-DQPSK with 2 bits per symbol. The signal constellation and data filtering are identical to PACS, but obviously the data rate and bandwidth are different.

8.5.5.6 W-CDMA Modulation

The operation of the W-CDMA system is similar to the operation of the CDMA system except that a PN code rate is higher to account for the wider bandwidth.

8.6 HANDOFFS

In analog cellular, the network measures the signal strength of the PS at a variety of RSs and processes a handoff to a channel on a different RS. In analog cellular, the handoff order is sent on the voice channel by blanking the voice and sending data. The PS then switches to the new channel where the network recognizes the presence of the PS and switches the network connection; thus the call continues. Sometimes the network can set up a bridge connection between the old and new RS and thus minimize the duration of the handoff. In general, depending on the system design, the handoff process (data transmission, channel switching, and network switching) can take 100–200 ms and produces a notice-

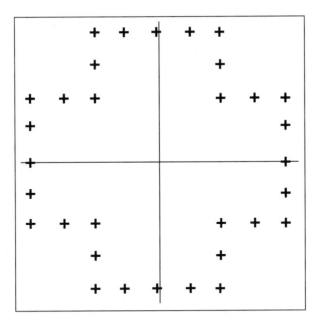

Fig. 8.46 PCS-2000 32-QAM Constellation

able click in the voice conversation. If the original voice channel is very noisy, the click will be imperceptible. However, if the handoff occurs at reasonable signal quality or is done to balance traffic at an RS, the user may hear the click. Too frequent handoffs may cause the user to have a poor impression of the service.

With digital systems, the handoff process is under control of the RS and the PS and can be designed to be imperceptible to the user.

Soft handoffs are common in CDMA where both RSs can transmit on the same frequency and Walsh code and the resultant signal is handled by the receiver as just one more multipath to be included in the decoded signal. Soft handoffs are imperceptible to the user. The soft handoff process requires that all RSs in a network be tightly synchronized so that, when handoffs occur, data synchronization is maintained. The CDMA systems support this function by using the Global Positioning System (GPS)* to provide a master clock. The IS-136 TDMA system does not currently support this function but may in the future.

Hard handoffs are the normal handoffs for an analog cellular system. TDMA systems typically process hard handoffs unless special means are taken to synchronize the RSs. CDMA systems forced to hand off to a different Wash function or a different frequency will have hard handoffs. The resultant break in connection causes speech or data to be lost and is perceptible to the user.

---

*The GPS system is a set of low-orbiting satellites that provides an atomic clock (corrected for relativity effects, satellite height, satellite location, and so on). GPS has seen significant civilian use throughout the world. The former Soviet Union provides a similar capability through its Glossnoss system.

Table 8.25 summarizes the support for each of the air interfaces for soft and hard handoffs.

The handoff process can be triggered by either the network or the PS (see Table 8.25 for a summary of handoff types by air interface).

☞ **Network-directed handoff trigger:** When the network controls the handoff, signal quality measurements (error rates, signal strength, and so on) are measured by the network. The network then initiates the process to handoff the PS to another RS or channel at an RS.

☞ **PS-directed handoff trigger:** When the PS controls the handoff process, it makes the signal quality measurements and initiates the handoff.

When handoffs occur, call information can potentially be lost; therefore, one network element will maintain the call information. That network element is called the anchor for the call. RSs and PCSCs can be anchors. The choice of the anchor is a network design issue.

Table 8.25 Air Interface Support for Handoff Types

System Hand-off Type	Analog	CDMA	PACS	PCS2000	TDMA	W-CDMA
Hard	X	X	X	NA[*]	X	NA[*]
Soft		X	NA[*]	NA[*]	[†]	X
PS triggered		X		X	X	X
Network triggered	X	X		X	X	

[*]Not available
[†]TDMA will support soft handoff in the future.

8.7 SUMMARY

For the North American PCS and cellular market, five digital air interfaces and one analog air interface supporting a common MAP have been defined by the standards bodies TR-46 and T1P1. This chapter discussed the operation of these air interfaces and the services they support.

8.8 PROBLEMS

1. The United States will be using six different air interfaces for PCS with two different service models (MAPs). Describe the advantages and disadvantages of this multiplicity of standards as compared to support for a single standard for the country. Make this list for

   • The user of a PCS telephone
   • The manufacturer of the PCS system and PSs
   • The operator of a PCS system

2. In section 8.4.3.1, we describe the call flow where the ISDN switch manages call records on a call. Describe the call waiting call flow where the RS manages the call record information.

3. Describe the type of information that will flow from the RS to the network and to another RS to support network-triggered handoffs.

4. Describe the type of information that will flow from the PS to the RS, to the network, and to another RS to support personal terminal-triggered handoffs.

8.9 REFERENCES

1. Committee T1—Telecommunications, "A Technical Report on Network Capabilities, Architectures, and Interfaces for Personal Communications," T1 Technical Report #34, May 1994.

2. Committee T1, "Stage 2 Service Description for Circuit Mode Switched Bearer Services," Draft T1.704.

3. "PCS-1900 Air Interface, Proposed Wideband CDMA PCS Standard," J-STD-007, T1 LB-461.

4. "PCS-1900 MHz, IS-136 Based, Mobile Station Minimum Performance Standards," J-STD-009, "PCS-1900 MHz, IS-136 Based, Base Station Minimum Performance Standards," J-STD-010, "PCS-1900 MHz, IS-136 Based, Air Interface Compatibility Standards," J-STD-011.

5. "Personal Access Communications System, Air Interface Standard," Draft J-STD-XXX, LB-477.

6. Recommendation G162, CCITT Plenary Assembly, Geneva, May–June 1964, Blue Book, Vol. 111, 52.

7. TIA Interim Standard, IS-41 C, "Cellular Radiotelecommunications Intersystem Operations."

8. TIA Interim Standard, IS-91, "Cellular System Mobile Station—Land Station Compatibility Specification."

9. TIA Interim Standard, IS-104, "Personal Communications Service Descriptions for 1800 MHz."

10. T1P1/94-088, "Draft American National Standard for Telecommunications—Personal Station-Base Station Compatibility Requirements for 1.8 to 2.0 GHz Code Division Multiple Access (CDMA) Personal Communications Systems," Draft J-STD-008, November 1994.

11. T1P1/94-089, "PCS2000, A Composite CDMA/TDMA Air Interface Compatibility Standard for Personal Communications in 1.8–2.2 GHz for Licensed and Unlicensed Applications," Committee T1 Approved Trial User Standard, T1-LB-459, November, 1994.

CHAPTER 9

European and Japanese Cellular Systems and North American PCS1900

9.1 INTRODUCTION

In this chapter we present an overview of the Global System for Mobile communications (GSM) as described in the European Telecommunication Standards Institute's (ETSI's) Recommendations. The chapter also addresses PCS1900, a derivative of the GSM for PCS application in North America. A brief description of the Japanese Digital Cellular (JDC) system is also given.

9.2 GSM PUBLIC LAND MOBILE NETWORK (PLMN)

ETSI originally defined GSM as a European digital cellular telephony standard. GSM interfaces defined by ETSI lay the groundwork for a multivendor network approach to digital mobile communication.

GSM offers users good voice quality, call privacy, and network security. Subscriber Identity Module (SIM) cards provide the security mechanism for GSM. SIM cards are like credit cards and identify the user to the GSM network. They can be used with any GSM handset, providing phone access, ensuring delivery of appropriate services to that user, and automatically billing the subscriber's network usage back to the home network.

Roaming agreements have been established between most GSM network providers in Europe, allowing subscribers to roam between networks and have access to same services no matter where they travel.

Of major importance is GSM's potential for delivering enhanced services requiring multimedia communication: voice, image, and data. Several mobile service providers offer free voice mailboxes and phone answering services to subscribers.

The key to delivering enhanced services is Signaling System Number 7 (SS7), a robust set of protocol layers designed to provide fast, efficient, reliable transfer and delivery of signaling information across the signaling network and to support both switched-voice and nonvoice applications. With SS7 on the enhanced services platform and integrating mailbox parameters, subscribers can be notified about the number of stored messages in their mailboxes, time and source of last messages, message urgency, and type of message—voice or fax. Future applications such as fax store-and-forward and audiotext can also use the platform's voice- and data-handling capabilities.

9.3 OBJECTIVES OF A GSM PLMN

A GSM PLMN cannot establish calls autonomously other than local calls between mobile subscribers. In most of the cases, the GSM PLMN depends upon the existing wireline networks to route the calls. Most of the time the service provided to a subscriber is the combination of the access service by a GSM PLMN and the service by some existing wireline network. Thus, the general objectives of a GSM PLMN with respect to service to a subscriber are:

☞ To provide the subscriber with a wide range of services and facilities, both voice and nonvoice, that are compatible with those offered by existing networks (e.g., PSTN, ISDN)

☞ To introduce a mobile RS that is compatible with ISDN

☞ To provide certain services and facilities exclusive to mobile situations

☞ To give compatibility of access to the GSM network for a mobile subscriber in a country that operates the GSM system

☞ To provide facilities for automatic roaming, locating, and updating of mobile subscribers

☞ To provide service to a wide range of mobile stations, including vehicle-mounted stations, portable stations, and handheld stations

☞ To provide for efficient use of the frequency spectrum

☞ To allow for a low-cost infrastructure, terminal, and service cost

9.4 GSM PLMN SERVICES

A telecommunication service supported by the GSM PLMN is defined as a group of communication capabilities that the service provider offers to the subscribers. The basic telecommunication services provided by the GSM PLMN are divided into three main groups: bearer services, teleservices, and supplementary services.

9.4.1 Bearer Services

These services give the subscriber the capacity required to transmit appropriate signals between certain access points (i.e., user-network interfaces).

The capabilities of the GSM bearer services are the following:

- ☞ Rate-adapted subrate information—circuit-switched asynchronous and synchronous duplex data, 300–9,600 bps
- ☞ Access to Packet Assembler/Disassembler (PAD) functions—PAD access for asynchronous data, 300–9,600 bps
- ☞ Access to X.25 public data networks—packet service for synchronous duplex data, 2,400–9,600 bps
- ☞ Speech and data swapping during a call—alternate speech/data and speech followed by data
- ☞ Modem selection—selection of 3.1-kHz audio service when interworking to an ISDN
- ☞ Support of automatic request for retransmission (ARQ) technique for improved error rates—transparent mode (no ARQ) and nontransparent mode (with ARQ)

Table 9.1 provides a summary of these services and compares them with services available with ISDN.

9.4.2 Teleservices

These services provide the subscriber with necessary capabilities including terminal equipment functions to communicate with other subscribers.

The GSM teleservices are:

- ☞ Speech transmission—telephony, emergency call
- ☞ Short message services—mobile terminating point-to-point, mobile-originating point-to-point, cell broadcast

Table 9.1 A Comparison of Bearer Services Supported by GSM and ISDN

Service	GSM	ISDN
Data services	x	x
Alternate speech/data	x	x
Speech followed by data	x	x
Clear 3.1-kHz audio	x	x
Unrestricted digital information (UDI)	x	x
Packet Assembler/Disassembler (PAD)	x	
3.1-kHz external to PLMN	x	
Others		x

☞ Message handling and storage services

☞ Videotex accesss

☞ Teletex transmission

☞ Facsimile transmission

A summary of the teleservices is given in Table 9.2. A comparison is made between the GSM and ISDN teleservices.

9.4.3 Supplementary Services

These services modify or supplement basic telecommunications services and are offered together with or in association with basic telecommunications services.

You should note that most of the supplementary services in GSM have been aligned with the North American supplementary services specified by ATSI. These are described in chapter 8.

The GSM supplementary services are:

☞ Number identification services

   ✗ Calling Number Identification Presentation (CNIP)

   ✗ Calling Number Identification Restriction (CNIR)

   ✗ Connected Number Identification Presentation (CNOP)

   ✗ Connected Number Identification Restriction (CNOR)

   ✗ Malicious Call Identification (MCI)

☞ Calling Offering Services

   ✗ Call Forwarding Unconditional (CFU)

   ✗ Call Forwarding mobile Busy (CFB)

   ✗ Call Forwarding No Reply (CFNRy)

Table 9.2 A Comparison of Teleservices Supported by GSM and ISDN

Service	GSM	ISDN
Circuit speech (telephony)	x	x
Emergency call	x	x
Short message point-to-point	x	x
Short message cell broadcast	x	x
Alternate speech/facsimile group 3	x	x
Automatic facsimile group 3 service	x	x
Voice-band modem (3.1-kHz audio)	x	x
Messaging teleservices	x	
Paging teleservices	x	
Others		x

- ✘ Call Forwarding mobile Not Reachable (CFNRc)
- ✘ Call Transfer (CT)
- ✘ Mobile Access Hunting (MAH)
- ☞ Call Completion Services
  - ✘ Call Waiting (CW)
  - ✘ Call Holding (HOLD)
  - ✘ Completion of Call to Busy Subscriber (CCBS)
- ☞ Multiparty Services
  - ✘ 3-Party service (3PTY)
  - ✘ Conference Calling (CONF)
- ☞ Community of Interest Services
  - ✘ Closed User Group (CUG)
- ☞ Charging Services
  - ✘ Advice of Charge (AoC)
  - ✘ Freephone Service (FPH)
  - ✘ Reverse Charging (REVC)
- ☞ Additional Information Transfer Service
  - ✘ User-to-User Signaling (UUS)
- ☞ Call Restrictions Services
  - ✘ Barring All Originating Calls (BAOC)
  - ✘ Barring Outgoing International Calls (BOIC)
  - ✘ BOIC except Home Country (BOIC-exHC)
  - ✘ Barring All Incoming Calls (BAIC)
  - ✘ Barring Incoming Calls when Roaming (BIC-Roam)

Table 9.3 summarizes the supplementary services and compares them to the supplementary services available with ISDN.

The GSM system offers an opportunity to a subscriber of moving freely through countries where a GSM PLMN is operational. Agreements are required between the various service providers to guarantee access to service offered to subscribers.

9.5 GSM Architecture

A series of functions are required to support the services and facilities in the GSM PLMN. The basic subsystems of the GSM architecture are: Base Station Subsystem (BSS), Network and Switching Subsystem (NSS), and Operational Subsystem (OSS).

The BSS provides and manages transmission paths between the mobile stations (MSs) and the NSS. This includes management of the radio interface between MSs and rest of the GSM system. The NSS has the responsibility of

Table 9.3 A Comparison of Supplementary Services Supported by GSM and ISDN

Service	GSM	ISDN
Call Number ID Presentation	x	x
Call Number ID Restriction	x	x
Connected Number ID Presentation	x	x
Connected Number ID Restriction	x	x
Malicious Call Identification	x	x
Call Forwarding Unrestricted	x	x
Call Forwarding Mobile Busy	x	x
Call Forwarding No Reply	x	x
Call Forwarding Mobile Not Reachable	x	x
Call Transfer	x	x
Call Waiting	x	x
Call Hold	x	x
Completion of Call to Busy Subscriber	x	x
3-Party Service	x	x
Conference Calling	x	x
Closed User Group	x	x
Multi-party	x	x
Advice of Charge	x	x
Reverse Charging	x	x
Flexible Alerting		x
Mobile Access Hunting	x	x
Freephone	x	
Barring All Originating Calls	x	x
Barring Outgoing Calls	x	x
Barring Outgoing International Calls	x	x
Barring All International Calls	x	x
Barring Outgoing International Calls—except Home	x	x
Barring Incoming Calls when Roaming	x	x
Do Not Disturb	Barring	x
Message Waiting Notification	SMS	x

Table 9.3 A Comparison of Supplementary Services Supported by GSM and ISDN (Cont.)

Service	GSM	ISDN
Preferred Language Service		x
Remote Feature Control		x
Selective Call Acceptance	Barring	x
Voice Privacy	Encryption	x
Priority Access & Channel Assignment		x
Password Call Acceptance		x
Subscriber PIN Intercept	Barring	x
Subscriber PIN Access		x
Voice Mail Retrieval		x
Others		x

managing communications and connecting MSs to the relevant networks or other MSs. The NSS is not in direct contact with the MSs. Neither is the BSS in direct contact with external networks. The MS, BSS, and NSS form the operational part of the GSM system. The OSS provides the means for a service provider to control them. Figure 9.1 shows the model for the GSM system.

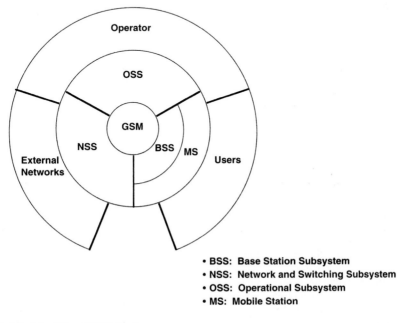

- **BSS: Base Station Subsystem**
- **NSS: Network and Switching Subsystem**
- **OSS: Operational Subsystem**
- **MS: Mobile Station**

Fig. 9.1 Model of the GSM System

In the GSM, interaction between the subsystems can be grouped into two main parts:

Operational part: external networks $\leftrightarrow$ NSS $\leftrightarrow$ BSS $\leftrightarrow$ MS $\leftrightarrow$ user

Control part: OSS $\leftrightarrow$ service provider

The operational part provides transmission paths and establishes them. The control part interacts with the traffic-handling activity of the operational part by monitoring and modifying it to maintain or improve its functions.

9.5.1 GSM Subsystems Entities

Figure 9.2 shows the functional entities of the GSM and their logical interconnection. A brief description of these functional entities is given below.

9.5.1.1 Mobile Station   The MS consists of the physical equipment used by the subscriber to access a PLMN for offered telecommunication services. Functionally, the MS includes a Mobile Termination (MT) and, depending on the services it can support, various Terminal Equipment (TE), and combinations of TE and Terminal Adaptor (TA) functions (TA acts as a gateway between the TE and MT) (see Figure 9.3). Various types of MS, such as a vehicle-mounted station, portable station, or handheld station, are used.

MSs come in five power classes that define the maximum RF power level that the unit can transmit. Table 9.4 provides the details of maximum RF power

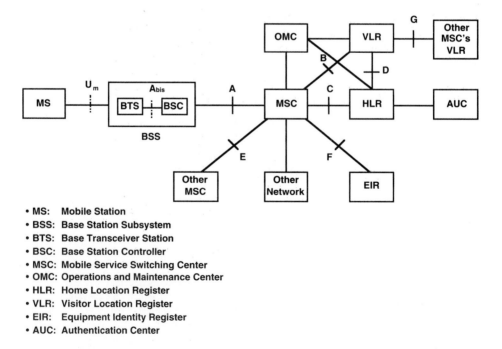

- **MS:** Mobile Station
- **BSS:** Base Station Subsystem
- **BTS:** Base Transceiver Station
- **BSC:** Base Station Controller
- **MSC:** Mobile Service Switching Center
- **OMC:** Operations and Maintenance Center
- **HLR:** Home Location Register
- **VLR:** Visitor Location Register
- **EIR:** Equipment Identity Register
- **AUC:** Authentication Center

Fig. 9.2 GSM Reference Model

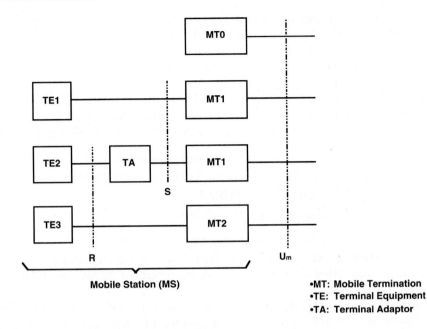

Fig. 9.3 Functional Model of a Mobile Station

for various classes. Vehicular and portable units can be either class I or class II, whereas handheld units can be class III, IV, and V.

Basically, an MS can be divided into two parts. The first part contains the hardware and software to support radio and man-machine interface functions and is available at retail shops to buy or rent. The second part contains terminal/user-specific data in the form of a smart card (SIM card), which can effectively be considered a sort of logical terminal. The SIM card plugs into the first part of the MS and remains in it for the duration of use. Without the SIM card, the MS is not associated with any user and cannot make or receive calls (except possibly an emergency call if the network allows). The SIM card is issued by the mobile service provider after subscription. The type of SIM-card mobility is analogous to terminal mobility, but it also provides a personal-mobility-like service within the GSM mobile network.

Table 9.4 Maximum RF Power for Mobile Stations

Class	Max. RF Power (W)
I	20
II	8
III	5
IV	2
V	0.8

An MS has a number of identities, including the International Mobile Equipment Identity (IMEI), the International Mobile Subscriber Identity (IMSI), and the ISDN number. The IMSI is embodied in the SIM. SIM is basically a smart card that contains all the subscriber-related information stored on the user's side of the radio interface.

International Mobile Station Identification   IMSI is assigned to an MS at subscription time. It uniquely identifies a given MS. IMSI will be transmitted over the radio interface only if necessary. IMSI contains 15 digits and has:

1. Mobile Country Code (MCC) – three digits (home country)
2. Mobile Network Code (MNC) – two digits (home GSM PLMN)
3. Mobile Subscriber Identification (MSIN)
4. National Mobile Subscriber Identity (NMSI)

Temporary Mobile Subscriber Identity   The TMSI is assigned to an MS by the Visitor Location Register (VLR). The TMSI uniquely identifies an MS within the area controlled by a given VLR. A maximum of 32 bits can be used for TMSI.

International Mobile Station Equipment Identity   The IMEI uniquely identifies the MS equipment. It is assigned by the equipment manufacturer. The IMEI contains 15 digits and carries:

1. Type Approval Code (TAC)—six digits
2. Final Assembly Code (FAC)—two digits
3. Serial Number (SNR)—6 digits
4. Spare (SP)—1 digit

Subscriber Identity Module   The SIM carries the following information:

☞ IMSI
☞ Authentication key (K_i)
☞ Subscriber information
☞ Access control class
☞ Cipher key (K_c)*
☞ TMSI*
☞ Additional GSM services*
☞ Location Area Identity (LAI)*
☞ Forbidden PLMN*

9.5.1.2 Base Station System   The BSS is the physical equipment that provides radio coverage to prescribed geographical areas, known as the cells. It contains equipment required to communicate with the MS. Functionally, a BSS consists of: a control function carried out by the Base Station Controller (BSC)

---

*Updatable by the network.

and a transmitting function performed by the Base Transceiver System (BTS). The BTS is the radio transmission equipment and covers each cell. A BSS can serve several cells because it can have multiple BTSs.

The BTS contains the Transcoder Rate Adopter Unit (TRAU). In TRAU, the GSM-specific speech encoding and decoding is carried out, as well as the rate adaptation function for data. In certain situations TRAU is located between the BSC and the Mobile Service Switching Center (MSC) to gain an advantage of a more-compressed transmission between the BTS and the TRAU. Interface between the BTS and BSC is A_{bis}. The interface between the MS and BSS is air interface (U_m).

9.5.2 Network and Switching Subsystem

The NSS includes the main switching functions of the GSM, databases required for the subscribers, and mobility management. Its main role is to manage the communications between the GSM and other network users. Within the NSS, the switching functions are performed by the MSC. Subscriber information relevant to provisioning of services is kept in the Home Location Register (HLR). The other database in the NSS is the Visitor Location Register (VLR).

The **Mobile Service Switching Center** performs the necessary switching functions required for the MSs located in an associated geographical area, called an MSC area. The MSC monitors the mobility of its subscribers and manages necessary resources required to handle and update the location registration procedures and to carry out the handoff functions. The MSC is involved in the interworking functions to communicate with other networks such as PSTN and ISDN. The interworking functions of the MSC depend upon the type of the network to which it is connected and the type of service to be performed. The call routing and control and echo control functions are also performed by the MSC.

The **Home Location Register** is the functional unit used for management of mobile subscribers. The number of HLRs in a PLMN varies with the characteristics of the PLMN. Two types of information are stored in the HLR: subscriber information and part of the mobile information to allow incoming calls to be routed to the MSC for the particular mobile. Any administrative action by the service provider on subscriber data is carried out in the HLR. The HLR stores IMSI, MS ISDN number, VLR address, and subscriber data (e.g. supplementary services).

The **Visitor Location Register** is linked to one or more MSCs. The VLR is the functional unit that dynamically stores subscriber information, such as location area, when the subscriber is located in the area covered by the VLR. When a roaming MS enters an MSC area, the MSC informs the associated VLR about the MS; the MS goes through a registration procedure that includes:

☞ The VLR recognizes that the MS is from another PLMN.

☞ If roaming is allowed, the VLR finds the MS's HLR in its home PLMN.

☞ The VLR constructs a Global Title (GT) from the IMSI to allow signaling from the VLR to the MS's HLR via the PSTN/ISDN networks.

☞ The VLR generates a Mobile Subscriber Roaming Number (MSRN) that is used to route incoming calls to the MS.

☞ The MSRN is sent to the MS's HLR.

The information included in the VLR is:

1. MSRN
2. TMSI
3. The location area in which the MS has been registered
4. Data related to supplementary service
5. MS ISDN number
6. IMSI
7. HLR address or GT
8. Local MS identity, if used

The NSS contains more than MSCs, HLRs, and VLRs. In order to setup a call for the GSM user, the call is first routed to a gateway switch, referred to as the Gateway Mobile Service Switching Center (GMSC). The GMSC is responsible for collecting the location information and routing the call to the MSC through which the subscriber can obtain service at that instant (i.e., the visited MSC). The GMSC first finds the right HLR from the directory number of the GSM subscriber and interrogates it. The GMSC has an interface with external networks for which it provides gateway function. It also has an interface with the SS7 signaling network for interworking with other NSS entities.

9.5.2.1 Operation and Maintenance Subsystem (OMSS)　　The OMSS is responsible for handling system security based on validation of identities of various telecommunications entities. These functions are performed in **AUthentication Center (AUC)** and **Equipment Identity Register (EIR)**.

The **AUC** is accessed by the HLR to determine whether an MS will be granted service.

The **EIR** provides MS information used by the MSC. The EIR maintains a list of legitimate, fraudulent, or faulty MSs.

The OMSS is also in charge of remote operation and maintenance of the PLMN. Functions are monitored and controlled in the OMSS. The OMSS may have one or more Network Management Centers (NMCs) to centralize PLMN control.

The **Operational and Maintenance Center (OMC)** is the functional entity through which the service provider monitors and controls the system. The OMC provides a single point for the maintenance personnel to maintain the entire system. One OMC can serve multiple MSCs.

9.5.3 Interworking and Interfaces

Necessary interfaces are required to achieve an optimum interworking between different entities of the GSM. The use of the CCITT SS7 between the

MSC and VLR and between the MSC and HLR, allows transmission of both call control signals and other information. The corresponding signaling capabilities are supported by the Mobile Application Part (MAP) in SS7 defined in the GSM Recommendations.

The interface labels on the GSM reference model (Figure 9.2—A, A_{bis}, B, C, D, E, F, G, and U_m) correspond to interfaces between network nodes in a GSM PLMN. Each interface is specified in the GSM Recommendations along with their corresponding procedures.

The GSM Recommendations in 09 Series cover interworking conditions between a PLMN and other networks.

9.5.4 GSM Service Quality Requirements

The GSM service quailty requirements are as follows:

☞ Time from switching to service ready: 4 sec in the home system and 10 sec in the visiting system
☞ Connect time to called network: 4 sec
☞ Release time to called network: 2 sec
☞ Time to alert mobile of inbound call: 4 sec in first attempt and 15 sec in final attempt
☞ Maximum gap due to handoff: 150 ms if intercell and 100 ms if intracell
☞ Maximum one-way speech delay: 90 ms
☞ Intelligibility of speech: 90%

9.6 GSM CHANNEL AND FRAME STRUCTURE

The bandwidth in the GSM is 25 MHz. The frequency band used for the uplink (i.e., transmission from the MS to the BS) is 890 to 915 MHz, whereas for the downlink (i.e., transmission from the BS to the MS) is 935 to 960 MHz. The GSM has 124 channels, each with a bandwidth of 200 kHz. For a given channel, the uplink (F_u) and downlink (F_d) frequency can be obtained from Eqs. (9.1) and (9.2), respectively:

$$F_u = 890.2 + 0.2\,(N-1)\ \text{MHz} \qquad (9.1)$$

$$F_d = 935.2 + 0.2\,(N-1)\ \text{MHz} \qquad (9.2)$$

where:
$N = 1, 2, \text{------}, 124.$

When the MS is assigned to an information channel, a radio channel and a timeslot are also assigned. Radio channels are assigned in frequency pairs—one for the uplink, F_u and other for the downlink, F_d. Each pair of radio channels supports up to eight simultaneous calls (see Figure 9.4). Thus, the GSM can support up to 992 simultaneous users with the full-rate speech coder. This number will be doubled to 1,984 users with the half-rate speech coder.

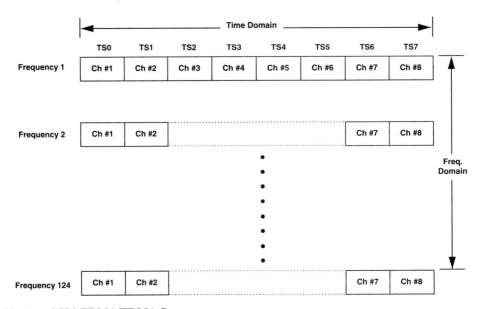

Fig. 9.4 GSM FDMA/TDMA Structure

9.6.1 Logical Channels

In the GSM, there are three types of logical channels: Traffic Channel (TCH), Control Channel(CCH), and Cell Broadcast Channel (CBCH). The TCHs are used to transmit user information (speech or data). CCHs are used to transmit control and signaling information. The CBCH is used to broadcast user information from a service center to the MS listening in a given cell area. It is a unidirectional (downlink only), point-to-multipoint channel used for a short-information message service. Some special constraints are imposed on the design of the CBCH because of the requirement that this channel can be listened in parallel with the Broadcast Control Channel (BCCH) information and the paging messages.

The TCH can be TCH/Full (TCH/F) or TCH/Half (TCH/H). The TCH/F allows the transmission of 13 kbs of speech or data at 12, 6, or 3.6 kbs. The TCH/H allows speech coded at a rate around 7 kbs or data at 6 or 3.6 kbs. The Control Channels (CCHs) consist of:

1. Broadcast Channel (BCH)
2. Common Control Channel (CCCH)
3. Dedicated Control Channel (DCCH)

The BCHs are point-to-multipoint, downlink-only channels. These channels consist of:

1. Broadcast Control Channel (BCCH)
2. Frequency Correction Channel (FCCH)
3. Synchronization Channel (SCH)

The BCCH is used to send cell identities, organization information about common control channels, cell service available, and so on. The FCCH is used to transmit a frequency correction data burst that contains a set of all "0." This gives a constant frequency shift of the RF carrier that can be used by the MS for frequency correction. The SCH is used to time synchronize the MSs. The data in this channel includes the TDMA frame number as well as the Base Station Identity Code (BSIC) required by MSs when measuring BS signal strength.

The CCCHs include:

1. Paging Channel (PCH)
2. Access Grant Channel (AGCH)
3. Random Access Channel (RACH)

The CCCHs are point-to-multipoint downlink-only channels that are used for paging and access. The PCHs are used to page mobile stations. The mobile stations need to listen for paging during certain times. The AGCHs are downlink-only channels used to assign mobiles to Stand-Alone Dedicated Control Channels (SDCCHs) for initial assignment. The RACHs are uplink-only channels used by mobile stations for transmitting their requests for dedicated connections to the system.

There are two types of DCCHs: SDCCH and Associated Control Channel (ACCH). The SDCCHs are bidirectional, point-to-point channels that are used for service request, subscriber authentication, ciphering initiation, equipment validation, and assignment to a TCH. The net SDCCH bit rate is about 0.8 kbs. The ACCHs are bidirectional, point-to-point channels that are associated with a given TCH or SDCCH. These channels are used to send out-of-band signaling and control data between the MS and the BS. Examples of their use are to send signal strength measurements from the MS to the BS or to send transmission timing information from the BS to the MS.

There are two types of ACCHs: Slow Associated Control Channels (SACCHs) and Fast Associated Control Channels (FACCHs). When the SACCH is sent along with TCH/F or TCH/H, it is then called SACCH/TF or SACCH/TH. It is also sent along with SDCCH/4 or SDCCH/8. It is then called SACCH/C4 or SACCH/C8.

The FACCH is used to send preemptive signaling on a full- or half-rate traffic channel. It is then called FACCH/F or FACCH/H. An example of its use is to send handoff messages.

Figure 9.5 shows the structure of the GSM logical channels. For more details, refer to the GSM 04.03, 05.01, and 05.02 series Recommendations and/or reference [12].

9.6.2 GSM Frame

The GSM multiframe is 120 ms. It consists of 26 frames of 8 time slots. The structure of a GSM hyperframe, superframe, multiframe, frame, and time slot is shown in Figure 9.6. A time slot carries 156.25 bits. The same format is used for the uplink and downlink transmission with various burst types as shown in Figure 9.7. In a normal burst, two user information groups of 58 bits account for

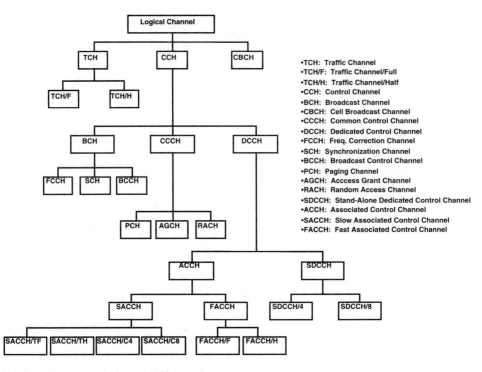

Fig. 9.5 Structure of Logical Channels

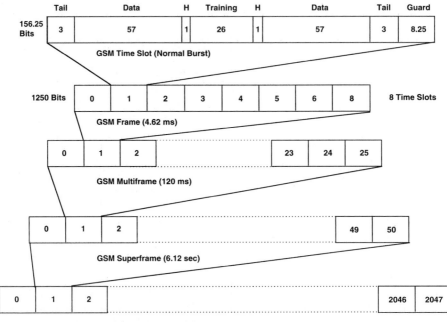

Fig. 9.6 Physical Structure for GSM Hyperframe, Superframe, Multiframe, Frame, and Time Slot

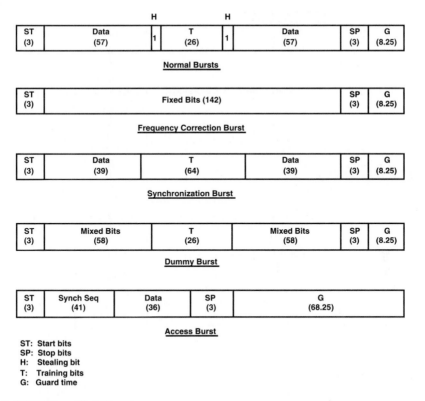

Fig. 9.7 GSM Time Slot Structure

most of the transmission time in a time slot (57 bits carry user data, while the H bit is used to distinguish speech from other transmissions). Twenty-six training (T) bits are used in the middle of the time slot. The time slot starts and ends with 3 tail bits. The time slot also contains 8.25 Guard (G) bits.

9.7 GSM SPEECH PROCESSING

Two major tasks are involved in transmitting and receiving information over a digital radio link: information processing and modulation processing. Information processing deals with the preparation of the basic information signals so that they are protected and converted into a form that the radio link can handle. Information processing includes transcoding, channel coding, encrypting, and multiplexing. Modulation processing involves the physical preparation of the signal to carry information on an RF carrier.

Each digital radio link process in the transmitting path has its peer in the receiving path (see Figure 9.8). The delay equalization process in the receiving path is required to compensate for the spread in time delays resulting from the multipath propagation. It can be the part of the demodulation process. However, it should be emphasized that it is required when delay spreads are significant

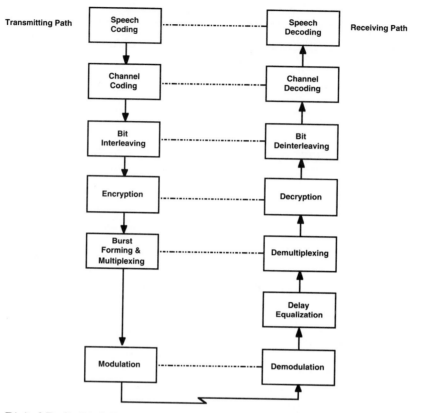

Transmitting Path ... **Receiving Path**

Fig. 9.8 Digital Radio Link Process

compared to the information bit period. In the GSM the transmitted bit period is about 37μsec, and delay spreads of about 5μsec are common. The delay spread problem becomes critical and difficult to solve with the higher transmission bit rate.

In the GSM, the analog speech from the mobile station is passed through a low-pass filter to remove the high-frequency contents from the speech. The speech is sampled at the rate of 8,000 samples per second, uniformly quantized to 2^{13} (8,192) levels and coded using 13 bits per sample. This results in a digital information stream at a rate of 104 kbs. At the base station, the speech signal is digital (64 kbs), which is first transcoded from the A-law 8-bit samples into 13-bit samples corresponding to a linear representation of the amplitudes. This results in a digital information stream at a rate of 104 kbs.

The 104-kbs digital signal stream is fed into the Regular Pulse Excited-Long-Term Prediction (RPE-LTP) speech encoder (see Figure 9.9) which then transcodes the speech into a 13-kbs stream. The full-rate speech encoder takes a 2,080 bit block from the 13-bit transcoder every 20-ms (i.e., 160 samples) and produces 36 "filter parameter" bits over the 20 ms period, 9 LTP bits every 5 ms, and 47 RPE bits every 5 ms (refer to Table 9.5). Thus, 260 bits are generated

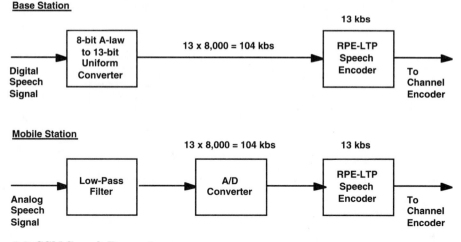

Fig. 9.9 GSM Speech Processing

Table 9.5 Details of Class I and Class II Bits

	Bits per 5 ms	Bits per 20 ms
Linear Prediction Coding (LPC) filter		36
Long Term Prediction (LTP) filter	9	36
Excitation signal	47	188
Total		260
Class I		182 (class Ia = 50, class Ib = 132)
Class II		78

every 20 ms. Of these 260 bits, 182 are classified as class I bits related to the excitation signal, whereas the remaining 78 are class II bits related to the parameters of Linear Prediction Coding (LPC) and LTP filters. The class I bits are further classified into class Ia (50 bits) and class Ib (132 bits). The class Ia bits are the most significant bits which are used to generate 3 Cyclic Redundancy Check (CRC) bits. The 3 CRC bits along with 4 tail bits are added to 182 class I bits before they are passed through the half-rate convolutional coder to produce twice as many bits output as there are bits input (i.e., $2 \times 189 = 378$). The 78 class II bits remain uncoded and are bypassed (see Figure 9.10). The total 456 bits (i.e., $378 + 78$) are fed to the bit interleaver. Since 456 bits are generated during 20 ms, the user data rate is $456/0.02 = 22.8$ kbs. This includes 13.0 kbs raw data and 9.8 kbs of parity, tail, and channel coding.

Of the 456 bits, 57 at a time are interleaved with 57 other bits from an adjacent data block to form a data burst of 114 bits. At this stage, 42.25 overhead bits

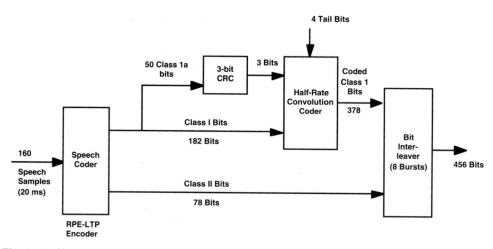

Fig. 9.10 Channel Coding for Full-Rate Speech in GSM

are added to the data burst to carry it into a time slot (see Figure 9.11). Bit inter-leaving is used to reduce the adverse effects of Rayleigh fading by preventing entire blocks of bits from being destroyed by a signal fade. Interleaved data is passed through the GMSK modulator where it is filtered by a Gaussian filter

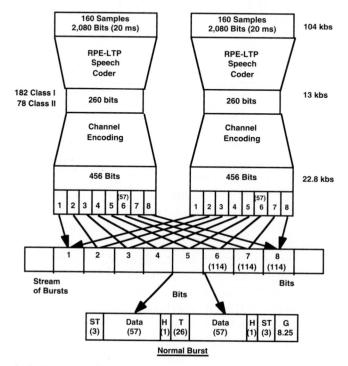

Fig. 9.11 Speech Coding in GSM

before applying it to a modulator. The modulated data passes through a duplexer switch where filtering is provided between the transmitted and the received signal. On the receiving side the signal is demodulated and de-interleaved before the error correction is applied to the recovered bits.

9.8 GSM CALL FLOW SCENARIOS

In this section, we discuss call flow scenarios used in the GSM. In these call flow scenarios, we assume that the MS enters the new MSC area and requires a location update procedure involving registration, authentication, ciphering, and equipment validation. The MS location registration update procedure is given in chapter 10. In this chapter we discuss the call flow scenarios involved with the call origination (i.e., MS to land call and MS to MS call), call termination (land to MS call), and handoff (i.e., inter-/intra-MSC).

9.8.1 Call Setup and Call Release

9.8.1.1 Call Setup with a Mobile   The procedure for a call setup with a mobile station is as follows (see Figure 9.12):

1. The MS sends a SETUP_REQ message to the MSC after it begins ciphering the radio channel. This message includes the dialed digits.

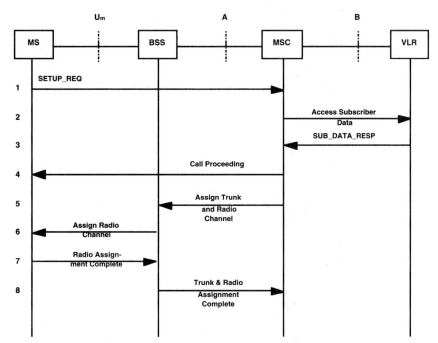

Fig. 9.12 Call Setup with a Mobile

2. Upon receiving the SETUP_REQ message, the MSC requests the VLR to supply the subscriber parameters necessary for handling the call. The message contains the called number and service indication.

3. The VLR checks for call-barring conditions. If the VLR determines that the call cannot be processed, the VLR provides the reason to the MSC. In this case, we assume that the procedure is successful and the call can be processed. The VLR returns a message SUB_DATA_RESP to the MSC containing the service parameters for the subscriber.

4. The MSC sends a message to the MS that the call is proceeding.

5. The MSC allocates an available trunk to the BSS currently serving the MS. The MSC send a message to the BSS supplying it with the trunk number allocated and asks to assign a radio traffic channel for the MS.

6. The BSS allocates a radio channel and sends the information to the MS over SDCCH.

7. The MS tunes to the assigned radio channel and sends an acknowledgment to the BSS.

8. The BSS connects the radio traffic channel to the assigned trunk on the MSC and deallocates the SDCCH. The BSS informs the MSC with a trunk and radio assignment complete message.

9.8.1.2 Call Setup with a Land Network   At this point a voice path is established between the MS and the MSC. The MS user hears silence since the complete voice path is not yet established. The last phase involves the MSC establishing a voice path from the MSC to Public Switched Telephone Network (PSTN) (see Figure 9.13).

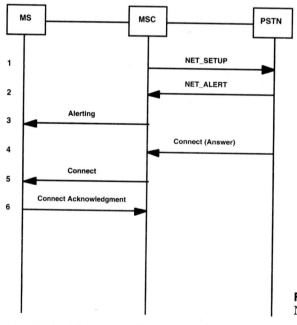

Fig. 9.13 Call Setup with Land Network

1. The MSC sends the NET_SETUP message to the PSTN to request the call setup. This message includes the digits dialed by the MS and details of the trunk that will be used for the call.
2. The PSTN sets up the call and notifies the MSC with a NET_ALERT message.
3. The MSC informs the MS that the destination number is being alerted. The MS hears the ringing tone from the destination local exchange through the established voice path.
4. When the destination party goes off hook, the PSTN informs the MSC.
5. The MSC informs the MS that the connection has been established.
6. The MS sends an acknowledgment to the MSC.

9.8.1.3 Call Release—Mobile Initiated   Under normal conditions, there are two basic ways a call is terminated: mobile initiated and network initiated. In this scenario, we assume that the mobile user initiates the release of the call (see Figure 9.14).

1. At the end of the call, the MS sends the CALL_DISC message to the MSC.
2. On receiving the CALL_DISC message, the MSC sends a NET_REL request message to the PSTN to release the call.
3. The MSC asks the MS to begin its clearing procedure using the CALL_REL message.

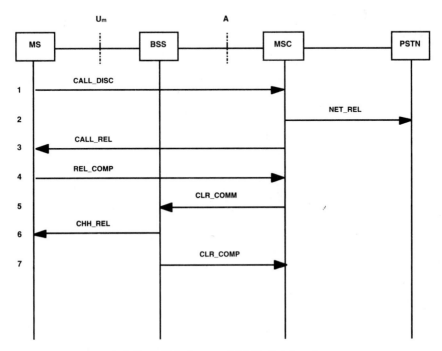

Fig. 9.14 Mobile to Land Call: Call Release—Mobile Initiated

4. After the MS has performed its clearing procedure, it informs the MSC through the REL_COMP message.

5. The MSC then sends the CLR_COMM message to the BSS to ask it to release all the allocated dedicated resources for a given Signaling Connection Control Part (SCCP) connection.

6. The BSS sends the CHH_REL message to the MS to release the traffic channel.

7. The BSS sends an acknowledgment message CLR_COMP to the MSC informing it that all allocated dedicated resources have been released.

9.8.1.4 Routing Analysis—Land to Mobile Call   In this scenario we assume that the MS is already registered with the system and has been assigned a TMSI. We also assume that the MS is in its home system. A land subscriber dials the directory number of the mobile subscriber (see Figure 9.15).

1. The PSTN routes the call to the MSC assigned this directory number. The directory number in the INC_CALL message is the Mobile Station ISDN Number (MSISDN).

2. The MSC sends the GET_ROUT message to the HLR to provide the routing information for the MSISDN.

3. The HLR returns the ROUT_INF message to the MSC. This message contains the Mobile Station Roaming Number (MSRN). If the MS is roaming within the serving area of this MSC, the MSRN returned by the HLR will

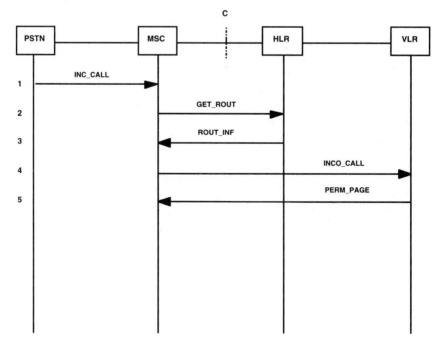

Fig. 9.15 Land to Mobile Call—Routing Analysis

most likely be the same as the MSISDN. In this scenario we assume that the MS is not roaming.

4. The MSC informs its VLR about the incoming call using a INCO_CALL message that includes MSRN.

5. The VLR responds to the MSC through a PERM_PAGE message that specifies Location Area Identification (LAI) and TMSI of the MS. If the MS is barred from receiving the calls, the VLR informs the MSC that a call cannot be directed to the MS. The MSC would connect the incoming call to an appropriate announcement.

9.8.1.5 Paging—Land to Mobile Call The following is the procedure for paging in a land to mobile call (refer to Figure 9.16).

1. The MSC uses the LAI provided by the VLR to determine which BSSs will page the MS. The MSC sends the PERM_PAGE message to each of the BSSs to perform the paging of the MS.

2. Each BSS broadcasts the TMSI of the MS in the page message (PAGE_MESS) on the PCH.

3. When the MS hears its TMSI broadcast on the PCH, it responds to the BSS with a CHH_REQ message over the common access channel, RACH.

4. On receiving the CHH_REQ message from the MS, the BSS allocates an SDCCH and sends the DSCH_ASS message to the MS over the AGCH. It is

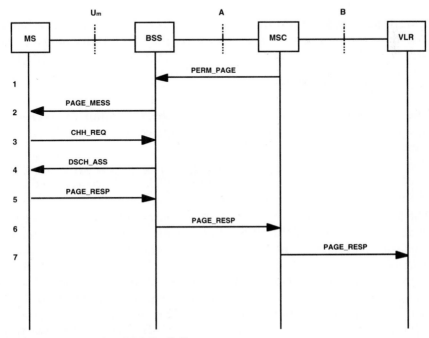

Fig. 9.16 Paging—Land to Mobile Call

over the SDCCH that the MS communicates with the BSS and MSC until a TCH is assigned.

5. The MS sends a PAGE_RESP message to the BSS over the SDCCH. The message contains the MS's TMSI and LAI.

6. The BSS forwards the PAGE_RESP message to the MSC.

7. The MSC informs its VLR that the MS is responding to a page.

At this point the MS goes through authentication, ciphering, equipment validation, call setup, and call release procedures. If the MS has already gone through the authentication, ciphering, and equipment validation procedures, then only call setup and call release (discussed earlier) are carried out.

9.8.2 Handoff

Basically there are two levels of handoffs: internal and external. If the serving and target BTSs are located within the same BSS, the BSC for the BSS can perform a handoff without the involvement of the MSC. This type of handoff is referred to intra-BSS handoff. However, if the serving and target BTSs do not reside within the same BSS, an external handoff is performed. In this type of handoff the MSC coordinates the handoff and performs the switching tasks between the serving and target BTSs. The external handoffs can be classified as: within the same MSC (i.e., intra-MSC) and between different MSCs (i.e., inter-MSC). In the following call flow scenarios we focus only upon the external handoffs. We discuss the intra-MSC and inter-MSC handoff.

9.8.2.1 Intra-MSC Handoff   When the MS determines that a handoff is required in an attempt to maintain the desired signal quality of the radio link (The signal quality is constantly monitored by the MS and BSS, and the BSS may optionally forward its own measurements to the MS), the following takes place (refer to Figure 9.17).

1. The MS determines that a handoff is required. It sends the STRN_MEAS message to the serving BSS. This message contains the signal strength measurements.

2. The serving BSS sends a HAND_REQ message to the MSC. This message contains a rank-ordered list of the target BSSs that are qualified to receive the call.

3. The MSC reviews the global cell identity associated with the best candidate to determine if one of the BSSs that it controls is responsible for the cell area. In this scenario the MSC determines that the cell area is associated with the target BSS. To perform an intra-MSC handoff, two resources are required: a trunk between the MSC and the target BSS, and a radio traffic channel in the new cell area. The MSC reserves a trunk and sends a HAND_REQ message to the target BSS. This message includes the desired cell area for handoff, the identity of the MSC-BSS trunk that was reserved, and the encryption key (K_c).

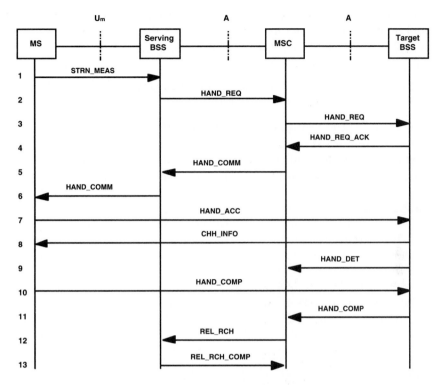

Fig. 9.17 Intra-MSC Handoff

4. The target BSS selects and reserves the appropriate resources to support the handoff pending the connection execution. The target BSS sends an acknowledgment to the MSC (HAND_REQ_ACK). The message contains the new radio channel identification.

5. The MSC sends the HAND_COMM message to the serving BSS. In this message the new radio channel identification supplied by the target BSS is included.

6. The serving BSS forwards the HAND_COMM message to the MS.

7. The MS retunes to the new radio channel and sends the HAND_ACC message to the target BSS on the new radio channel.

8. The target BSS sends the CHH_INFO message to the MS.

9. The target BSS informs the MSC when it begins detecting the mobile handing over.

10. The target BSS and the MS exchange messages to synchronize/align the MS's transmission in the proper time slot. On completion, the MS sends the HAND_COMP message to the target BSS.

11. At this point the MSC switches the voice path to the target BSS. Once the MS and target BSS synchronize their transmission and establish a new sig-

naling connection, the target BSS sends the MSC the HAND_COMP message to indicate that the handoff is successfully completed.

12. The MSC sends the REL_RCH message to the serving BSS to release the old radio traffic channel.

13. At this point the serving BSS frees up all resources with the MS and sends the REL_RCH_COMP message to the MSC.

Note that GSM Recommendations require that the "open interval gap" during a handoff will not exceed 150 ms for 90% of the handoffs. The "open interval gap" starts when the MS retunes to the new radio channel and ends after synchronization without any loss in voice/data transmission in the BSS or MSC.

9.8.2.2 Inter-MSC Handoff In this scenario we assume that a call has already been established. The serving BSS is connected to the serving MSC and the target BSS to the target MSC. The inter-MSC handoff procedure is as follows (see Figure 9.18):

1. Same as in the intra-MSC handoff (step 1).

2. Same as in the intra-MSC handoff (step 2).

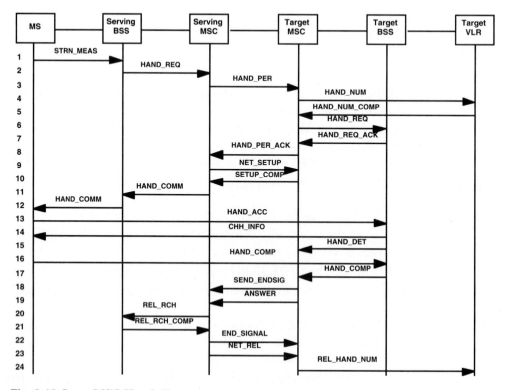

Fig. 9.18 Inter-MSC Handoff

3. When a call is handed over from the serving MSC to the target MSC via PSTN, the serving MSC sets up an inter-MSC voice connection by placing a call to the directory number that belongs to the target MSC. When the serving MSC places this call, the PSTN is unaware that the call is a handoff and follows the normal call routing procedures and delivers the call to the target MSC.

4. The target MSC sends a HAND_NUM message to its VLR to assign the TMSI.

5. The target VLR sends the TMSI in the HAND_NUM_COMP message.

6. Same as step 3 in the intra-MSC handoff.

7. Same as step 4 in the intra-MSC handoff.

8. The target MSC sends the HAND_PER_ACK message to the serving MSC indicating that it is ready for the handoff.

9. The serving MSC sends the NET_SETUP message to the target MSC to set up for the call.

10. The target MSC acknowledges this message with a SETUP_COMP message to the serving MSC.

11. Same as step 5 in the intra-MSC handoff.

12. Same as step 6 in the intra-MSC handoff.

13. Same as step 7 in the intra-MSC handoff.

14. Same as step 8 in the intra-MSC handoff.

15. Same as step 9 in the intra-MSC handoff.

16. Same as step 10 in the intra-MSC handoff.

17. Same as step 11 in the intra-MSC handoff.

18. At this point the handoff has been completed, the target MSC sends the SEND_ENDSIG message to the serving MSC.

19. The MS retunes to the new radio channel. A new voice path is set up between the MS and the target BSS. The target MSC sends an ANSWER message to the serving MSC.

20. Same as step 12 in the intra-MSC handoff.

21. Same as step 13 in the intra-MSC handoff.

22. The serving MSC sends the END_SIGNAL message to the target MSC.

23. The serving MSC releases the network resources and sends the NET_REL message to the target MSC.

24. The target MSC sends the REL_HAND_NUM message to its VLR to release the connection.

9.9 MSC PERFORMANCE

The MSC performance will meet the GSM Recommendations, Series 2.08 and 3.05.

The reliability objectives of the MSC as per GSM Recommendations, Series 3.05 and 3.06 are:

☞ Cutoff call or call release failure rate probability: $P \leq 0.0002$

☞ Probability of incorrect charging, misrouting, no tone, or other failures: $P \leq 0.0001$

☞ Mean Accumulated Intrinsic Down Time (MAIDT) for one termination, or MAIDT (1) $\leq$ 30 minutes/year

☞ Probability of losing HLR/VLR messages: $P \leq 0.0000001$

The service availability of an MSC is expressed in terms of the frequency or duration of loss of service. The loss of service to particular circuits, groups of circuits, subsystems, or the complete MSC is determined by the faults in the MSC.

The average cumulative duration of service denial due to faults affecting more than 50% of the circuits will not exceed three minutes during the first year of operation and two minutes during each subsequent year. On the average, a fault that causes more than 50% of the established calls to be disconnected prematurely will occur less than once a year.

9.10 NORTH AMERICAN PCS1900

Figure 9.19 shows the functional model that has been derived from the T1P1 reference model shown in Figure 8.2 and discussed in chapter 8. Several physical scenarios can be developed using the functional entities shown in Figure 9.19. Figure 9.20 shows the Functional Entity (FE) grouping in which the physical interface between the RS and the Switching System Platform (SSP) carries both the call control and mobility management messages.

Radio Terminal Function (RTF) FE: It is the subscriber unit (SU). The only physical interface is to the RS using the air interface.

Radio Control Function (RCF) FE and Radio Access Control Function (RACF) FE: These are included in the RS. Combining these FEs onto the same platform allows air-interface-specific functions (such as those that would impact handoff) to be isolated from the other interfaces. Operations Systems (OS) information, including performance data and accounting records, is generated, collected, and formatted on this platform. There is only one physical interface to SSP to carry both the call control and mobility management signaling.

Service Switching Function (SSF)/Call Control Function (CCF) FE: It is contained in SSP and provides interfaces to operator services, E911, international calls, and network repair/maintenance centers. Physical interfaces for this collection include: to the RS, to the mobility management platform, to the IP, and to other SSPs and external networks.

Specialized Resource Function (SRF) FE and data InterWorking Function (IWF): They are contained in the Internal Peripheral (IP). Physical interfaces for this collection include one to the SSP and another to the mobility management platform.

Individually the SSF/CCF FE and CCF FE represent interswitch and internetwork functional entity collections and physical interfaces.

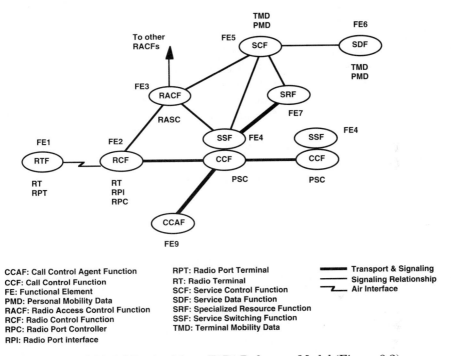

CCAF: Call Control Agent Function
CCF: Call Control Function
FE: Functional Element
PMD: Personal Mobility Data
RACF: Radio Access Control Function
RCF: Radio Control Function
RPC: Radio Port Controller
RPI: Radio Port interface

RPT: Radio Port Terminal
RT: Radio Terminal
SCF: Service Control Function
SDF: Service Data Function
SRF: Specialized Resource Function
SSF: Service Switching Function
TMD: Terminal Mobility Data

━━━ Transport & Signaling
─── Signaling Relationship
⌁ Air Interface

Fig. 9.19 Functional Model Derived from T1P1 Reference Model (Figure 8.2)

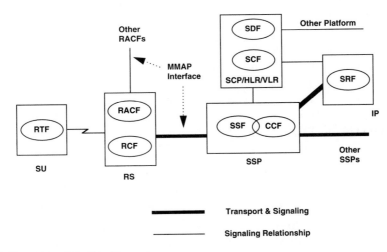

━━━ Transport & Signaling

─── Signaling Relationship

Fig. 9.20 Functional Entity Groupings

As shown in Figure 9.20, the only interface to the RS is from the Switching Control Point (SSP). There is no direct physical path between the RS and the SCP/HLR/VLR. All operations to or from the RS pass through the SSP, whether or not the SSP consumes or produces the operation.

The proposed North American PCS1900 standard is an adaptation of the ETSI DCS1800 standard that was initially developed for the digital cellular band, i.e., GSM 900. It consists of 200-kHz radio channels shared by eight time slots, one per terminal. The standard supports a frequency duplex arrangement for forward and reverse links. The PCS1900 system supports a fixed-rate RPE based on a speech coder that operates at 13 kbs.

The North American types of handoff are network initiated and Mobile-Assisted Handoff (MAHO). In case of the network-initiated handoff, both "hard" and "soft" handoff are supported. The PCS1900 standard defines support for MAHO and a form of network-initiated handoff that applies only to "hard" handoff. For PCS1900 systems to function as an integral part of the North American PCS environment, handoff needs will be supported between PCS1900 and North American systems.

PCS1900 supports voice privacy through the encryption capabilities (refer to chapter 10 for details). The encryption (voice privacy) is an air interface capability that can be controlled by the network operator rather than a service that may not be controlled by the network operator but may also be offered as a service to the end user. The GSM encryption is only an air interface function and does not depend on the GSM MAP function or the home network.

The authentication algorithm (refer to chapter 10 for details) in the PCS1900 uses IMSI as input. The terminal possesses a "key" that is the same "key" known by the home network. The network computes a signature that is specific for an end user. This signature is used to authenticate the end user through the duration of the service. This authentication scheme has its strength in the authentication algorithm. However, there is no mechanism to recognize clones.

To satisfy the PCS needs and requirements for accessibility and seamless service, air interface transparency must exist. Transparency implies that an end user can have access to service regardless of the access method.

In the initial phase of PCS, multiple air interface may exist, and therefore "dual mode" or "dual spectrum" terminals may be used. The aim is to attain some level of interoperability with the existing North American networks. If interoperability does not exist between PCS1900 air interface and analog AMPS 800-MHz air interface, the ubiquity of service is precluded. The PCS1900 air interface may access the network that provides GSM services. The AMPS analog air interface may have access to the IS-41 services. Access to two different types of networks and services may preclude ubiquity of service from an end-user prospective.

9.11 JAPANESE DIGITAL CELLULAR (JDC) SYSTEM

The JDC system uses three-channel TDMA. Two frequency bands are reserved: 800-MHz band with 130-MHz of duplex separation and 1.5-GHz band with 48

MHz of duplex separation. The 800-MHz band is used first. The 1.5-GHz band will be used later on. The modulation scheme is π/4-QPSK, and the interleaved carrier spacing is 25 kHz. The speech CODEC (coder-decoder) was selected based on evaluations of the speech quality, complexity, and delay of several speech CODECs. The selected CODEC uses 11.2-kbs Vector Sum Excited Linear Prediction (VSELP) including channel coding.

In the 800-MHz band, the uplink transmission (from the MS to the BS) frequency is 940 to 956 MHz and the downlink transmission (from the BS to the MS) frequency is 810 to 826 MHz. Since a 25-kHz channel bandwidth is used, this provides 640 carriers and 3 channels per carrier. Thus, a total of 1,920 channels is available. The number of channels will be doubled as the half-rate speech coders are introduced.

9.11.1 JDC Channel and Frame Structure

The logical channel structure is shown in Figure 9.21. The channels are divided into the Traffic Channel (TCH) and Control Channel (CCH). The TCH is used to transmit user information and the CCH for control information. There are two types of CCH: Common Access Channel (CAC) shared by many users and User Specific Channel (USC) dedicated to a user. The CAC is further divided into three types. The Broadcast Control Channel (BCCH) provides the MS with system information that contains the MS related data such as maximum transmission power of the MS, location identity code to register the user's location, and

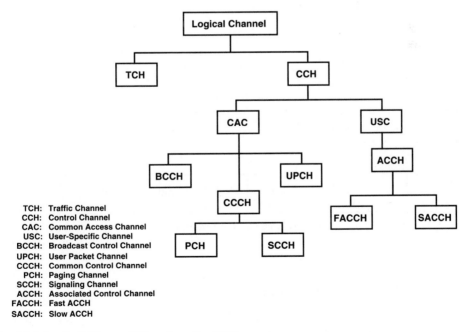

Fig. 9.21 Logical Channel Structure for JDC

the information related to CCH structure, such as the number of CCHs available in the cell. The Common Control Channel (CCCH) is a point-to-multipoint duplex channel used for transmission of signaling information. It comprises the Paging Channel (PCH) and the Signaling Control Channel (SCCH). The PCH transmits paging information to the MS. The SCCH is used to communicate between the network and the MS for signaling information other than the paging message. The User Packet Channel (UPCH) is a point-to-multipoint duplex channel used to transmit user packet data.

The USC is the CCH associated with the TCH. It defines the Associated Control Channel (ACCH) that is used to transmit the signaling data required during communications. There are two types of control channels: Slow Associated Control Channel (SACCH) and Fast Associated Control Channel (FACCH). The SACCH has a slow data transmission rate, whereas the FACCH is used for a high data-transmission-rate channel using the TCH bits.

A Radio-Channel-Housekeeping Channel (RCH) is also provided. It is a layer-1 channel without the layered signaling protocol. It is used for control functions that require real-time response such as transmission power control and status notification.

The JDC frame structure is given in Figures 9.22 and 9.23. The frame is 20 ms long and has three time slots. Each frame carries 840 bits; this corresponds to

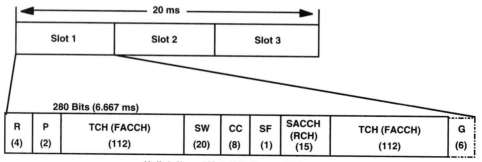

Uplink (from MS to BS) Transmission

Downlink (from BS to MS) Transmission

G: Guard Time
R: Ramp Time
P: Preamble
SW: Synchronization Word
CC: Color Code

SACCH: SACCH bits
FACCH: FACCH bits
RCH: Housekeeping bits
SF: Steal Flag
CAC: CAC bits
E: Collison Control bits

Fig. 9.22 JDC Frame Structure—Traffic Frame

First Unit: 280 Bits (6.667 ms)

R (4)	P (48)	CAC (66)	SW (20)	CC (8)	CAC (116)	G (18)

(a) Uplink (from BS to MS)

Second Unit: 280 Bits (6.667 ms)

R (4)	P (2)	CAC (112)	SW (20)	CC (8)	CAC (116)	G (18)

(b) Uplink (from BS to MS)

280 Bits (6.667 ms)

R (4)	P (2)	CAC (112)	SW (20)	CC (8)	CAC (112)	E (22)

(c) Downlink (from BS to MS)

G: Guard Time
R: Ramp Time
P: Preamble
SW: Synchronization Word
CC: Color Code

SACCH: SACCH bits
FACCH: FACCH bits
RCH: Housekeeping bits
SF: Steal Flag
CAC: CAC bits
E: Collison Control bits

Fig. 9.23 JDC Frame Structure—Control Frame

a data rate of 42 kbs (i.e., $840/(20 \times 10^{-3})$). Thus, the data rate per user is 14 kbs (i.e., 280/0.02). The information data rate per user is 11.2 kbs (i.e., 224/0.02). The difference of 2.8 kbs is allocated between preamble, ramp, guard bits, color code, steal flag, and so forth. Figures 9.24 and 9.25 show the JDC physical channel for traffic and JDC physical channel for control, respectively. A comparison between the GSM and JDC system is given in Table 9.6.

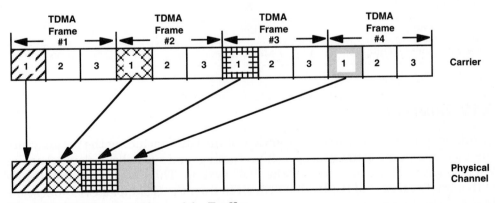

Fig. 9.24 JDC Physical Channel for Traffic

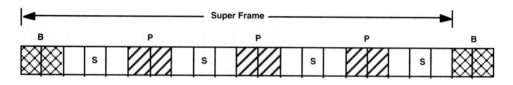

B: BCCH (Broadcast Control Channel)
S: SCCH (Signaling Control Channel)
P: PCH (Paging Channel)

Fig. 9.25 JDC Physical Channel for Control

Table 9.6 GSM and JDC System Comparison

	GSM	JDC
Access Method	TDMA/FDMA	TDMA/FDMA
Frequency range (MHz)	890–915 uplink 935–960 downlink	940–956 uplink 810–826 downlink 1447–1489 uplink 1429–1441 downlink 1501–1513 uplink 1453–1465 downlink
Channel bandwidth (kHz)	200	25
Modulation	GMSK	π/4-DQPSK
Bit rate (kbs)	270.83	42.0
Voice channel coding	RPE-LTP/Convolutional 13 kbs	VSELP/Convolutional 11.2 kbs
Voice frame (ms)	4.6	20
Interleaving (ms)	40	26.667
Slot/frame	8	3
No. of channels	124	640
Associate control channel	Extra frame	Same frame

9.12 SUMMARY

In this chapter, we presented an overview of the GSM including the architecture, channel, and frame structure; speech processing; and typical call flow scenarios. We also gave a brief overview of the JDC system. The chapter was concluded with a brief overview of the PCS1900 derived from the GSM for PCS applications in the North America.

9.13 PROBLEMS

1. What are the four basic subsystems in the GSM architectures? Briefly specify their roles.

2. What are basic objectives of the GSM PLMN?

3. Discuss briefly the channel structure used in GSM.

4. What are the differences in the modulation techniques used in GSM and North American IS-136 (TDMA) systems?

5. Discuss basic differences of JDC and GSM systems.

6. Compare the three digital cellular systems used in the world, i.e., GSM, JDC, and North American IS-136 (TDMA).

9.14 REFERENCES

1. GSM Specification Series 1.02–1.06, "GSM Overview, Glossary, Abbreviations, Service Phases."

2. GSM Specification Series 2.01–2.88, "GSM Services and Features."

3. GSM Specification Series 3.01–3.88, "GSM PLMN Functions, Architecture, Numbering and Addressing, Procedures."

4. GSM Specification Series 4.01–4.88, "MS-BSS Interface."

5. GSM Specification Series 5.01–5.10, "Radio Link."

6. GSM Specification Series 6.01–6.32, "Speech Processing."

7. GSM Specification Series 7.01–7.03, "Terminal Adaptation."

8. GSM Specification Series 8.01–8.60, "BSS-MSC Interface, BSC-BTS Interface."

9. GSM Specification Series 9.01–9.11, "Network Interworking, MAP."

10. Japanese Digital Cellular Standards—Digital Mobile Network Inter-node (DMNI) Specification—DMNI 3.1.

11. Japanese Digital Cellular Standards—Research & Development Center for Radio Systems (RCR) Specification—RCR STD—27a.

12. Mouly, M. and Pautet, M., *The GSM System for Mobile Communications*, Palaiseau, France, 1992.

Security and Privacy in Wireless Systems

10.1 INTRODUCTION[*]

Although radio has existed for almost 100 years, most of the population uses wired phones. Only over the last 10 years have large numbers of people used wireless or cordless phones. With this exposure, users of wireless phones and the news media have challenged two bedrocks of the telecommunications industry: privacy of the conversation and billing accuracy.

The current concepts of privacy of communications and correctness of billing are based on the telephone company's ability to route an individual pair of wires to each residence and office in the country. Thus, when a call is placed on a pair of wires, the telephone company can correctly associate the call on a wire with the correct billing account. Similarly, since there is a pair of wires from a home to the telephone company central office, no one can easily listen to the call.

For most people, a wiretap is an abstract concept that only concerns someone who is involved in illegal activities. Anyone in the telephone industry could easily refute this with examples of wiretaps, but most people do not worry.

Communications on shared media[†] can be intercepted by any user of the media. When the media are shared, anyone with access to the media can listen to

---

[*]Section 10.1 and 10.2, Fig. 10.1, 10.2, and 10.3 adapted from reference [9]. Copyright IEEE. Used with permission.

[†]Examples of shared media are: party-line telephone service, radio communications, cable TV, local area networks, wide area networks, and the Internet. Only radio communications are discussed here, but the concepts apply equally well to the other media.

or transmit on the media. Thus, communications are no longer private. In shared media, the presence of a communication request does not uniquely identify the originator, as it does in a single pair of wires per subscriber. In addition, all users of the network can overhear any information that an originator sends to the network and can resend the information to place a fraudulent call. The participants of the phone call shown in Figure 10.1 may not know that their privacy is compromised. When the media are shared, privacy and authentication are lost unless some method is established to regain it. Cryptography provides the means to regain control over privacy and authentication.

There have been attempts to control privacy and authentication though noncryptographic means. These have failed thus far. The designers of the original cellular service implemented authentication of the mobile telephone using a Number Assignment Module (NAM) and an Electronic Serial Number (ESN). The NAM would be implemented in a Programmable Read Only Memory (PROM) for easy replacement when the phone number changed. The ESN would be implemented in a "tamper-resistant module" that could not be changed without damaging the cellular telephone. In practice, many manufacturers implement the NAM and the ESN in either battery-backed Random Access Memory (RAM) or Electrically Erasable PROM (EEPROM). The manufacturer and the installer place the data in the phone via external programming from either the keypad on the phone or a set of programming leads associated with the battery eliminator and feature connector on the phone.

Similarly, the designer assumed that privacy of cellular communications would occur because 900-MHz scanners would be too difficult and too expensive

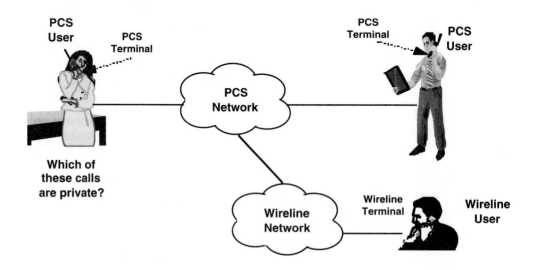

Fig. 10.1 PCS Privacy

to build. When those scanners became easily available, Congress passed the Electronic Communications Privacy Act in 1986, and in 1992 the FCC banned the importation and manufacture of scanners covering cellular phone bands. In practice, the laws do not help since there are millions of scanners in existence today. Furthermore, cellular test equipment is easy to build or buy, and most cellular phones can be placed in a maintenance mode that allows them to monitor any channel. Any cellular phone can be easily converted to a cellular scanner.

To provide the proper privacy and authentication for a PCS PS, some cryptographic system is necessary.

Some of the cryptographic requirements are in the air interface between the PS and the RS. Other requirements are on databases stored in the network and on information shared between systems in the process of handoffs or giving service for roaming units.

In this chapter we examine the requirements needed for strong privacy and authentication of wireless systems, and then we examine how each of the cellular and PCS systems supports these requirements. The chapter first discusses four levels of voice privacy. We then identify requirements in the areas of privacy, theft resistance, RS performance, system lifetime, physical requirements as implemented in PSs, and law enforcement needs. We will examine four different methods that are in use to meet these needs.

Throughout the chapter, as we describe the operation of cellular and PCS systems, we have deliberately left some areas vague. We believe that security through secrecy or the assumption that only insiders will know or understand the system, in general, does not provide security. However, if we explained everything, some security experts would consider us foolish. In addition, details of cryptographic systems are subject to the International Traffic in Arms Regulation (ITAR) of the U.S. government and cannot be exported outside the United States without permission. We have abstracted the material presented in this chapter from the public sections of the cellular and PCS standards, and the material in this chapter can be exported. Detailed information on the cryptographic system is covered by ITAR and is not included here.

10.2 SECURITY AND PRIVACY NEEDS OF A WIRELESS SYSTEM

10.2.1 Privacy Definitions

When most people think of privacy, they think of either of two levels:

1. None
2. Privacy good enough for military users

However, as we describe here, there are really four levels of privacy that need to be considered.

☞ **Level 0: None.** With no privacy enabled, anyone with a digital scanner could monitor a call.*

☞ **Level 1: Equivalent to wireline.** As discussed in the introduction, most people think wireline communications are secure. Anyone in the industry knows that they are not, but the actions to tap a line often show the existence of the tap. With wireless communications, the tap can occur without the knowledge of anyone. Therefore, the actions to tap a wireline call must be translated into a different requirement for a wireless system.

The types of conversations that would be protected with this level of security are the routine everyday conversations of most people. These types of communications would be personal discussions that most people would not want exposed to the general public—for example, details of a recent operation or other medical procedure, family financial matters, mail orders using a credit card, family discussions, requests for emergency services (911), discussion of vacation plans (thus revealing when a home will be vacant).

The cryptographic system must be designed so that information about one conversation does not compromise any other conversations from the same or different participants. Thus, a cryptographic system that would protect individual conversations for a year or more before it could be defeated with current technology would provide a secure enough system for most people. Once a particular conversation was broken, the same effort would be needed to break other conversations.

☞ **Level 2: Commercially secure.** This level would be useful for conversations where the participants in the conversation discuss proprietary information—for example, stock transactions, lawyer-client discussions, mergers and acquisitions, contact negotiations.

A cryptography system that allows industrial activities to be secure for 10–25 years would be adaquate. Once a particular conversation was broken, the same effort would be needed to break other conversations.

☞ **Level 3: Military and government secure.** This is the level that most people think of when cryptography is discussed. This would be used for the military activities of a country and nonmilitary government communications. The appropriate government agencies would define requirements for this level.

---

*Although these scanners do not exist today, they would be built as PCS services become available. Cellular scanners did not exist in the 1970s but were built when the use of the 800-MHz band increased. Even if the FCC bans the manufacturing and importing of the scanners, as they have for cellular-capable scanners, there are enough electronics magazines (*Radio Electronics, Popular Electronics, Nuts and Volts*) published each month that plans for the digital scanner would soon appear.

In addition, PSs placed in the maintenance mode could be used to monitor calls. Manufacturers need maintenance modes to test PSs as they are manufactured or repaired. The literature to determine the maintenance modes is difficult, if not impossible, to keep out of the hands of people without the need to know, yet at the same time be readily available for legitimate purposes.

10.2.2 Privacy Requirements

This section discusses the privacy needs of a wireless telephone user. Figure 10.2 is a high-level diagram of a PCS system showing areas where hackers can compromise privacy. Therefore designers must pay attention to each of these areas to maintain privacy. A user of a PCS PS needs privacy in the following areas:

☞ **Privacy of call setup information.** During call setup, the Personal Station will communicate information to the network. Some of the information that a user could send is: calling number, calling card number, type of service requested. The system must send all this information in a secure way.

☞ **Privacy of speech.** The system must encrypt all spoken communications so that hackers cannot intercept the signals by listening on the airwaves.

☞ **Privacy of data.** The system must encrypt all user data communications so that hackers cannot intercept the data by listening on the airwaves.

☞ **Privacy of user location.** A user should not transmit any information that enables a listener to determine the user's location. The usual method to meet this need is to encrypt the user ID. Three levels of protection are needed:

    ✗ Radio link eavesdropping

    ✗ Unauthorized access by outsiders (hackers) to the user location information stored in the network at the VLR and HLR

    ✗ Unauthorized access by insiders to the user location information stored in the network. The third level is difficult to achieve, but not impossible

☞ **Privacy of user identification.** When a user interacts with the network, the user ID is sent in a way that does not show the user ID. This prevents analysis of user calling patterns based on user ID.

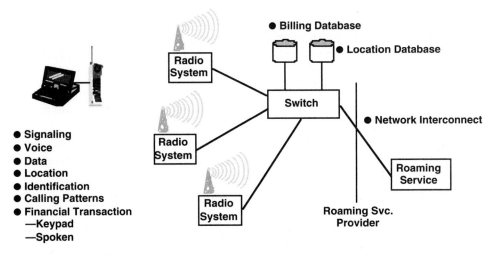

Fig. 10.2 Privacy Requirements

☞ **Privacy of calling patterns.** No information must be sent from a PS that enables a listener of the radio interface to do traffic analysis on the PCS user. Typical traffic analysis information is:

1. Calling number
2. Frequency of use of the PSs
3. Caller identity (previously discussed)
4. Privacy of financial transactions

If the user transmits credit card information over any channel, the system must protect the data. Users may order items from mail order houses via a telephone that is wireless. Users may choose to speak their credit card numbers rather than dialing them via a key pad.

Users may access bank voice response systems, where they send account data via tone signaling. Users may access calling card services of carriers and may speak or use tone signaling to send the card number.

All these communications need to be private. Since the user can send the information on any channel—voice, data, or control—the system must encrypt all channels.

10.2.3 Theft Resistance Requirements

The system operator may or may not care if a call is placed from a stolen PS as long as the call is billed to the correct account. The owner of a PS will care if the unit is stolen.

The terminal design should reduce theft of the PS by making reuse of a stolen PS difficult. Even if the PS is registered to a new legitimate account, the use of the stolen terminal should be stopped. The terminal design should also reduce theft of services by making re-use of a stolen PS unique information difficult. Requirements needed to accomplish the reduction in theft are:

☞ **Clone-resistant design.** In the current wireless systems, cloning of PSs is a serious problem; methods must be put in place to reduce or eliminate fraud from cloning. To accomplish fraud reduction, PS unique information must not be compromised by any of the following means:

1. Over the air: Someone listening to the radio channel should not be able to determine information about the PS and then program it into a different PS:
2. From the network: The databases in the network must be secure. No unauthorized people should be able to obtain information from those databases.
3. From network interconnect: Systems will need to communicate with each other to verify the identity of roaming PSs. A fraudulent system operator could perpetrate fraud by using the security information about roaming PSs to make clone PSs.
4. The communications scheme used between systems to validate roam-

ing PSs should be designed so that theft of information by a fraudulent system does not compromise the security of the PS.

5. Thus, any information passed between systems for security checking of roaming PSs, must have enough information to authenticate the roaming PS. It must also have insufficient information to clone the roaming PS.

6. From users cloning their own PS: Users can perpetrate fraud on the system. Multiple users could use one account by cloning PSs. The requirements for reducing or eliminating this fraud are the same as those to reduce repair and installation fraud described below.

☞ **Installation and repair fraud.** Theft of service can occur when the service is installed or when a terminal is repaired. Multiple PSs can be programmed with the same information (cloning). The cryptographic system must be designed so that installation and repair cloning is reduced or eliminated.

☞ **Unique user ID.** More than one person may use a handset. It is necessary to identify the correct person for billing and other accounting information. Therefore, the user of the system must be uniquely identified to the system.*

☞ **Unique PS ID.** When all security information is contained in a separate module (smart card), the identity of the user is separate from the identity of the PS. Stolen PSs can then be valuable for obtaining service without purchasing a new (full price) PS. Therefore, the PS should have unique information contained within it that reduces or eliminates the potential for stolen PSs to be reregistered with a new user.

10.2.4 Radio System Requirements

When a cryptographic system is designed, it must function in a hostile radio environment characterized by bit errors caused by:

☞ **Multipath fading and thermal noise.** The characteristics of the radio channel affect the choice of cryptographic algorithms. The radio signals will take multiple diverse routes from the PS to the RS. The effect of the multiple diverse routes is to cause fading that can be severe and cause burst errors. Although the system may be interference limited, there may be conditions when the limiting factor on performance is thermal noise. The choice of cryptographic modes must include both of these channel characteristics.

☞ **Interference.** The PCS systems may initially share radio spectrum with other users. The modulation scheme and the cryptographic system must be designed so that interference with shared users of the spectrum does not compromise the security of the system.

---

*This may require a separate security module that plugs into the PS.

☞ **Jamming.** Although usually thought about only in the context of military communications, civilian systems can also be jammed. As wireless communication becomes ubiquitous, jamming of the service may be a useful means of social protest. It can also be a method of breaking the security of the system. Therefore, the cryptographic system must work in the face of jamming.

☞ **Support for handoffs.** When the call handoff occurs to another radio port in the same or adjacent PCS system, the cryptographic system must maintain synchronization.

10.2.5 System Lifetime Requirements

It has been estimated that computing power doubles every two years. An algorithm that is secure today may be breakable in 5–10 years.[*] Since any system being designed today must work for many years after design, a reasonable requirement is that the procedures must last at least 20 years. Thus, the algorithm design must consider the best cracking algorithms available today and must have provisions for being upgraded in the field.

10.2.6 Physical Requirements

Any cryptographic system used in a PS must work in the practical environment of a mass-produced consumer product. Therefore, the cryptographic system must meet the following requirements:

☞ **Mass production.** It can be produced in mass quantities (millions per year).

☞ **Exported and/or imported.**[†] The algorithm must be capable of being exported and imported. Two problems are solved with export and import restrictions lifted:

    ✗ It can be manufactured anywhere in the world.

    ✗ It can be carried on trips outside the United States.

As an alternative, if an import/export license for the algorithm cannot be obtained, the following restrictions must apply:

    ✗ Either only U.S. manufacturing or two-stage manufacturing:

    ✗ All PSs must be made in the United States or all PSs made outside the United States will have final assembly in the United States.

    ✗ All PSs must be impounded on leaving the United States.

---

[*]The original cellular security model was developed in 1974 by AT&T. It is just now in the process of being upgraded, and cellular phones using it will exist for several more years.

[†]Recently, the U.S. government has indicated that it is willing to permit export of secure wireless telephones and that it will only restrict the export of network equipment. If this actually happens, the material in this item is needed only for completeness.

☞ **Basic handset requirements.** Any cryptographic system must have minimum impact on the following PS requirements:

✗ Size

✗ Weight

✗ Power drain

✗ Heat dissipation

✗ Microprocessor speed

✗ Reliability

✗ Cost

☞ **Low-cost level 1 implementation.** The level 1 implementation would be the expected baseline for most PCS systems. Therefore, level 1 implementation must be low cost. Designers obtain low-cost solutions by implementations that can be done either in software or in low-cost hardware. Software solutions are attractive. Often PSs have spare read only memory (ROM), RAM, and central processing unit (CPU) cycles in the microprocessors.

10.2.7 Law Enforcement Requirements

When a valid court order is obtained, current analog telephones (either wired or wireless) are relatively easy to tap by the law enforcement community. The same requirements described in this chapter to ensure privacy and authentication of wireless PCS communications make it more difficult to execute legitimate court wiretap orders.

The law enforcement community can wiretap PSs after properly obtaining court orders. When an order is obtained, there are several ways a PCS system operator can meet the needs of the order. Any method used must not compromise the security of the system.

Figure 10.3 shows possible approaches to tapping the call. The tap can be done over the air or at the central switch.

This discussion assumes that only the radio portion of the link is encrypted and the call appears in the clear in the wired portion of the network. If end-to-end encryption is used, other means must be considered to obtain the information since the call never appears in the clear except at the end points.

10.2.7.1 Over-the-Air Tap When the tap is done over the air, a wiretap van will be needed. Since PCS is a cellular type of system, the van must be driven to inside the cell where the call is placed. A centrally located RS would receive interference from PSs in many cells or would not be able to receive a low-power PSs at all.

In a large-cell PCS system, wiretap stations could be deployed in each cell, but, in a small cell system, the number of tap points would be too high. Therefore, a wiretap van would be needed and would have to be driven to the correct cell where the call is placed.

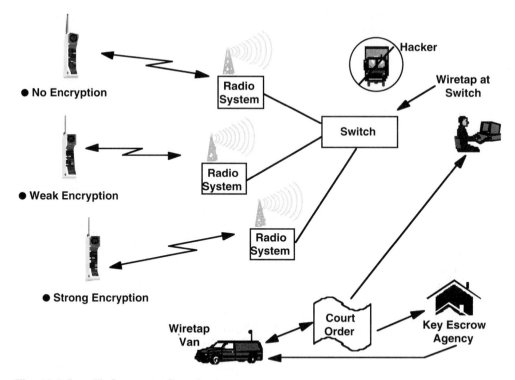

Fig. 10.3 Law Enforcement Requirements

After the van is driven to the correct cell, it needs to be close to the PS. A van might have an antenna that is a maximum of 6–10 feet high versus a RS antenna height of 25 to 100 feet or more. Thus, the van must be closer to the PS than a cell radius. A quick rule of thumb for the wiretap van is that, if the PS is visually observable, then the wiretap van can receive the PS's transmission.

If a wiretap van is used, then the transmissions of the PS must be decrypted. The following options are possible:

☞ **No encryption:** This approach makes tapping the easiest; if no encryption is used, anyone can listen in to a call over the airwaves. Thus, law enforcement personnel can listen to and record a call. Unfortunately, so can anyone else.

☞ **Breakable algorithms:** If the algorithm is weak enough, law enforcement agencies can break the algorithm when permitted to do so by an appropriate court order. Unfortunately, given the proliferation of desktop personal computers (PCs), any algorithm that can easily be broken by the law enforcement community will also be quickly broken by anyone else.

☞ **Strong encryption:** Strong encryption makes it difficult, if not impossible, for the wiretap van to decrypt the transmission. One method to resolve this dilemma is to use a key escrow system where all cryptographic keys would be available from an appropriate key escrow agency. With a court order, the information could be obtained by law enforcement agencies so that they could listen to and record a call.

Procedures must be in place and a set of trusted key escrow agencies chosen so that the key could not be obtained through fraud.

10.2.7.2 Wiretap at Switch Since all PCS calls must be routed through a central switch, those calls that use radio-link-only encryption can be tapped at the central switch under a court order.

This is the preferred method for low-power wireless calls. This method leaves it to the user and the system provider to have appropriate levels of security in the wireless portion of the call.

10.3 METHODS AND PROCEDURES OF PROVIDING PRIVACY AND SECURITY IN WIRELESS SYSTEMS

As we described in chapters 8 and 9, the North American and European cellular and PCS systems support a variety of air interface protocols. They are:

☞ the Advanced Mobile Phone System (AMPS)

☞ the IS-95 Code Division Multiple Access (CDMA)

☞ the Global System for Mobile (GSM) Communications

☞ the Personal Access Communications System (PACS)

☞ the PCS-2000 protocol

☞ the IS-54 (now IS-136) Time Division Multiple Access (TDMA) protocol

☞ the Wideband CDMA (W-CDMA) System

The AMPS protocol has been modified but is essentially the same as was standardized in 1979. The CDMA and TDMA protocols are similar for cellular and PCS, and the goal is for them to be identical. The GSM protocol is used in Europe for digital cellular and in North America with some protocol modifications for the U.S. market and a change to the PCS frequencies. The other protocols are new for PCS. There are other variations considered for cellular, but they have not seen widespread use.

Across these seven protocols, there are four security models that are being used or proposed for cellular/PCS phones in the United States and Europe.[*]

---

[*]As noted in the introduction to this chapter, some details of the security of each of the systems are deliberately not presented. Details of cryptographic systems are governed by ITAR and cannot be exported outside of the United States without permission. The material presented in this chapter is abstracted from the public sections of the cellular and PCS standards and can be exported.

1. **MIN/ESN:** The original AMPS system used a 10-digit Mobile Identification Number (MIN) and a 32-bit ESN. All data is sent in the clear. Data is shared between systems on bad MINs, ESNs, and MIN/ESN pairs. When a mobile telephone roams into a system, first the bad list is checked, and then a message is sent to the home system to validate the MIN/ESN pair. The intersystem communications are sent via SS7 using a protocol called IS-41.

2. **Shared Secret (Key) Data (SSD):** The TDMA and CDMA cellular protocols use SSD stored in the network and the cellular telephone. At subscription time, a secret key is stored in the phone and in the network. AMPS plans to support the SSD in the future. The PCS versions of CDMA, PACS, CDMA/TDMA, TDMA, and W-CDMA all support SSD. The intersystem communications are sent via SS7 using IS-41.

3. **Security Triplets (Token Based):** GSM uses its own unique algorithm and does not share secrets between cellular or PCS systems. It uses a token-based authentication scheme. The Omnipoint PCS-2000 system will also support GSM authentication. When a mobile telephone roams into a system, a message is sent to the home system asking for sets (3–5 typically) of triplets (unique challenge, response to the challenge, and a voice privacy key derived from the challenge). Each call that is placed or received uses one triplet. After all triplets are used up, the visited system must send a new message to get another set of triplets. The intersystem communications use SS7 and a different protocol than IS-41. In Europe CCITT SS7 is used; in the United States, ANSI SS7 is used.

4. **Public Key:** PACS will optionally support public key encryption. The key length is not yet defined. The complete system operation is also not yet defined. Public key systems do not need communications to the home system to validate the mobile telephone. The intersystem communications are still needed to validate the account and get user profile information. These intersystem communications have not yet been defined.

10.3.1 MIN/ESN Authentication

When cellular was invented in the United States, the only mobile telephones were on the Improved Mobile Telephone System (IMTS) (see chapter 1 for a description of IMTS). IMTS had problems with fraud; the last four or five digits of the line number of the IMTS telephone were programmed into the phone using either switches or jumpers. Since all IMTS telephones (from a given manufacturer) used the same key to open the lock on the phone, anyone with access to the key could change their phone number. Later fraud was perpetrated by small boxes on the dashboard of the automobile that allowed any number to be programmed into the phone. When an IMTS phone seized an idle channel, its sent its identity to the network. If the line number was the same as a subscriber, the call would be billed to that subscriber. If the line number did not match, the customer was assumed to be roaming into the system and the operator would come on line to request your full 10-digit telephone number. Clearly the system could be defrauded. Fraud was limited by the low availability of the service.

The cellular system improved security by storing the full 10-digit number in the phone and adding a 32-bit serial number to identify the phone. The FCC rules required that the serial number module be tamper resistant, and attempts to change the serial number would render the phone inoperative. The designers of the system naively assumed that manufacturers would design their phones to be meet the FCC rules. They were also aware that anyone with a microcomputer system could alter the programming of the phone, but the number of people with that capability was very small in 1974 when the cellular security model was invented. No one anticipated the desktop PC explosion of the 1980s and 1990s. Similarly, since 900-MHz technology was expensive and hard to build in the 1970s, no one anticipated the explosion of desktop and handheld scanners that would cover the cellular phone band. Therefore, no attempts were made to encrypt the signaling or the voice communications. The expense of the equipment would limit the availability of equipment to monitor telephone calls.

With the original security model, when a user placed or received a cellular call, the phone transmitted its MIN and ESN to the network. The network first checked a "bad" list of stolen units. The bad list could have MINs, ESNs, or MIN/ESN pairs. If the phone was not on the bad list, then the call was processed. The process has been updated in recent years to standardize intersystem communications using the IS-41 protocol over the SS7 network. Thus, in theory, all calls can be validated in real time. In practice, too many systems still do not perform authentication of the phone.

As an improvement to this approach, some systems require that a user enter a PIN before placing calls. The main advantage of the PIN is that it can be changed in the network when it is compromised, and the user can continue to have the same phone number. Cellular phones that are cloned must have their phone number (MIN) to stop the fraudulent use.

In the 1990s, the MIN/ESN approach with no voice privacy is woefully inadequate for use in a wireless system. Cellular scanners, while banned by the FCC as of April 1994, are still plentifully available on the used market. Equipment exists to decode the cellular data stream and control a scanner for the tracking and monitoring of cellular phone calls, and many cellular phones can be reprogrammed via hardware or software. Thus, there is absolutely no security in North American analog cellular systems.

10.3.2 Shared Secret Data Authentication

When the TDMA protocol was designed, a new authentication scheme was developed. This scheme, called Shared Secret Data (SSD), uses a common authentication key in the mobile telephone unit and in the network. At the time the telephone unit is placed in service, a 64-bit A-key is entered into the unit and into the network in a database called the Home Location Register (HLR). From the A-key, SSD-A and SSD-B are derived and used to authenticate the telephone set and establish a voice privacy key.

All PSs are also assigned an ESN at the time of manufacturing. They are also assigned a 15-digit International Mobile Subscriber Identity (IMSI),[*] that is unique worldwide, an A-key, and other data at the time of service installation. When the PS is turned on, it must register with the system. When it registers, it sends its IMSI and other data to the PCS/cellular network. The VLR in the visited system then queries the HLR for the security data and service profile information. The VLR then assigns a Temporary Mobile Subscriber Identity (TMSI) to the PS. The PS uses the TMSI for all further accesses to that system. The TMSI provides anonymity of communications since only the PS and the network know the identity of the PS with a given TMSI. When a PS roams into a new system, some air interfaces use the TMSI to query the old VLR and then assign a new TMSI; other air interfaces request that the PS send its IMSI and then assign a new TMSI. The call flows for registration are described in a later section of this chapter.

On the control channel, the radio system transmits a random number, RAND, that is received by all PSs. When a PS accesses the system, it calculates AUTHR, an encrypted version of RAND, using a derivative of the A-key called SSD-A. It then transmits to the network the desired message with its authentication. The network does the same calculation and confirms the identity of the PS. All communications between the PS and the RS are encrypted to prevent decoding of the data and using the data to clone other phones. Furthermore, each time a PS places or receives a call, a call counter is incremented (CHCNT). The counter is also used for clone detection since clones will not have a call history identical to the legitimate phone.

Procedures have been designed to permit a system to challenge an individual PS with a unique challenge and to update the SSD. Call flows for the global and unique challenge are described in the following sections. Call flow for the SSD update are not described.

10.3.2.1 Shared Secret Key Registration

Registration is the means by which a PS informs a service provider of its presence in the system and its desire to receive service from that system. The PS may initiate registration for several different reasons. Each of the air interfaces supports different types of registration and are controlled through parameters transmitted by the network on the forward control channels.

A PS registering on an access channel may perform any of the following registration types:

☞ **Distance-based registration:** when the distance between the current cell and the cell where the mobile last registered exceeds a threshold.

☞ **Geographic-based registration:** whenever a PS enters a new area of the same system. A service area may be segmented into smaller regions, called

---

[*]Most cellular phones and some early PCS phones may use MIN instead of IMSI. For the discussions in this chapter, only IMSI will be considered.

"location areas," which are a group of one or more cells. The PS identifies the current location area via parameters transmitted on the forward control channel. Location-based registration reduces the paging load on a system by allowing the network to page only in the location area(s) where a PS is registered.

☞ **Parameter change registration:** when specific operating parameters in the PS are changed.

☞ **Periodic registration:** when the system sets parameters on the forward control channel to indicate that all or some of the PSs must register. The registration can be directed to a specific PS or a class of PSs.

☞ **Power down registration:** when the PS is switched off. This allows the network to deregister a PS immediately upon its power down.

☞ **Power up registration:** when power is applied to the PS, used to notify the PCS/cellular network that the PS is now active and ready to place or receive calls.

☞ **Timer-based registration:** when a timer expires in the PS. This procedure allows the database in the network to be cleared if a registered PS does not reregister after a fixed time interval. The time interval can be varied by setting parameters on the control channel.

The following are the call flows for the registration of all PSs listening to a control channel (see Figure 10.4):

1. The PS determines that it must register with the system.
2. The PS listens on the control channel for the global challenge, RAND.
3. The PS sends a message to the RS with IMSI, RAND, and other parameters, as needed, in the PS Registration Request.
4. The RS validates RAND.
5. The RS sends an ISDN REGISTER message to the PCSC.
6. The PCSC receives the REGISTER message and sends a message to the serving VLR.
7. If the PS is not currently registered to the serving VLR, the VLR sends an REGistration NOTification (REGNOT) message to the user's HLR containing the IMSI and other data as needed.
8. The PS's HLR receives the REGNOT message and updates its database accordingly (stores the location of the VLR that sent the REGNOT message).
9. The PS's HLR sends an IS-41 REGistration CANCel (REGCANC) message to the old VLR where the PS was previously registered so that the old VLR can cancel the PS's previous registration.
10. The old VLR returns a Confirmation message that includes the current value of CHCNT.
11. The user's HLR then returns a REGNOT Response message to the (new) visited VLR and passes along information that the VLR needs (e.g., PCS

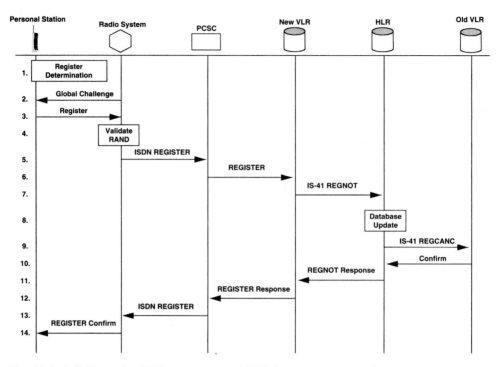

Fig. 10.4 Call Flows for PS Registration of All PSs Listening to a Control Channel

user's profile, interexchange carrier ID, shared secret key for authentication, and current value of CHCNT). If the registration is a failure (due to invalid IMSI, service not permitted, nonpayment of bill, or other), then the REGNOT message will include a failure indication.

12. Upon receiving the REGNOT message from the user's HLR, the VLR assigns a TMSI and then sends a Registration notification Response message to the PCSC.

13. The PCSC receives the message, retrieves the data and sends an ISDN REGISTER message to the RS.

14. The RS receives the REGISTER message and forwards it to the PS to Confirm Registration.

Some air interfaces support the sending of the old TMSI when a PS registers in a new system. When a PS sends its old TMSI, the process flow is similar except the new VLR communicates with the old VLR to obtain the IMSI before an HLR query can be done (see problem 2 at the end of this chapter).

10.3.2.2 Shared Secret Key Global Challenge   All PSs accessing the RS must respond to the global challenge as part of their access. The global challenge response is an integral part of the system access (origination, page response, registration, and so on). We describe it separately here for clarity. The following are

the call flows for a global challenge of all PSs listening to a control channel (see Figure 10.5):

1. The RS continuously broadcasts RAND that changes periodically.
2. The PS calculates its specific response to the challenge (AUTHR) and includes it and RAND (or a shorted version, RANDC) within a Service Request (registration, origination, page response, or data burst message).
3. The RS compares RAND or RANDC with a short list of most recently sent RANDs.
4. If RAND (RANDC) is valid, the RS sends a PCSAP message to the PCSC with TMSI (or MIN or old TMSI), RAND, AUTHR, and other data as needed.
5. The PCSC sends an Authentication Request message to the VLR with TMSI (or MIN or old TMSI) and RAND and requests that the VLR perform the same calculation as done by the PS.
6. The VLR checks its database for TMSI (or MIN or old TMSI). If the data is in not in the VLR, the VLR queries the HLR for the data. When an old TMSI is supplied, the VLR may need to query the old VLR for the location of the HLR and the IMSI. When the data is in the database at the new VLR, it calculates the value of AUTHR and looks up the value of CHCNT.
7. The VLR returns a message to the PCSC.

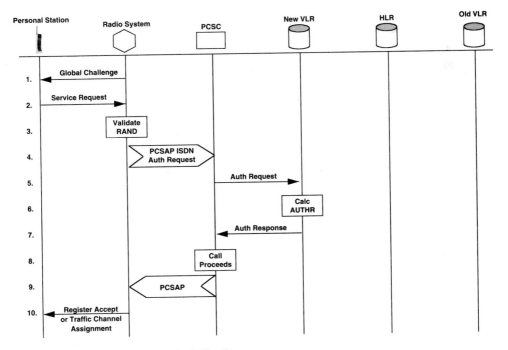

Fig. 10.5 Call Flows for a Global Challenge

8. The PCSC compares the values of AUTHR and CHCNT from the PS and from the VLR. If they match, then a service request message is formed and call processing proceeds as described in other sections. If they do not match, then a service reject message is formed.

9. The PCSC then sends the appropriate PCSAP message to the RS.

10. The RS forwards the accept or reject message to the PS. For Registration messages this is a Register Accept message. For pages or originations, this is a Traffic Channel Assignment.

10.3.2.3 Shared Secret Key Unique Challenge   The unique challenge can be sent to a PS at any time. It is typically initiated by the PCSC in response to some event (registration failure and after a successful handoff are the most typical cases). The following is the call flow for a unique challenge (see Figure 10.6):

1. The PCSC decides to perform a unique challenge

2. The PCSC sends a PCSAP message to the RS with TMSI (or MIN or IMSI if the PS is not registered) and RANDU.

3. The RS forwards the Unique Challenge message to the PS.

4. The PS calculates its specific response to the unique challenge (AUTHU) and sends to the RS a unique challenge order response message that may include TMSI (or MIN or IMSI), AUTHU, and other data as needed.

5. The RS forwards the message to the PCSC in the PCSAP message.

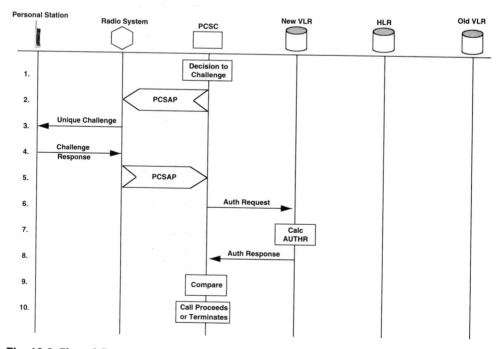

Fig. 10.6 Shared Secret Key Unique Challenge

6. The PCSC sends an Authentication Request message to the new VLR with TMSI (or MIN or IMSI) and RAND (or RANDC) and requests that the VLR perform the same calculation as done by the PS.

7. The VLR checks its database for TMSI (or MIN or IMSI). If the data is not in the VLR, the VLR queries the HLR for the data. When the data is in the database at the VLR, it calculates the value of AUTHU.

8. The VLR returns a message to the PCSC.

9. The PCSC compares the AUTHU from the PS and from the VLR.

10. The PCSC decides to continue or interrupt call processing. If the two AUTHUs match, then the PCSC continues call processing. If they do not match, then the PCSC optionally may take action (e.g., terminate a call in progress, deregister the PS, or other).

10.3.3 Token-Based Authentication

The designers of the GSM system wanted a security system that was under control of the service provider and did not require the sharing of secret data between systems. They were also designing the system for Europe where each system is the size of a country. Thus, intersystem communications needs were low. Only when a PS roamed into another country would an intersystem message be needed. The token-based system using security triplets meets this need.

In GSM-based systems, the triplets consist of a pseudorandom number RAND; its corresponding response, SRES, generated by the authentication algorithm; and a temporary encryption key, K_c, used for data, signaling, and voice privacy. The triplets are requested by the visited system from the home system. These triplets are computed and stored in the PS and in the home authentication center and in the visited VLR. The comparison of the SRES values is done in the visited VLR.

Authentication is performed after the user identity is known by the network and before the channel is encrypted.

Each system operator can choose its own authentication method. The PS and the HLR each support the same method and have common data. The concept of a global challenge as in the SSD systems does not exist. Each PS sends a registration request; then the network sent a unique challenge. The PS calculates the response to its challenge and sends a message back to the network. The VLR contains a list of triplets (unique challenge, response to the challenge, and privacy key); the network compares the triplet with the response it received from the PS. If the response matches, the PS is registered with the network. The just-used triplet is discarded. After all triplets are used, the VLR must query the HLR for a new set. Each query typically results in three to five triplets.

Anonymity is handled by an IMSI/TMSI approach that is similar to the SSD support for IMSI/TMSI. However, unless a failure occurs, once a PS is assigned a TMSI, it uses it until a new TMSI is assigned. When a PS roams into a new area, the old TMSI is sent and a new one is assigned.

The GSM system does not support a call history counter; thus no clone detection is possible. Security of the PS is maintained via a Subscriber Identity Module (SIM) that can be removed from the PS. The SIM card is a smart-card-based system. Smart cards are microprocessor-based secure systems mounted on a card that looks like a credit card. The card is resistant to tampering and cloning.

10.3.3.1 Token-Based Registration   When a PS registers with the PCS network, it sends its TMSI and Location Area Indicator (LAI). The LAI informs the system where to find the old VLR. The network then queries the old VLR for data and uses the data to authenticate the PS. The new VLR then communicates with the HLR to update the location of the PS. The HLR sends a registration cancellation message to the old VLR.

The operation of the token-based system is slightly different from the SSD system in that the segmentation of call processing between RS, PCSC, and VLR is defined differently. Therefore the call flows do not, in most cases, distinguish where call processing is done. The call flows for token-based registration (see Figure 10.7) are:

1. The PS sends a Registration message to the PCS system with the old TMSI and old LAI.
2. The new PCS system queries the old VLR for data.
3. The old VLR returns security-related information (e.g., unused triplets and location of HLR).
4. The new system issues a challenge to the PS.

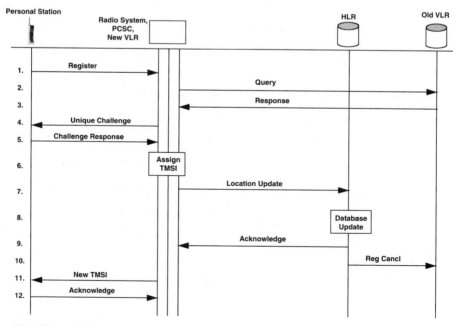

Fig. 10.7 Token-Based Registration

5. The PS responds to the challenge.

6. The new system assigns a new TMSI.

7. The new system sends a message to the HLR with location update information.

8. The HLR updates its location database with the new location of the PS.

9. The HLR acknowledges the message and may send additional security-related data (additional security triplets).

10. The HLR sends a Registration Cancellation message to the old visited system.

11. The new system sends an encrypted message to the PS with the new TMSI.

12. The PS acknowledges the message.

Note that steps 7–10 and 11–12 can occur in either order.

If the old VLR is not reachable, for any reason, then the network will request that the PS send its IMSI to the network, and communications with the HLR will then occur.

10.3.3.2 Token-Based Challenge   Since token-based systems must query the HLR for additional triplets when they are used, provisions are made to reuse the triplets. In those areas of the world where encryption of the radio link is not permitted or during times of network overload when encryption is disabled, the reuse of triplets can ultimately result in a security breach since it may be possible for other PSs to send a previously used challenge response pair and falsely gain access to the network. As token-based systems are more widely deployed, this type of fraud may be seen.

The security-related information consisting of the triplets of RAND, SRES, and K_c are stored in the VLR. When a VLR has used a token to authenticate a PS, it either deletes the token or marks it as used. When a VLR needs to use a token, it uses a set that is not marked as used in preference to a set that is marked as used. If all sets are used, then the VLR may reuse a set that is marked as used. The system operator defines how many times a token may be reused in the VLR. When a token is used the maximum number of times, it is deleted.

When a VLR successfully requests tokens from the HLR or an old VLR, it discards any tokens that are marked as used. When a HLR receives a request for tokens, it sends any sets that are not marked as used. Those sets shall then be deleted or marked as used. The system operator defines how many times a set may be reused before being discarded. When the HLR has no tokens, it will query the authentication center for additional tokens.

The token-based challenge is integrated into the various call flows (e.g., registration, handoff). It is described separately here, for clarity.

Whenever a PCS network must challenge a PS, it will use one token from its set and use the following call flow (see Figure 10.8):

1. The network transmits a nonpredictable RAND to the PS.

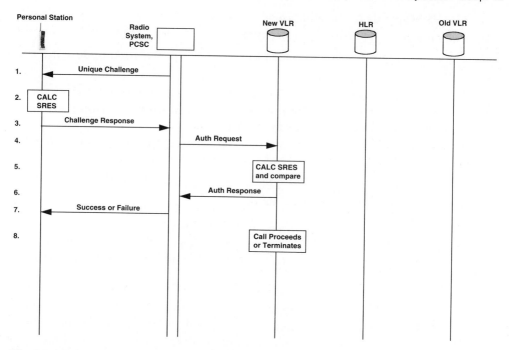

Fig. 10.8 Token-Based Unique Challenge

2. The PS computes the signature (SRES) of RAND using the encryption algorithm and the user authentication key (K_i).
3. The PS transmits the SRES to the network.
4. The PCSC sends a message to the VLR requesting an authentication.
5. The VLR tests SRES for validity
6. The VLR returns the status to the PCSC.
7. The PCSC sends a message to the PS with a success or failure indication.

10.3.4 Public-Key-Based Authentication

The preceding cryptographic systems use secret data that is stored in the PS and the network. The same key is used for encrypting and decrypting the data. With public key systems, two keys are used, one for encrypting and one for decrypting. The two keys are mathematically related to each other with the property that knowing one key does not divulge the other key. The two keys are called the "public key" (USERPUB) and "private key" (USERPRIV) of the user. The PCS network also has a public key (NETPUB) and a private key (NETPRIV).

The following description of public key cryptography uses combination locks and boxes to describe how public key cryptography is done.[*][1]

---

[*]The material is copyright 1994 by AT&T and used with permission.

Imagine that Alice wants to leave a paper message for Bob (historically, Alice and Bob are the actors used in describing cryptographic protocol) in some public place in such a way that Bob can assure himself that only Alice could have left the message. Through prearrangement, Alice tells Bob the combination of a lock that she has had made for her exclusive use, and they agree on a suitable public location where messages will be left (similar to the lockers at a bus terminal). At some future time, Alice can deposit a message in the lockbox, locking it with her personal lock, and inform Bob that the message is ready to be retrieved. Bob can then unlock the combination lock with the combination that Alice has provided and be quite confident that no one else has surreptitiously inserted a false message.

The same type of arrangement can be used if Alice wants to send Bob a message in such a way that they can both be sure that a third party cannot read the message. In this case, Bob provides the combination lock to Alice (in an unlocked state), but does not reveal the combination to her. Alice can then lock her message in the locker, being assured that no one other than Bob knows the combination and can retrieve the message.

Instead of using the combination locks, paper messages, and lockers described in the analogy above, the Rivest-Shamir-Adleman (RSA) [3] algorithm uses hard mathematical algorithms to compute message exchanges that model the exchanges above.

Analogous to the lock and its combination, the RSA algorithm relies on two cryptographic keys, intimately related to each other but each underivable from the other. To sign a message m, Alice would reveal her public key, e_a, but not her secret key, d_a. Signing a message then requires only that the message be encrypted with d_a. Anyone can verify the signature by using e_a to decrypt the message. Correspondingly, RSA could be used to send messages secretly if Bob gave Alice his public key, e_b, to encrypt the message and used his private key, d_b, to decrypt the message.

Mathematically, the two keys used are multiplicative inverses of each other in a finite field of size n, where n is the product of two large prime numbers p and q. Both p and q are kept secret but n can be published. Anyone knowing p and q can pick an e, relatively prime to $(p-1)*(q-1)$, and can compute d, the multiplicative inverse of e, in the field relatively easily. Anyone else who knows only n and d, will have great difficulty in computing e, perhaps needing to expend as much effort as trying to factor n into the two prime numbers it is composed of.

Encrypting (or signing) a message m requires that the originator calculate a cipher, c, where

$$c = m^e \ (\text{mod } n) \qquad (10.1)$$

Decrypting (or verifying the signature) requires that the recipient calculate

$$m = c^d \ (\text{mod } n) \qquad (10.2)$$

Of course, since e and d are multiplicative inverses,

$$(m^e)^d = m^{ed} = m^1 = m \qquad (10.3)$$

and the original message is recovered.

The public and private keys are related by mathematical relationships that make it difficult to calculate the private key when the public key is known. As computing power increases, it will be necessary to lengthen the key length of the public key system. The mathematical calculations for encryption and decryption are processor intensive and are difficult to do on the 8-bit microprocessors in PSs.

When a PCS system uses public key cryptography, it publishes its NETPUB, or ea-NET, and all PSs have NETPUB stored in memory. The public key of the PS is also known to the network. The public key system is used to authenticate the PS and to communicate with the network. Full anonymity is guaranteed since only the network can decrypt messages. Voice and data privacy is done by using the public key system to exchange in a secure transmission the secret key needed for privacy.

10.3.5 Air Interface Support for Authentication Methods

The various air interfaces used for PCS and cellular in Europe and North America support one or more of the different authentication methods. Only the older AMPS system supports MIN and ESN as the authentication method. All of the digital cellular systems in North America and all of the PCS systems being proposed for North America, except for GSM, support SSD. PACS will support public key authentication as a future option. GSM supports only token-based authentication. The PCS-2000 system supports token-based authentication as an option when it is supporting the GSM MAP. Table 10.1 summarizes this information.

Table 10.1 Summary of Authentication Systems for PCS and Cellular in Europe and North America

| Air Interface | Type of Authentication | | | | Type of Voice Privacy Supported |
	MIN/ESN	SSD	Token-Based	Public Key	
AMPS	X	X			none
CDMA		X			strong
GSM			X		strong
PACS		X		X	strong
PCS-2000		X	X		strong
TDMA		X			weak[*]
W-CDMA		X			strong

[*]The PCS version of TDMA will have stronger voice privacy.

10.3.6 Summary of Security Methods

Each of the four security methods meets the security needs for a wireless system in different ways. The older AMPS system has poor security. The digital systems using either SSD or tokens meet all of the security needs of wireless systems except for full anonymity. The public-key-based security system meets all the requirements, including anonymity, but is not yet fully designed. The following is a summary of the support for security requirements for the PCS and cellular systems in North America and Europe (see Table 10.2).

Privacy of communications is maintained via encryption of signaling message and voice and data for the digital systems. AMPS sends all data in the clear and so has no privacy unless the user adds it to the system.

10.3.6.1 Billing Accuracy   Since AMPS phones can be cloned from data intercepted over the radio link, billing accuracy for AMPS is low to none. For the other systems, when authentication is done, billing accuracy is high. If a system operator gives service before authentication or even if authentication failure occurs, then billing accuracy will be low.

10.3.6.2 Privacy of User Information   Privacy of user information is high for the public key system, moderate for the SSD and token-based systems (since sometimes IMSI is sent in the clear) and low for the AMPS system.

10.3.6.3 Theft Resistance of PS   PS theft resistance is high for over-the-air transmission for all systems except AMPS. Since the token-based system doesn't support a call history count, it has a lower resistance to cloning than the SSD or public key systems. AMPS phones using MIN/ESN have no resistance to cloning but will support SSD in the future. The resistance to stealing data from network interconnects or from operations support systems in the network will depend on the system design.

10.3.6.4 Handset Design   Except for some public key systems, all of the authentication and privacy algorithms easily run in an standard 8-bit microprocessor used in PSs.

10.3.6.5 Law Enforcement   The AMPS system is relatively easy to tap at the air interface. The digital systems will require a network interface since privacy is maintained over the air interface. The reference models and network requirements are currently being updated to meet the needs of the law enforcement community doing legal wiretaps.

10.4 SUMMARY

This chapter described the requirements that any cryptographic system should meet to be suitable for use in a ubiquitous wireless network. We then examined

Table 10.2 Summary of Support for Security Requirements for PCS and Cellular in Europe and North America

Feature	MIN/ESN (AMPS)	SSD	Token-Based	Public Key
Privacy of communications				
• Signaling	None	High: messages are encrypted	High: messages are encrypted	High: messages are encrypted
• Voice	None	High: voice is encrypted	High: voice is encrypted	High: voice is encrypted
• Data	None	High: data is encrypted	High: data is encrypted	High: data is encrypted
Billing Accuracy	None: phones can be cloned	High: if authentication is done.	High: if authentication is done.	High: if authentication is done.
Privacy of user information				
• Location	None	Moderate: using IMSI/TMSI	Moderate: using IMSI/TMSI	High: public key provides full anonymity
• User ID	None	Moderate: using IMSI/TMSI	Moderate: using IMSI/TMSI	High: public key provides full anonymity
• Calling patterns	None	High: using TMSI and encryption	High: using TMSI and encryption	High: public key provides full anonymity
Theft resistance of PS				
• Over the air	None	High	High	High
• From network	Depends on system design	Depends on system design	Depends on system design	Depends on system design
• From interconnect	Depends on system design	Depends on system design	Depends on system design	Depends on system design
• Cloning	None	High	Medium	High
Handset design	Algorithms run in microprocessor of handset	Algorithms run in microprocessor of handset	Algorithms run in microprocessor of handset	Microprocessor speed may be fast enough for some algorithms
Law enforcement needs	Easily met on the air interface (if van is nearby to PS) or at the switch	Must wiretap at the switch	Must wiretap at the switch	Must wiretap at the switch

four security models and described how they met the requirements. Certain details were left out to maintain security and meet U.S. government export requirements on cryptographic systems.

10.5 PROBLEMS

1. Describe how you would design a PS so that the security data stored in the terminal is tamper resistant.

2. Computing power is estimated to double every 2 years or less. If a cryptographic algorithm takes 365 days to break using a readily available computer today (e.g., a desktop microcomputer), in how many years can the algorithm be broken in 1 hour? In how many years can it be broken in 1 second and therefore be included in a digital scanner design?

3. Repeat the previous problem if a weaker algorithm is used that today can be broken in 30 days.

4. Describe the call flows that would be necessary for a shared secret key registration when the old VLR must first be queried for the IMSI before the correct HLR can be queried.

5. Describe the call flows that would be necessary for a token-based registration when the old VLR cannot be queried for the IMSI and the network must request that the PS send its IMSI before messages can be exchanged with the HLR. If you have access to the GSM standards or reference [2], compare your answer with the standards.

10.6 REFERENCES

1. D'Angelo, D. M., B. McNair, and J. E. Wilkes, "Security in Electronic Messaging Systems." *AT&T Technical Journal* 73, no. 3, May/June 1994.

2. JTC(AIR)/94.03.25-257R1, "Minimum Requirements for PCS Air Interface Privacy and Authentication."

3. Report of the Joint Experts Meeting on Privacy and Authentication for PCS, Phoenix, Arizona, November 8–12, 1993.

4. Rivest, R. L., A. Shamir, and L. Adleman, "A Method for Obtaining Digital Structures and Public-Key Crypto Systems." *Commun. ACM* 21, no. 2 (February 1978): 120–7.

5. TIA Interim Standard, IS-41 C, "Cellular Radiotelecommunications Intersystem Operations."

6. TR-46 P&A ad hoc/94.04.17.01 R5, "TR-46 PCS Privacy and Authentication, Volume 1, Common Requirements," Version 6, November 1994.

7. TR-46 P&A ad hoc/94.04.17.02 R4, "TR-46 PCS Privacy and Authentication, Volume 2, PCS1900 Based Requirements."

8. TR-46 P&A ad hoc/94.05.17.02 R3, "TR-46 PCS Privacy and Authentication, Volume 3, Shared Secret Data Requirements."

9. Wilkes, J. E., "Privacy and Authentication Needs of PES." *IEEE Personal Communications* 2, no. 4 (August 1995).

Network Management for PCS and Cellular Systems

11.1 INTRODUCTION

In the preceding chapters we discussed the elements of a cellular and PCS system that are needed to provide service to a subscriber. In this chapter we discuss those elements of the system that keep the system operating on a day-to-day basis. In the telecommunications business, these systems are referred to the Operations, Administration, Maintenance, and Provisioning (OAM&P) systems. With these systems, cellular and PCS service providers monitor the health of all network elements, add and remove equipment, test software and hardware, diagnose problems, and bill subscribers for services. In the past each new network element had a corresponding operations support system that provided the management capabilities. Each of these systems had a different user interface, used a different computing platform, and typically managed one type of network element. As service providers moved into a mixed vendors' environment, they could no longer afford different systems for each network element. Therefore, the standards groups in Committee T1M1 and CCITT (International Telecommunications Union, or ITU) defined the interfaces and protocols for operations support systems under the umbrella of Telecommunications Management Network (TMN).

Service providers will deploy PCS into an existing cellular competitive environment. Therefore, they must offer cost-effective PCS services. While management functions are necessary for the smooth operation of a system, they are a cost of doing business and do not directly improve the bottom line of a business. Therefore service providers must manage the cost of these systems. Considerations of initial cost of the systems, personnel costs, and operational costs must be examined. A key component to the successful deployment of the PCS systems and their cost-effective operation will depend upon the operator's ability to effectively manage the new equipment and services in a mixed vendors' environment. For PCS to succeed, the OAM&P costs for PCS networks and services must be lower than those of either the existing wireless or wireline networks. TMN will meet this requirement.

We will discuss, in this chapter, the goals and the requirements for PCS management. We will then briefly discuss the PCS-specific management standards before presenting information on the TMN. We will then describe the TMN and Open System Interconnection (OSI) management. Since TMN is based on Object-Oriented Design (OOD) we will provide a brief discussion of OOD. With these requirements, we will present the information model that can be used to manage the PCS network.

11.2 GOALS FOR PCS MANAGEMENT

The original cellular systems were purchased from equipment providers as a complete and proprietary system. The interfaces between network elements were proprietary to each vendor, and network elements from one vendor did not work with network elements from another vendor. Included with the systems were management systems that were specific to the vendor's hardware and software. Both cellular and PCS are migrating to a mixed vendor environment with open interfaces. Thus, a service provider can purchase switching hardware from one company, radio equipment from another company, and network management equipment from a third company.

In this new environment, specific goals for network management software and hardware are necessary. These network management goals for PCS are:

- ☞ **Operation in mixed vendor environment:** The network management software, hardware, and data communications must support network elements from different vendors.

- ☞ **Multiple solutions:** Service providers should be able to purchase network management solutions from multiple and competitive vendors. This will ensure the lowest cost and richest feature set for the systems.

- ☞ **Support for mixed air interfaces:** In chapters 8 and 9, we described in excess of six air interfaces that will be deployed in various countries. Since service providers may have different radio technologies in different cities that are managed by a common network management function, support for mixed air interfaces is necessary.

- ☞ **Use of existing resources:** Cellular companies and existing wireline companies should be able to manage their PCS business with minimum additions of new data communications, computing platforms, and so on. If possible, minimal additions should be made to existing distribution networks to implement the required control or reporting system.

☞ **Support for multiple and interconnected systems:** When a service provider has multiple systems deployed, the different management systems should interact effectively and efficiently to deliver service by sharing a common view of the network.

☞ **Support for sharing of systems and information across multiple service providers:** Multiple service providers must interact to share information on billing records, security data, subscriber profiles, and so forth and to share call processing, i.e., intersystem handoffs. The network management systems, therefore, should allow flexible telecommunication management relationships among multiple service and network providers of PCS. Flexible management relationships include complex network and service provider arrangements consistent with individual providers being separate business entities. Wireless access providers might also want management access through intermediate networks.

☞ **Support for common solutions between end users and service providers:** Many services require a joint relationship between the service provider and the end user. End-to-end data communication is the most common example of the need for a joint relationship. When service providers and end users are operating in this mode, their management functions should be interconnected. The solutions should allow flexible telecommunication management relationships between service providers and end users.

11.3 MANAGEMENT REQUIREMENTS FOR PCS

A PCS network is a telecommunications network where the users and the terminals are mobile rather than fixed. In addition, there are radio resources that must be managed and do not exist in a wireline network. Furthermore, the network model can be a cellular model where the radio resources and the switching resources are owned and managed by one company. A second newer model has the wireline company operating the standard switching functions and a wireless company operating the radio-specific functions. With these new modes of operation, network management requirements for PCS include standard wireline requirements and new requirements specific to PCS.

The new PCS-specific requirements should be developed for the following areas:

☞ **Management of radio resources:** PCS allows terminals to connect to the network via radio links. These links must be managed independent of the ownership of the access network and switching network. The two networks may consist of multiple network service providers or one common service provider. The multiple operator environment will require interoperable management interfaces.

☞ **Personal mobility management:** Another important aspect of PCS is mobility management. Users are no longer in a fixed location but may be anywhere in the world. They may be using wireless or wireline terminals to place and receive calls. In some cases the network will determine the location automatically; in other cases the user may need to report his or her location to the network. This will increase the load on the management systems as end users manage various decision parameters about their mobility; for example they may request different services based on the time of day or terminal busy conditions.

☞ **Terminal mobility management:** The primary focus of PCS is to deliver service via wireless terminals. These terminals may appear anywhere in the worldwide wireless network. Single or multiple users may register on a wireless terminal. The terminal management function may be integrated with the user management function or may be separate from the user management function. Different providers may operate their system in a variety of modes. The management functions must support all modes.

In chapter 8, we identified two reference models for PCS. We left the discussion of the Operations Systems (OSs) to this chapter where the primary focus will be to address the Operations Systems functional block and the q reference point (see Figure 8.2). We also discussed management functions related to other network elements and other reference points. The q reference point in Figure 8.2 is one of many possible connections, which can logically be extended to any Network Element (NE) function.

Telecommunications engineers have studied the management function in depth and segment it into eight areas: accounting management, configuration management, fault management, materials management, performance management, planning for growth, security management, and work force management. In the rest of this chapter, we will specify the requirements for the PCS management function in six of the above eight areas. Materials management (inventory control) and work force management (personnel practices and management) are not discussed here since they have seen less attention in standards bodies, do not directly affect the telecommunications aspects of the PCS service, and are not directly specific to the telecommunications industry.

11.3.1 Accounting Management

Accounting management for PCS involves the real-time generation and collection of usage data at network elements and the processing and the distribution of the usage data to applications such as billing. Accounting management for PCS is more complex than accounting management for wireline services because usage data that belongs to the same service may be generated and/or collected in several networks simultaneously. Also, systems may be required to provide additional flexibility to transmit data to the appropriate billing system.

The requirements for PCS accounting management include: a framework architecture for accounting management, data element definitions and relevance to the applications, the association between logical PCS network resources and data elements, and interprovider accounting specifications. Accounting management is a distributed network function that measures usage of network resources by a subscriber or by the network itself (e.g., for audit purpose) and generates formatted records containing this usage information. Users of the formatted accounting management records include the billing Operations Systems, other Operations Systems, and the subscribers. The purpose of accounting management data is to provide a bill to the subscriber of a service. Since the user of the service may be an employee of a company or another family member in a family, the subscriber and the user may not be the same. Systems may need to generate two sets of bills to account for the separation between subscriber and user.

The accounting process (Figure 11.1) has three subprocesses:

1. The **call detail recording** process is responsible for the creation of detailed records for each accountable event in the network.

2. The **rating process** is responsible for collecting the call detail records that belong to a particular transaction, combining them into service transaction records, and rating the transaction to determine the bill.

3. The **billing process** is responsible for collecting the service transaction records and selecting from them those that belong to a particular subscriber over a particular time period to produce billing.

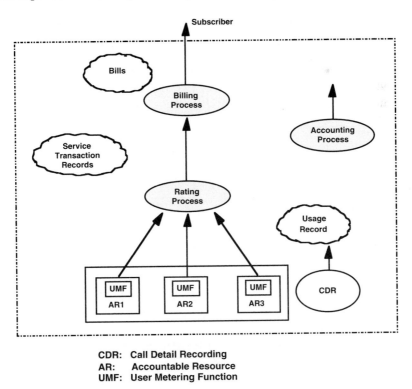

CDR: Call Detail Recording
AR: Accountable Resource
UMF: User Metering Function

Fig. 11.1 PCS Accounting Management of Resource Utilization

The following requirements should be met for PCS accounting management:

1. **Multiple providers:** The usage environment will consist of multiple service and network providers. The set of providers will change more rapidly than in today's environment.

2. **Distributed collection of usage information:** Since many services will be "network services," it will be, in many cases, no longer possible for a single node (such as a switch) to generate a complete record of a call.

3. **Multitude of charging strategies:** The usage information collected for billing must be flexible to support a variety of charging strategies.

4. **Rapid deployment of diverse services:** The accounting management structures must allow for the timely introduction of additional formats as services and technologies evolve.

5. **Multiple use of usage data:** The usage information will be used by more than one process. Anticipated uses are fraud detection, network engineering, and customer care, among others.

6. **Integration with existing network:** In the present environment, many switches produce automatic message accounting (AMA) records, while other switches produce charging records of various formats. It must be possible to take advantage of the accounting infrastructure that already exists.

7. **Global breadth:** It is desirable that standards developed for PCS accounting management should allow interwork on a global scale.

Figure 11.2 shows the network reference model for Data Message Handler (DMH) that is defined in TIA/EIA IS-124 [2]. We will now discuss the functional entities and reference points of this architecture as it applies to PCS accounting management.

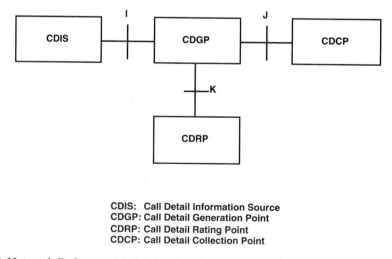

CDIS: Call Detail Information Source
CDGP: Call Detail Generation Point
CDRP: Call Detail Rating Point
CDCP: Call Detail Collection Point

Fig. 11.2 Network Reference Model for the Data Message Handler

The network model represents four basic functional entities:

1. Call Detail Information Source (CDIS)
2. Call Detail Generation Point (CDGP)
3. Call Detail Rating Point (CDRP)
4. Call Detail Collection Point (CDCP).

In the reference model, each functional entity is a logical processing unit rather than a physical processing unit.

The call detail information source may be a PCS switching center, a radio port controller, a radio access system controller, a terminal mobility controller, a personal mobility controller, or any other source of call detail information. It could even be another call detail collection point. The call detail information source provides call information without additional services or call detail information storage.

The call detail generation point is responsible for collecting call detail information and encoding the messages for delivery to a call detail collection point. The call detail generation point may edit the collected call detail information, as necessary, to perform operations including reducing redundant information, discarding inconsistent records, eliminating information not needed by collection point, merging related call information, or adding rating information. The call detail generation point is also responsible for addressing messages to the proper call detail collection point. The call detail generation point should be capable of storing certified call detail information until it is received and reconciled by the call detail collection point.

The call detail collection point is the recipient of call detail records. It may use or process call details. Examples of services it supports include subscriber billing, intersystem charging, net settlement, and subscriber fraud detection.

The call detail reduction point calculates and returns the billing information based on call detail information. In a Public Switched Telephone Network (PSTN) environment, this function is normally referred to as tariffing and would most frequently be associated with the billing system.

11.3.2 Configuration Management

PCS configuration management supports the management of resources and services in the PCS network. The resources are: radio channels, data links, base stations, the switch, the Home Location Register, the Visited Location Register, traffic and signaling links, among others. Service configuration supports the network and subscriber data necessary to customize services to a specific user.

Standard telecommunications network management for nonwireless specific network elements will be provided as part of a complete PCS system. In addition, we assume that providers will configure their radio resources to maximize coverage of an area's traffic patterns. This includes location of users for call placement and delivery, user mobility, and local radio control after calls are connected. PCS-specific resource configuration management has:

1. PCS network elements implemented as shared resources to reduce the cost of PCS networks

2. PCS network elements installation, replacement, provisioning, and management using minimum manual effort

3. The management interface providing the ability to name network elements independently of the service provider

4. The management interface allowing the rapid provisioning of network element names within the control of administering organization

5. The frequency assignment algorithms and related parameters specified by industry standards

6. Messages to support self-configuration management capabilities of radio ports and other network elements

7. PCS network elements supporting remote provisioning capabilities and self-inventory

8. The capabilities for PCS service providers to share access to Network Elements, that allow radio network management, terminal mobility management, or personal mobility management

9. A standard mechanism to accomplish remote software downloads to network elements

10. A mechanism for the configuration and reconfiguration of software, including functions for remote distribution, activation, maintenance, and administration, of software by a management system

11. A mechanism for quickly deploying software to handle special situations such as traffic overload algorithms and hardware failure not detected or envisioned earlier

12. Messages for remote control from an OS for selection of frequency assignment algorithms and related parameters

13. A mechanism for the assignment of nonfrequency-unique channel characteristics of network elements

14. Messages to manage a variable set of equipment designs that support different air interfaces

15. A system to allow multiple providers to exchange data on a timely basis and maintain data integrity about customer services and network interfaces

PCS-specific service configuration management will have:

1. PCS networks supporting some user control of personal profiles such as feature options preferences and related user behaviors

2. User profile editing that are under user's control

11.3.3 Fault Management

PCS service and network providers will need bidirectional testing and alarm surveillance capabilities. Control capabilities will be negotiated and exercised as providers deem appropriate. Although most providers will want to operate stand-alone systems and not permit other providers to have access to their equipment, it may be impossible to diagnose problems across networks unless the systems are interconnected. For example, intersystem handoffs that fail because of network problems will not be resolved unless communication between neighboring systems is implemented. While the communications can be verbal, problem resolution will occur faster when the networks are interconnected by operations support systems. However, service providers cannot expect carte blanche control of other network provider resources to test service operation. Network providers cannot arbitrarily invoke PCS services to test network connectivity. Fault management of PCS includes alarm surveillance, testing, and fault localization and correction.

Alarm surveillance will need to support:

1. Fault management messages to report alarms and failure conditions for various reference points (Figure 8.2).

2. Fault management messages to report alarms and failure conditions for radio network elements, PCS switching center elements, and mobility management elements.

3. Specific mechanisms for detection of failures before they affect customer service.

4. Messages to support selected alarms through a broadcast technique. This would allow either upstream or downstream surveillance (distribution link or air interface) to a multiple surveillance operations system.

5. Specification and storage of the records needed to record the existence of a failure and its retention for later analysis.

6. Specification of the alarms resulting from the following failures conditions: loss of power, equipment failure, loss of communications links, continuous transmission, and abnormal environmental conditions (e.g., temperature, battery voltage, doors, vibration).

7. Measurements to indicate that proactive maintenance should be initiated.

8. Alarm indications for software failure (e.g., database error, service logic abnormalities or exception conditions, failed registrations, or failed profile updates).

9. Alarm notifications for equipment failures of all network elements.

After alarms are generated, then the system must automatically remove failed equipment from service and perform diagnostics on the failed equipment. Alternately, maintenance personnel must be able to manually request diagnostics. In either case, the diagnostics will do the following:

1. Provide for the verification of the operation of the radio equipment, the elements of the PCS switching center, and the auxiliary network elements, such as the mobility management functions of the Home and Visited Location Registers

2. Provide for the verification of operation of connecting links (voice, data, and signaling) between all internal and external network elements

3. Provide for the verification of the proper operation of services delivered to the user

4. Support in-service testing for those critical network elements that can not be removed from service

After the systems determine the fault, it may be necessary to further localize the fault to a subsystem and correct that fault. The fault localization and correction capabilities will use:

1. Specific mechanisms for multiple cooperating business entities to exchange fault localization information in order to isolate the cause of a reported fault

2. Capabilities to allow the correction of fault either through repair of the equipment, replacement of the equipment, or reinitialization of the equipment.

3. Equipment change-out procedures that either process with no break in service or use automatic service recovery when the replacement system is brought on line.

4. Automatic correction of software-based errors. Systems will correct errors with error-handling software routines.

5. Automatic correction of data element errors by reloading them from backup databases.

11.3.4 Performance Management

PCS service and network providers need bidirectional performance monitoring and control capabilities. Control capabilities are negotiated and exercised as providers feel appropriate. Service providers may want notification of network conditions that impact the service perceived by customers. Network providers may want confirmation that they are providing adequate capacity. Performance measurement requirements fall in one of three categories: traffic and signaling, quality of service, and availability.

Traffic and signaling measurements are taken to determine traffic load on the radio interface, including both user and signaling traffic; occupancy, and throughput load measurements and congestion indications within network nodes; and traffic load on circuits connecting the network elements or providing connections to other networks, including both user and signaling traffic. The purpose of such data collection is to enable traffic management to reconfigure the PCS network or modify its operation to adjust to unused traffic patterns and to provide traffic capacity data for purposes of planning network growth.

Quality of service measurements are intended to represent the performance experienced by the user. Measurements of connection establishment, connection retention, connection quality for voice and data services, and billing integrity are considered to define the quality of service. The purpose of this data collection is to identify degradation of service quality, points of failure, or failure trends.

Availability measurements are taken to allow PCS providers and peer network providers to exchange signaling traffic measurements. They are also taken to allow PCS signaling measurements to identify the signaling traffic in separate categories (registration, alerting, call delivery, user administration, and other), and to identify signaling traffic destinations separately.

Performance measurements also include counts of successful handoffs. Handoffs may be initiated by either terminals or the network. It is important to note that a handoff does not always occur because of end user mobility. Handoffs can occur as the radio environment around the end user changes.

PCS-specific performance controls are necessary to moderate control traffic. For example, without performance controls, handoff requests and terminal registrations could congest signaling channels. Another example is to "gap" originating traffic to avoid exception processing in a PCS network concentration point. This includes support to control traffic across various interfaces.

11.3.5 Planning Management

The service provider needs to plan for the orderly growth of the PCS system as new users are brought into the system. We discuss the planning process in detail in chapter 13 as we describe how to design and grow a wireless system. The service provider will need to use measurement data to engineer the growth of the system.

11.3.6 Security Management

PCS security management deals with anonymity, bearer and signaling channel privacy, security of subscriber data, subscriber authentication, and inter-operability between PCS service providers. We have described the security of PCS in detail in chapter 10.

11.4 OAM&P STANDARDS FOR PCS AND CELLULAR NETWORK

The PCS operation, administration, maintenance and planning (OAM&P) standard developed in the ATIS T1M1 committee is based upon OSI management and provides a mechanism for easily adding new technology and vendor extensions to the standard, while still maintaining backward compatibility. The PCS OAM&P standards are based upon T1P1 reference model discussed in chapter 8 and use the Telecommunications Management Network (TMN) framework developed within the International Telecommunications Union (ITU) to define the OAM&P interface. One key component in TMN is the application of the OSI management standard.

The GSM standards developed by ETSI consist of parts (series) 1 through 15. Series 12 of these standards describes the OAM&P functions of a GSM network and provides a GSM-specific TMN model for managing GSM networks. These standards have been brought to the U.S. ATSI T1P1 and TIA TR-46 committees to support PCS1900 (the 1900-MHz variant of GSM 900 MHz) in North America. The GSM model is considerably more detailed than the T1M1 model and includes technology-specific parameters such as those used by the GSM handoff algorithms. It also deals with administration of subscriber profiles, management of security, and collection of usage information. The GSM OAM&P standards are custom-tailored to manage GSM networks and therefore do not currently fit within the multiple-vendor, multiple-service provider model for operations systems in North America. The T1M1 model could be extended to handle GSM; however, these extensions would take time to develop. Many of the approaches used in the T1M1 standard (such as maintaining a physical and logical split) have been taken from GSM Series 12 specifications. Since the GSM standards are specific to a GSM network, much of the GSM Series 12 information model is not sufficiently generic to apply to PCS as a whole.

We have previously described another OAM&P effort, the Data Message Handler (DMH) standard developed by TIA committee TR-45. The DMH deals only with billing and provides a means of exchanging usage and rating data across jurisdictions. While not a TMN standard, it is also based upon an OSI model. This standard has been developed to exchange data between existing cellular networks. The standard also deals with operating systems (OS)-to-OS transfers, and there is no overlap between this standard and the T1M1 or GSM standards.

11.5 TELECOMMUNICATIONS MANAGEMENT NETWORK

The development of OAM&P functions and protocol standards for PCS follows the concepts developed in CCITT Recommendation M.3000-series which defines the Telecommunications Management Network (TMN).

TMN as applied to PCS:

1. Uses a common set of standard telecommunication management functions for managing generic resources to save the expense of providing a unique PCS functional management architecture.

2. Segments the common characteristics of telecommunications equipment into managed objects, new object definitions (if required), and new attribute definitions for existing object definitions. This segmentation provides a consistent data model for PCS resources and services.

3. Helps define customer access to management information with standard managed object definitions for resources, services, and customers.

The parts of the TMN architecture applicable for managing PCS consist of:

✗ A logical layered architecture with five layers relevant to PCS: the business management layer, the service management layer, the network management layer, the element management layer, and the network element layer.

✗ A physical architecture to define management roles for operations systems, communication networks, and network elements.

✗ A functional architecture that defines fault management, performance management, configuration management, accounting management, and security management.

4. TMN provides an organized architecture (see Figure 11.3) to achieve the interconnection between various types of OSs and telecommunication equipment by using standard protocols and interfaces. Telecommunication equipment includes switching systems, transmission systems, multiplexers, and so on. When managed by an operations system, the equipment is called a "network element."

11.5.1 Layered Management Architecture

The important layers of the TMN management architecture are described in the following paragraphs (see Figure 11.4).

The **Business Management Layer (BML)** has the responsibility for the total enterprise and is the layer at which agreements between network operators are made. This layer normally carries out goal-setting tasks rather than goal achievement but can become the focal point for action in cases where executive action is called for. This layer is part of the overall management of the enterprise, and many interactions are necessary with other management systems.

The **Service Management Layer (SML)** is concerned with the contractual aspects for services that are provided to customers or available to potential new customers. This layer has five principle roles: interfacing with customers, interacting with service providers, interacting with the network management

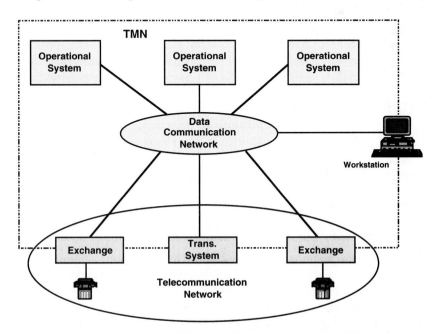

Fig. 11.3 General Relationship of a TMN to a Telecommunications Network

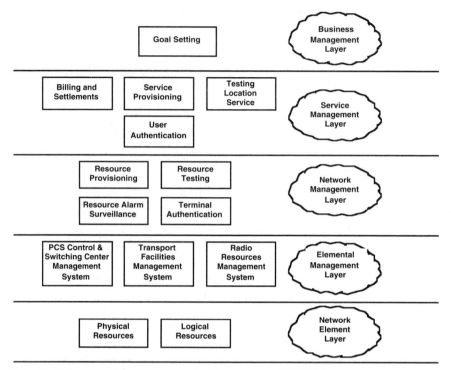

Fig. 11.4 TMN Layered Management Architecture and PCS Management

layer, maintaining statistical data (e.g., quality of service), and interacting between services.

The **Network Management Layer (NML)** has responsibility for management of all the network elements, as presented by the element management layer, both individually and as a set. It is not concerned with how a particular element provides services internally. At this layer, functions addressing the management of a wide geographic area are located with complete visibility of the whole network along with a vendor-independent view. The NML has three principle roles: the control and coordination of network view of all NEs within its scope or domain, the provision, cessation or modification of network capabilities for the support of services to customers, and interaction with SML on performance, usage, availability, and so on.

The **Element Management Layer (EML)** manages each network element on an individual basis and supports an abstraction of functions provided by the network element layer. The EMLs has a set of element managers that are individually responsible for some subset of NEs. Each element manager has three main roles: control and coordination of a subset of NEs, provide a mediation function to permit the network management layer to interact with network elements, and maintain statistical, log, and other data about elements.

The **Network Element Layer (NEL)** consists of logical and physical resources to be managed.

Information models for different levels of abstraction provides the required flexibility in developing implementations supporting the Network Element Layer, Element Management Layer, Network Management Layer, Service Management Layer and Business Management Layer functions. The initial PCS standard supports the specification of information models based on levels of abstraction to the Element Management Layer-Network Management Layer and Element Management Layer-Network Element Layer reference points.

11.5.2 The Physical Architecture

The general TMN architecture (see Figures 11.5 and 11.6) is described in CCITT Recommendation M.3010. Functionally, TMN is the means to transport and process information that relates to the management of telecommunication networks. It includes: the Operations Systems Function (OSF) blocks, the Mediation Function (MF) blocks, and the Data Communication Function (DCF) blocks.

The operations systems function (OSF) block performs the operations system functions. The mediation function (MF) block routes and/or acts on information that flows over the TMN q interfaces and may threshold, buffer, store, filter, condense, adapt, or process the information. The data communication function

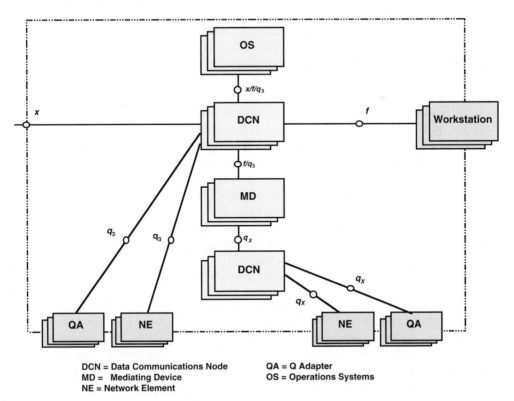

DCN = Data Communications Node   QA = Q Adapter
MD = Mediating Device   OS = Operations Systems
NE = Network Element

Fig. 11.5 TMN Physical Architecture

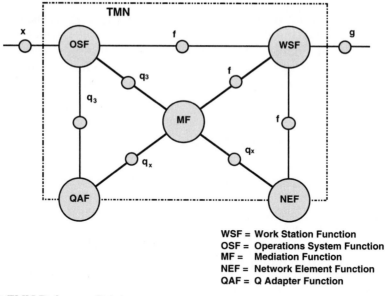

WSF = Work Station Function
OSF = Operations System Function
MF =   Mediation Function
NEF = Network Element Function
QAF = Q Adapter Function

Fig. 11.6 TMN Reference Points

(DCF) block provides information transport mechanisms, including routing and relaying functions. The data communications function is a means to transport information related to telecommunications management between management function blocks. Data communications are provided by layers 1 through 3 of the OSI reference model.

TMN also includes the Network Element Function (NEF) blocks, Q Adapter Function (QAF) blocks, and Workstation Function (WSF) blocks to support the management function. The Q Adapter (QA) includes the device that connects network elements or operations systems with non-TMN-compatible interface to a TMN q-type interface. The workstation system performs Work Station Functions (WSFs), translating information at an "f" reference point to a displayable format at a "g" reference point (see Figure 11.6).

The interconnections of network elements, operations systems, work stations, and mediation devices via a data communications network are provided by standard interfaces to insure interoperability of systems interconnected to accomplish the given OAM&P functions. The TMN standard interfaces are:

1. q3-Interface: This interface supports the full set of OAM&P functions between operations systems and network elements and between operations systems and mediation devices. The q3-Interface may also be capable of supporting the sets of OAM&P functions between operations systems.

2. q_x-Interface: This interface supports a full set or a subset of OAM&P functions between mediation devices and network elements, between mediation devices, and between a network element with mediation function and another network element.

3. x-Interface: This interface supports the set of OS-to-OS functions between TMNs or between a TMN and the x-Interface on any other type of management network.

4. f-Interface: This interface supports the set of functions for connecting workstations to operations systems, mediation devices, or network elements through a data communication network.

5. g-Interface: This interface is located outside of the TMN and is between users and the WSF.

6. m-Interface: This interface is located outside the TMN and is between QAFs and non-TMN-managed entities.

11.6 OSI MANAGEMENT

The TMN information architecture uses the OSI management principles. OSI management deals with a management protocol, a set of templates to specify management information, and a toolbox of management capabilities that can be used over an interface. The protocol defined within OSI management, called Common Management Information Protocol (CMIP), is an object-oriented protocol that sends messages about objects.

OSI management communications requires connection-oriented transport and relies on the OSI application layer environment. Managers and agents are the peer applications that use the services of a Common Management Information Service Element (CMISE) to exchange managed information. CMISE provides Service Access Points (SAPs) to support controlled association between managers and agents. Associations are used to exchange managed information queries and responses, to handle event notifications, and to provide remote invocations of managed object operations. To support these services, CMISE utilizes the services of OSI's Association Control Service Element (ACSE) and the Remote Operations Service Element (ROSE).

OSI management is based on object-oriented design principles defined in CCITT Recommendation X.700-series and the corresponding ISO standards (ISO 10040 and ISO 10165-4). The key aspects within OSI Management are: managed objects, manager/agent, management function, and CMISE/ROSE. Figure 11.7 shows the OSI layers with CMISE/ROSE. We will now discuss each of these aspects.

11.6.1 Managed Objects

Managed objects are conceptual views of the resources that are being managed or may be used to support certain management functions. The managed object may also represent a relationship between resources (e.g., a path) or combination of resources (e.g., a network). PCS-managed objects include entities such as "transceivers" or "handoff algorithms." Typical messages that use Common Management Information Protocol (CMIP) might include such operations as

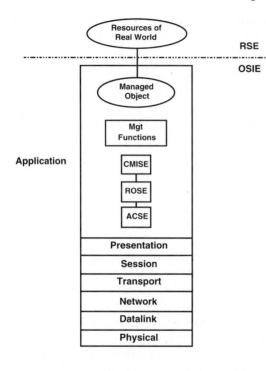

Application

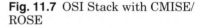

Fig. 11.7 OSI Stack with CMISE/ROSE

"request transceiver A to perform a diagnostic test B" or "set parameter X of handoff algorithm Y to value Z." Managed object definitions are a key aspect of the specification. Object oriented principles used to describe managed objects classes allow hierarchical modeling, inheritance, containment, and so on. A managed object is defined by: superclasses of the managed object; visible attributes of the object (data that can be read or written); management operations that can be applied to an managed object (i.e., requests that can be made to be a object); behavior shown by an managed object in response to management operations or in reaction to other types of stimuli—these can be internal (e.g., threshold crossing) or external (e.g., interaction with other objects); notification emitted by the managed object (alert that an object can send to inform the managing system of some event).

These characteristics (except for the superclasses) are collected in modular packages. Managed objects are modeled as collections of packages. Packages may be mandatory or conditional. Mandatory packages contain the characteristics that are essential and must be possessed by every instance that is a member of that class. Conditional packages are nonessential and may be present in some class members, but not all. The conditional packages are initiated at the same time as the object is created.

Package information is specified using templates that are in machine-readable form and can be directly compiled by the machine. Using this specification, the message can be generated automatically with commercially available software. The specification of the managed objects are called an "information model."

11.6.1.1 Attributes   Attributes are named characteristics of an managed object. A value is associated with each attribute of an managed object. Such values may be read or modified through internal system or network activity or through management actions. The modifications that may be performed on such attributes are subject to the internal constraints imposed by object and attribute behavior.

Attributes may either be implemented as stored values that can be read or modified, or they may be interpreted as being references to methods that will return values in response to a GET statement and will put the computational process in the right state in response to a SET statement.

11.6.1.2 Behavior   The behavior of an managed object specifies dynamic characteristics of an object and its attributes, notifications, and actions. Behavior includes the semantics of attributes and describes the way in which management operations affect the object and its attributes. It also describes the internal events that occur and result in observable changes in the object. This includes conditions under which attribute values may change and under which notifications may be emitted by the object.

11.6.1.3 Inheritance   Object-oriented modeling allows specialization of one managed object class from another. A class, derived from another class, is called a subclass of that class; the parent class is referred to as its superclass. A subclass may be specialized from a superclass by modifying the characteristics of the superclass. A class derived by means of specialization is said to inherit the characteristics of its superclass. Managed objects must obey the inheritance and compatibility rules. The relationship between subclasses and superclasses results in a hierarchy referred to as the "inheritance tree."

For example, inheritance can be used to allow the derivation of a "CDMA transceiver" managed object from a generic "transceiver" managed object. The CDMA transceiver object inherits all the attributes, actions, and notifications from the generic transceiver and adds new information relevant only to CDMA transceivers.

This subclassing mechanism allows an easy way to extend generic managed objects to meet the needs of a specific technology. This mechanism can also be used to derive vendor-specific managed objects from generic objects.

11.6.1.4 Allomorphism   Allomorphism, a concept similar to polymorphism in object-oriented programming languages, allows one managed object to masquerade as a more basic managed object. This provides a way to achieve backward compatibility over the interface, as a newer version of a managed object can present itself as the older version of the managed object.

11.6.1.5 Manager/Agent Because the environment being managed is distributed, network management is a distributed application involving the exchange of management information between management processes for monitoring and controlling the various physical and logical network resources. For management associations OSI management uses the manager/agent concept.

A managing system acting as a manager issues management operation requests and receives notifications from the system being managed, which plays the agent role. The agent is responsible for maintaining the managed objects. The agent provides selection functions to locate the managed object records accessed by the GET/SET/ACTION SAPs of CMISE. The agent also provides event detection and forwarding notifications to managing entities enrolled (through management information tree records) to receive them. A CMISE entity provides service access points to support communications with the agent. It dispatches/receives CMIP Protocol Data Units (PDUs) to/from other service elements such as ACSE and ROSE. These protocol data units are exchanged through a connection-oriented transport. CMIP protocol data units can be viewed as carriers of requests and replies generated by respective CMISE primitives. For example, a CMISE M-GET accessed by a manager generates a CMIP GET-REQUEST protocol data unit to the agent and respective GET/SET-RESPONSE, protocol data units from the agent.

11.6.1.6 CMISE/ROSE The management exchanges between manager and agent take place through the use of CMISE that provides a set of (seven) generic services that a manager can use to create or delete managed objects, GET/SET their attributes, or invoke actions on them. The services defined for CMISE in ISO/IEC 9595 are: M-GET, M-SET, M-CREATE, M-DELETE, M-ACTION, M-EVENT-REPORT, and M-CAN-CELGET. The agent, after servicing these requests, issues responses through CMISE. CMISE is also used by the agent to spontaneously emit notifications. CMISE is merely a vehicle to carry management information and operations. The actual syntax/semantics of the information is specific to the managed object and/or management function and transparent to CMISE. CMISE uses the services of ROSE to carry its protocol data units. ROSE provides general transaction-oriented services.

11.6.1.7 Management Functions Management functions in general provide functionality to add value beyond the services available from CMISE. These functions are generic in nature and apply to many different managed objects.

We illustrate them through some examples. The object management function defines services to report creation and deletion and changes in attribute values of managed objects. The event report management function allows a manager to control the transmission of event reports from managed objects independently from the definition of these objects—e.g., if an event report is generated within a system, the forwarding mechanism specifies to which destination the event should be routed. The alarm reporting function supports an alarm notification service. The log control function allows a manager to control the repository of event reports generated within the agent. For example, events coming from a specific managed object may be stored in a log and queried later.

11.7 PCS INFORMATION MODEL

It is beyond the scope of this book to discuss information models, managed objects and other details in a great depth. If you are interested in these topics, we suggest you consult references [5] and [6] for more details. We briefly introduce the concept of an information model.

The information model specifies what OAM&P information can be exchanged between the OS and the managed nodes. The information models are shown in Figure 11.8, Figure 11.9, and Figure 11.10, using modified entity-relationship (ER) notations. The boxes represent managed objects defined in the model, and the diamonds represent relationships between objects. These relationships give much of the structure of the model.

Figure 11.8 gives an overview of the overall information model. The information model contains many managed objects "borrowed" from other TMN standards. The model also contains managed objects to represent hardware and software that will vary from vendor to vendor. These managed objects are therefore modeled separately from the functionality they provide (which should be common across vendors). In the ER diagrams (Figures 11.9 and 11.10), functional objects are represented by boxes with rounded corners.

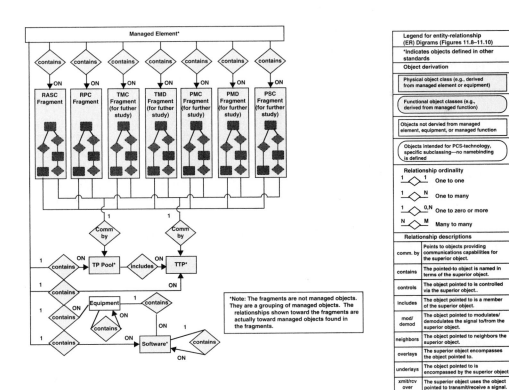

Fig. 11.8 Overview of PCS Information Model Fragments

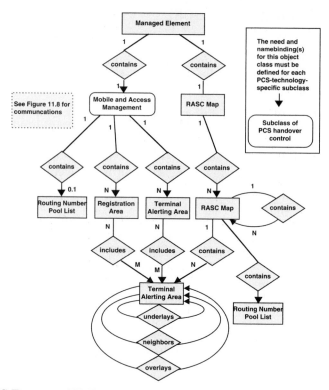

Fig. 11.9 RASC Fragment ER Diagram

The various support objects are not shown in the ER diagrams. Support objects are managed objects that do not represent PCS functionality or resources, but are part of the management infrastructure. These include objects such as scanners (report generators), event-forwarding discriminators (alarm routing), or current data (set of performance counters). The support objects are also borrowed from other standards.

The information model is generally broken into "fragments" to reduce its complexity. Figures 11.9 and 11.10 show fragments of the Radio Access System controller (RASC) and Radio Port Controller (RPC). The RASC fragment contains information needed for location management. The RPC fragment contains information required to manage the wireless access system.

11.8 SUMMARY

In this chapter we have introduced the concepts of managing a telecommunications network in general and a PCS network specifically. We described the goals for PCS network management where new services and technologies place increasing demands on operations systems. We then reviewed the standard man-

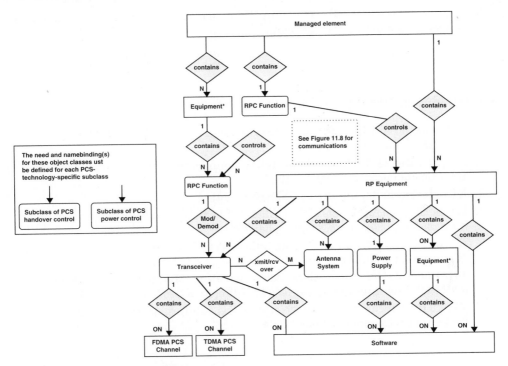

Fig. 11.10 RPC Fragment ER Diagram

agement categories for all telecommunications networks (accounting, configuration, fault, performance, planning, and security) and described how they are performed for PCS. Since standards are important for any service provider operating in a mixed-equipment and multiple-interacting provider environment, we then reviewed work on the Telecommunications Management Network standard. We also described the OSI management model that forms the foundation for TMN. Finally we gave a brief introduction to the PCS information model to show how material in this chapter can be used to build operations systems. For more information, consult the references at the end of this chapter.

11.9 PROBLEMS

1. What are the goals for PCS network management?

2. What are the layers of the TMN management architecture? Specify the functions of these layers.

3. What are the main functional blocks in the TMN physical architecture? Specify their roles.

4. Name the elements of TMN functional architecture. Specify briefly the role of each element.

5. What services of CMISE are used in OSI management? Name the application layer protocols used in OSI management.

6. What is an information model? Discuss its role in PCS network management.

11.10 REFERENCES

1. Aidarous, S., and T. P. Levyak, eds., *Telecommunications Network Management into the 21st Century,* IEEE Press, 1994.

2. EIA/TIA IS-124, "Cellular Radio Telecommunications Intersystem Non-signaling Data Communications (DMH)," 1994.

3. ETSI Technical Specifications, GSM 12 Series.

4. Hayes, Stephen, "A Standard for OAM&P of PCS Systems," *IEEE Personnel Communications Magazine,* 1, no. 4 (December 1994).

5. Klerer, M., "System Management Information Modeling" *IEEE Communications Magazine* 4, no. 5 (May 1993).

6. T1M1.5/94-001R2, "Operations, Administration, Maintenance, and Provisioning (OAM&P) Interfaces Standards for PCS," Proposed Draft Standard, 1994.

7. T1P1 TR-34 Technical Report, "Network Capabilities, Architecture and Interfaces for Personnel Communications," 1994.

Interworking in Wireless Systems

12.1 INTRODUCTION

The reference architectures for U.S.-based and European-based systems (see Figures 8.1, 8.2, and 9.1) show an interworking function for PCS systems. This chapter discusses the needs of that interworking function plus other interworking needs so that users of wireless PCS systems can place and receive calls to other networks, both wireless and wireline.

For a wireless system to have full access to the wireline network and other wireless systems, it must interwork in the following areas:

- ☞ Speech coding
- ☞ Data transmission
- ☞ Signaling, numbering and routing
- ☞ Authentication and security
- ☞ Basic services
- ☞ Supplementary services
- ☞ Roaming between similar systems
- ☞ Roaming between dissimilar systems
- ☞ Emergency calling
- ☞ Billing

We will first discuss interworking of speech, data signaling, and security, since without them no further capabilities are possible. Then we will discuss interworking for services (basic and supplementary) and for roaming.

The various standards bodies have identified solutions for speech coding interworking and signaling systems interworking. Some of the systems can interwork for some data services and for some authentication and security services.

The interworking functions do not cleanly segment themselves into the sections of this chapter. Interworking affects all aspects of the PCS system and cannot be accomplished in isolation. Thus, there is some overlap in the material presented in the sections of this chapter. We have attempted to minimize the overlap, but unfortunately it is not possible to avoid it completely.

As we described in previous chapters, while Europe is in the process of implementing one Pan-European system—GSM, the United States plans to implement multiple systems. This chapter will focus on the problems and solutions to interworking between systems that are deployed in the United States. Within the United States there are two services-oriented MAPs—one for PCS1900 (GSM) and one for IS-41-based systems. There are also in excess of eight air interface specifications that will see deployment in the United States (AMPS, CDMA, DECT, PACS, PCS1900, PCS2000, TDMA, and W-CDMA). This chapter will discuss the key issues when PSs place or receive off-network calls and when they roam into other networks. The issues are complex and may not be fully resolved until the end of the decade.

12.2 SPEECH CODING INTERWORKING

In chapters 8 and 9, the speech coding algorithms for each of the air interfaces was described. Except for some older analog switches, the phone network transmits speech over digital links at 56/64 kbs with Pulse Code Modulation (PCM). The analog speech is digitized in either the telephone set or the switching equipment (central office or PBX). Two speech coding rules are used for the PCM—μ-law and A-law. The North American-based systems use μ-law and the European systems use A-law. The 64-kbs channels are then aggregated into 24 channels with a rate 1.544 Mbs or 32 channels with a rate of 2.048 Mbs. North America uses 24 channels in a T1 system and Europe uses 32 channels. When a system is used in a country, then the PCS switch must support the correct version of speech coding and the correct T1 data rate for that particular country. Only the PCS2000 air interface supports PCM as an option. None of the other air interfaces use 64-kbs PCM, since frequency spectrum is a valuable and expensive resource. Various rates from 8–32 kbs are used with various speech coding algorithms (see Table 12.1). When a user on a PCS system makes a voice call to another user on the PCS system, both PSs will use the same speech coding algorithm, and no speech coding interworking is needed. Some, but not all, systems may convert to PCM at the base station anyway, for uniformity.

When a PCS system user makes a voice call to the wireline network, the voice coding system must be converted to 64-kbs PCM. When a PCS system user

Table 12.1 Speech Coding Used in PCS

System	Coding Type	Coding Rate (kbs)
CDMA	QCELP	14.4
European Wireline	PCM (A-law)	64
GSM	RPE-LTP	13
Japanese	VSELP	11.2
North American Wireline	PCM (μ-law)	64
PACS	ADPCM	32
PCS2000	ADPCM	32
	PCM	64
	PCS HCA	8
TDMA	VSELP	8
W-CDMA	ADPCM	32

makes a voice call to another PCS system, the speech will be converted to 64-kbs PCM for transmission over standard transmission facilities.

When two PCS systems use the same speech coding system and have a significant amount of traffic, then the efficiency of the transmission facilities can be improved. For example, if two PCS systems that use 32-kbs ADPCM communicate, then each 64-kbs transmission facility between the switches can carry two voice calls rather than one.

12.3 DATA INTERWORKING

Low-speed data is transmitted over the telephone network today using voice-band modems that range from 1,200 baud to 28,800 baud. Some of the uses for this data are:

☞ Accessing electronic mail
☞ Accessing remote computers
☞ Transferring files
☞ Transmitting facsimiles
☞ Making transaction services (e.g., credit card validation)

Transmission at higher rates is typically used for video conferencing, mainframe to mainframe communications, and other uses that would not initially be carried over a wireless mobile link.

Except for those PCS systems supporting PCM or ADPCM, the speech coding systems for transmission of voice have been optimized for the voice transmission and have not been optimized for the transmission of nonvoice signals such as voice-band modems. Therefore, both the PS and the PCS system must have interworking capabilities to provide the wide range of services currently using voice-band modems. Over the air interface, the data must be transmitted digitally since that is the only option available. If the air interface supports a data rate higher than the basic data rate of the voice-band modem, then interworking is possible. Interworking is not possible if the required data rate is higher than the needed data rate. For example, a raw date rate of 28.8 kbs is not possible over a 16-kbs data link. Those systems supporting PCM or ADPCM can interwork with voice-band modems without a data interworking platform. However, error performance may suffer because of the high error rate of the radio channel.

For the remainder of this section, we will focus on those systems using speech coding systems not capable of supporting voice-band data and show the interworking functions needed to support data. The interworking function can support the data needs of the voice-band modem only when the air interface data rate is higher than the modem speed. Figure 12.1 shows one method for interworking by having a modem in the PS that communicates with a modem in the auxiliary device (fax machine, PC, credit card validation terminal, among others) via an industry-standard telephone jack (e.g., RJ-11). This method requires no modifications to the auxiliary device but does require a different modem in the PS for each data service. Figure 12.2 shows a second method where a digital port (e.g., RS-232) appears on the PS. With this method, the auxiliary device must have an option for bypass of the built-in modem and therefore a direct digital connection. It should be noted, however, that many devices (e.g., fax machines and transactions telephones) do not currently support a direct RS-232 connection in place of the internal modem.

Figure 12.3 shows the architecture of the PCS network side of the interworking function. In the network, a modem pool (see Figure 12.4) must be installed for each service that needs interworking.

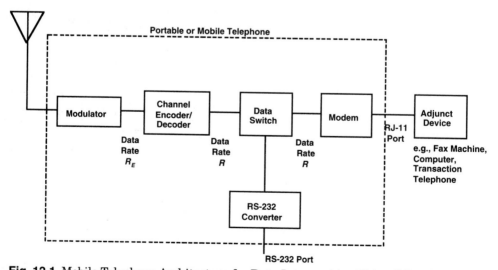

Fig. 12.1 Mobile Telephone Architecture for Data Interworking Using RJ-11 Port

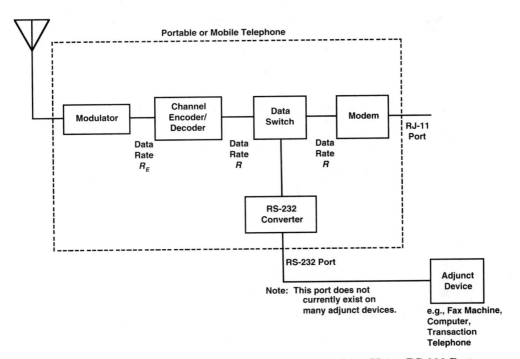

Fig. 12.2 Mobile Telephone Architecture for Data Interworking Using RS-232 Port

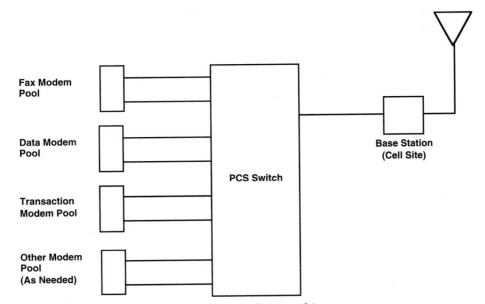

Fig. 12.3 PCS Network Architecture for Data Interworking

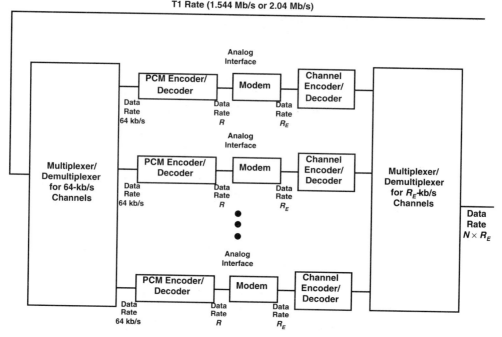

Fig. 12.4 Modem Pool for Data Interworking

The radio link has a high bit error rate, and the data transmission must account for these errors by using bit interleaving, error-correcting codes, and re-transmission of the data as necessary. Figure 12.5 shows these concepts. Bit interleaving is the process where data bits are not sent in the order in which they are generated but are placed in a buffer and permuted in a known pattern (see Figure 12.6 for an example). With bit interleaving, a burst of errors on the radio link will appear as isolated bit errors in the received data. Obviously, the buffer length and the distance between bits must be higher than the expected burst error length on the radio path. Error-correcting codes are used to identify and correct errors on data links. With error-correcting codes, spare transmission data is used to send additional data that tells the receiver if the received data is correct or in error. For more details of error-correcting codes, see appendix A and the references in that appendix. In summary, the data from the auxiliary device

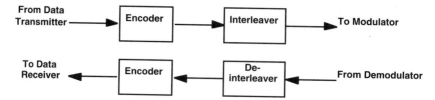

Fig. 12.5 Block Diagram of Encoder/Decoder

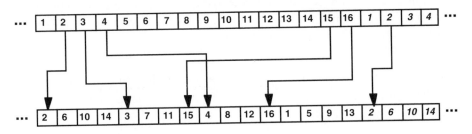

Fig. 12.6 Example of Bit Interleaving

is converted to a data rate for transmission over the air interface and then converted back to the form that the service normally uses. When the user wants to use one of the services, the network must be notified of the service request and the appropriate modem placed in the connection.

Table 12.2 describes the data interworking capabilities of each of the air interfaces used for cellular and PCS in North America.

Table 12.2 Supported Data Services

System	Data	FAX	Transactions
AMPS	1.2–28.8 kbs with modem adapter	with modem adapter	with modem adapter
CDMA	1.2–14.4 kbs	*	*
PACS	Voice-band data—0–4.8 kbs circuit data—28.8 kbs, packet data—192 kbs	with modem adapter	with modem adapter
PCS1900	300 bps–9,600 bps asynchronous and synchronous, X.25	CCITT Group 3	—
PCS2000	320 bps–256 kbs	*	*
TDMA	*	*	*
W-CDMA	voice band, 0–144 kbs, X.25	Group 3	with modem adapter

*This data service is not defined in the 1995 standards.

12.4 SIGNALING, NUMBERING, AND ROUTING INTERWORKING

Whenever the call must leave the PCS switch, there must be a signaling path to the destination switch so that a traffic (voice or data) path can be established between the two switches. The establishment of the correct traffic path is a routing function, and the communications needed between various switches to establish that traffic path is a signaling function. For signaling, the originating switch must support standard signaling interfaces.

Signaling between switches can use a variety of means:

☞ Dual-Tone Multifrequency (DTMF) appropriate for the national network
☞ Multifrequency (MF) tones appropriate for the national network
☞ Signaling System 7 (SS7)

Although some cellular systems support either DTMF or Multifrequency signaling, the current standard for intersystem communications is SS7. Two dialects of SS7 are used in the world: CCITT SS7 and ANSI SS7. ANSI SS7 is used in North and South America, Japan, and some other places; CCITT SS7 is used in Europe.

The PCS switch must support the dialect of SS7 for the area of the world where the system is deployed. If calls from one switch have destinations at locations where an incompatible dialect of SS7 exists, the international wireline gateway switch in the country where ANSI SS7 is used will perform the protocol conversion needed.

The signaling network is organized differently in different parts of the world. In North America, the United States in particular, the telephone network is privately owned rather than owned by the government as it is in most of the world. In the United States, there are local telephone companies and three major inter-exchange companies (AT&T, MCI and Sprint). Each of these companies maintains its own signaling network, and extensive controls are in place so that problems on one network cannot affect other networks. The signaling network is used extensively to route calls efficiently. Differences in the size of countries, ownership of the telephone network and the signaling network result in different philosophies in different parts of the world. Interworking between these systems is not just a protocol conversion problem, but must address issues of network stability, message priority, and use of signaling transfer points, among other things. The export of one signaling system to an area of the world where a different one is in use must be carefully done.

When a call originates by a PS, then the originating switch must perform an analysis of the dialed digits to determine the correct interworking functions to route the call and establish a talking path for voice calls or a data path for data calls.

In North America, the dialing plan is defined by the North American Number Plan (NANP) and uses 10 digits in the following form:

☞ Area code: 3 digits. The area code is also known as the Number Plan Area (NPA). Most NPAs are geographic and identify the area of North America where the telephone exists and are independent of the carrier that routes the call. Previously, NPAs had the form of N(0/1)X, where the first digit, N, is a digit from 2 to 9, the center digit can be only 0 or 1, and the last digit, X, is any digit. For example, 212 is New York City, 714 is Anaheim, California, and 702 is the state of Nevada in the United States. In 1995, the center-digit restriction was lifted, and all NPAs can have the form of NXX.

Changes have been introduced to support wireless users and users desiring nongeographic numbers. The 700 NPA is a nongeographic NPA, but is carrier dependent. Thus, when routing calls to numbers that start with 700, the carrier that owns the number must be identified since multiple carriers can use the same 700 number. Thus, 700–555–1212 would reach a different destination depending on the carrier. Other nongeographic numbers that will see use are 200 and 500 numbers.

☞ Central Office (CO) code: 3 digits. The CO code has the form NXX.

☞ Line number: 4 digits. The line number has the form of XXXX.

Dialing of telephone numbers consists of dialing a 7-digit number where the NPA is assumed to be the same as that of the calling telephone or, if the NPA is different from that of the calling telephone, by dialing 1+NPA+NXX+XXXX. The leading 1, for calls out of the current area code, informs the switch to do 10-digit analysis rather than 7-digit analysis to determine routing. The absence of the 1, for calls within the area, informs the switch to start analysis after 7 digits rather than waiting for the interdigit time-out; thus calls are routed quicker. It is the long-term goal of the NANP to have full 10-digit dialing on all calls.

In the United States, while almost any phone can reach almost any number, sometimes it is necessary to dial a carrier access code to correctly route the call. Often a customer of a telephone may wish to route the call over a carrier that is different from the one that is the default for the telephone. In that case, the carrier access code must be dialed. The most common area where this occurs is at public telephones. For those customers dialing numbers in the 700 NPA, the access code must be dialed to correctly route the call.

In other parts of the world, the telephone number is of variable length from 6 to 15 digits and consists of:

☞ Country code: 2 or 3 digits. The country code is in the form of NX or NXX depending on the country. For example: 39 is Italy, 81 is Japan, 91 is India, 44 is the United Kingdom, 967 is Yemen, and 213 is Algeria.

☞ City code: 0, 1, or 2 digits of the form X or XX. City codes are not unique but are country dependent. For example: Tokyo is 3; London is 44; Reykjavik, Iceland, Dublin, Ireland, and Baghdad, Iraq, are all 1; Algeria does not have a city code.

☞ Local telephone number: 4 to 7 digits of the form XXXX to XXXXXXX.

Calls to different regions of the world are identified by an outside-of-the-region prefix. In North America, the prefix is 011 (for "sent-paid" calls).

Thus, digit analysis for routing of calls is a complex task and requires extensive entries in a routing database. Each switch has a database of routing tables. A query of the database is made with the smallest number of digits needed to determine a route to the correct switch. When new switches are added to a network, the routing is handled by additional entries to the database. This is a standard part of the provisioning process and is covered in chapter 11 on OAM&P. No new switch software needs to be developed to change routing tables.

The database either can be associated with the switch or can be remote from the switch and installed in a Switching Control Point (SCP). Switches communicate with SCPs via the SS7 network. Often, a switch can do only partial routing and must pass the routing information off to another switch or SCP to do further routing. This may occur two or three times until the proper route is found.

When a call must be routed, the first step is to determine if the number is a geographic or nongeographic number or if the determination cannot be made. Then the following routing conditions can occur:

1. **Geographic number:** This is the case of a call to a wireline user or a wireless user where no (mobility) routing information can be determined from the number.

   ✘ The destination is a wireline destination in the local serving area, and a connection to a local exchange carrier must be established,

   ✘ The destination is to a wireline destination outside of the local serving area, and a connection to an interexchange carrier must be established.

2. **Nongeographic number:** These numbers can be 700 or 800 numbers where the call must be routed to the correct carrier or 500 numbers of wireless or wireline users who report their location and need calls routed to them as they move around the world (or a country). The following routing conditions exist:

   ✘ A database query to the local database determines that the call is to a wireless user (either home or roaming) that is currently reporting a location served by the same switch and thus no interworking is needed.

   ✘ A database query is made to the wireless user's HLR and the HLR returns routing information to route the call to the correct PCS switch where the user is currently reporting his or her location. See the sections on roaming services and registration services for more details.

   ✘ A database query is made at the wireline user's HLR, and the HLR returns routing information to route the call to the correct wireline or wireless switch where the user is currently reporting his or her location. A directory number where the user can be found must be returned.

   ✘ The number can be routed only by another carrier (e.g., 700 or 800 numbers), and the call must be routed to that carrier as the first step in a multistep routing process.

☞ Unknown routing information: The switch cannot determine if the destination is a wireless or wireline user and must therefore route the call to either a local or interexchange carrier according to the methods in step 1 above.

We covered call flows for geographic and nongeographic routing in more detail in chapter 8.

12.5 SECURITY AND AUTHENTICATION INTERWORKING

A complete description of the security needs for a wireless system is discussed in chapter 10, including the methods in use for each of the various air interfaces. This section will examine the issues related to interworking of the various security systems.

When a user enters a new system and requests service, the service provider has three methods of operation. The provider can deny service to the user; unfortunately, this approach does not permit the provider to make money. The second method is to give service for free to all users; unfortunately, this approach also does not permit the service provider to make money. The third method is to obtain information about the subscriber's profile from the home system and determine a method to authenticate the user. If the service provider fails to obtain the profile and authentication information before giving service, the provider is operating using the second method (i.e., free service mode) by default.

In chapter 10 we discussed the four security mechanisms that have been proposed for wireless systems.

1. MIN/ESN systems
2. SSD systems
3. Token-based secret key systems
4. Public key systems

Each of these systems has different needs for communicating information between systems to properly perform security functions for roaming users.

As we described in chapter 10, the MIN/ESN system has significant security flaws and will not be discussed further. Since the public key system is for future study, we will not consider it here. The remainder of this section will discuss issues related to interworking between SSD and token-based systems.

When a PS registers with a system, security information must be obtained from the user's HLR and populated in the current system's VLR. When the home system and the visited system use the same security mechanism, then the interworking proceeds smoothly and is described in chapter 10 under the registration call flows for both systems.

However, when the visited and the home systems have different security mechanisms, a more demanding interworking function is needed. The following sections describe the operation of the PS and the PCS network when various assumptions are made about the support for different authentication procedures.

For the following discussions, we will assume that the PS has hardware and software to support multiple air interfaces and that the service provider in the home system has populated the PS with the proper authentication information.

12.5.1 PS Support for One Authentication Method

If no changes are made in system designs, each PS will support one authentication method—the one in use in its home system. Thus, when the PS roams into a visited system, it can only interwork with authentication systems identical to its home system. If the authentication systems are different, service may not be possible. For example, let a PS whose home system uses SSD roam into a system that uses token-based authentication. The following steps occur:

1. The PS listens for the global challenge (RAND) and does not find it.
2. If the PS supports unauthenticated registration messages, then it sends an unauthenticated registration. If not, then no service is possible!
3. The network receives the unauthenticated registration message and issues a specific challenge (RAND) to authenticate the PS.
4. At the same time, the PCS network sends a message to the user's HLR requesting N sets of triplets (challenge, response, voice privacy key).
5. The PS returns with its response to the challenge, AUTHR.
6. The HLR not supporting token-based system either returns a "protocol denied" message with reason code, not supported, or the HLR returns the SSD so that the VLR can calculate its own triplets.
7. The VLR does not support calculation of its own triplets, so it issues a "registration denied" message to the PS.
8. The PS activates its "No Service" indicator, and the user does not have service.

For the case where the user's home system supports token-based authentication and the visited system supports SSD, the following steps occur.

1. The PS sends a registration message to the network.
2. The network receives the unauthenticated registration message and issues a specific challenge (RAND) to authenticate the PS.
3. At the same time, the PCS network sends a message to the user's HLR requesting the SSD.
4. The PS returns with its response to the challenge, AUTHR.
5. The HLR not supporting SSD either returns a "protocol denied" message with reason code, not supported, or the HLR returns a set of N-triplets.
6. The VLR does not know how to calculate the SSD from the triplets (because good security codes make that nearly impossible), so it issues a "registration denied" message to the PS.
7. The PS activates its "No Service" indicator, and the user does not have service.

In summary, systems supporting SSD will return an authentication key, and the visited system will use that information to challenge and authenticate the visiting PS. However, if the home system supports token-based secret data, a request for an authentication key will result in a "protocol not supported" response, and the visited system then must deny service to the user. An alternative to denying service is to give service (with reduced capabilities) on a "trust" basis without authenticating the visiting PS. Clearly, in those systems where fraud is known to be high, giving service on a "trust" basis will lead to even higher fraud and is not an acceptable business practice. If the visited system uses token-based secret data and a query is made to a home system that supports SSD or public key data, the same result will occur.

12.5.2 PS Support for Multiple Authentication Methods

Let us look at the case where the PS supports all three authentication mechanisms. This is the case that was examined by T1P1 with the standardization of a smart card as the authentication storage mechanism for the PS. Let us look at the case from the previous section where a PS whose home system supports SSD roams into a system that supports token-based authentication. The following steps occur:

1. The PS listens on the control channel and determines the method of authentication used in the visited system (token-based).
2. The PS sends an unauthenticated registration message to the system.
3. The network receives the unauthenticated registration message and issues a specific challenge (RAND) to authenticate the PS.
4. At the same time, the PCS network sends a message to the user's HLR requesting N sets of triplets (challenge, response, voice privacy key).
5. The PS returns with its response to the challenge, AUTHR.
6. The HLR not supporting the token-based system either returns a "protocol denied" message with reason code, "not supported," or the HLR returns the SSD so that the VLR can calculate its own triplets.
7. The VLR does not support calculation of its own triplets, so it issues a "registration denied" message to the PS.
8. The PS activates its "No Service" indicator, and the user does not have service.

Thus, even though the PS supports the "correct" authentication method for the home system, that is not sufficient to obtain service. Similar protocol denials occur for any other pairs of dissimilar authentication methods. Thus, for support of roaming PSs, a network interworking function is needed.

12.6 BASIC SERVICES INTERWORKING

In chapter 8, we discussed the following basic services defined by standards committee T1P1:

- ☞ Automatic registration
- ☞ Terminal authentication and privacy (using private key cryptography)
- ☞ Terminal authentication and privacy (using public key cryptography)
- ☞ User authentication and validation
- ☞ Automatic personal registration
- ☞ Automatic personal deregistration

☞ Personal registration

☞ Personal deregistration

☞ Call origination

☞ Call delivery

☞ Call clearing

☞ Handoff

☞ Emergency (E911) calls

☞ Data acquisition

☞ Roaming

The registration and authentication services have been discussed in the section on security and authentication, and the data services have been discussed in the data services section. Emergency calling and roaming will be handled later in this chapter. Call origination, call delivery, and call termination will be discussed in this section.

12.6.1 Call Origination, Delivery, and Clearing

When a call is originated by a PS, if the destination is on the same switch as the PS, then no interworking is needed. If the call must leave the switch, then the PCS switch must interwork with other networks in order to place the call.

Whenever the call must leave the switch, there must be a signaling path and a voice or data path. The voice and data interworking are discussed in other sections. For signaling, the originating switch must support standard signaling interfaces. Although a variety of signaling systems exist, the current standard is SS7. Two dialects of SS7 are used in the world: CCITT SS7 and ANSI SS7. ANSI SS7 is used in North and South America, Japan, and some other countries; CCITT SS7 is used in Europe.

The PCS switch must support the dialect of SS7 for the area of the world where the system is deployed. If calls from one switch have destinations at locations where an incompatible dialect of SS7 exists, the international gateway switch in the country where ANSI SS7 is used will perform the protocol conversion needed for the signaling for the call.

The issues of dialing plans and signaling interworking have been discussed in section 12.4.

The last step for basic services interworking is for the service provider to assign to each PS a directory number that is consistent with the dialing plan for its home area of the world. The databases in the network must then be updated with routing information to the PCS switch. With those steps, the system can interwork for basic services. We have described call flows in chapters 8 and 9.

12.6.2 Supplementary Services

Supplementary services are often considered as part of the ISDN but can be defined for any communications service. GSM defines one set of services, and the U.S.-based systems define a different set. Some of these services have similar names in each system; unfortunately they may operate differently in each system. The PCS1900 MAP is evolving to support the same operation as the IS-41 MAP. Some of these services require interworking with other networks. Some are completely defined within their own networks. Table 12.3 compares the various supplementary services, the network that supports them, and the degree of interworking needed. To some extent, all supplementary services require off-network interworking since the user's service profile must be obtained from the home system. For Table 12.3, we fill in the box for off-network interworking only if additional signaling between systems is necessary. We have described each of these services in chapters 8 and 9; consult those chapters for an explanation of the services.

Table 12.3 Supplementary Services Interworking

Service	PCS1900 MAP	IS-41 MAP	Off-Network Interworking	On-Network Interworking
Advice of charge information	•		•	
Advice of charge charging	•		•	
Automatic recall		•	•	
Automatic reverse charging		•	•	
Barring of calls	•		•	•
Call hold and retrieve	•	•		•
Call forwarding:				
Unconditional	•	•	•	•
On PS busy	•	•	•	•
On PS no reply	•	•	•	•
On PS not reachable	•			
Call transfer	•	•	•	•
Call waiting	•	•		•
Calling line ID presentation	•	•	•	
Calling line ID restriction	•	•	•	
Closed user group	•		•	•
Conference calling	•	•	•	
Connected line ID presentation	•			
Connected line ID restriction	•			
Do not disturb	•	•	•	
Flexible alerting		•	•	
Freephone	•			

Table 12.3 Supplementary Services Interworking (Continued)

Service	PCS1900 MAP	IS-41 MAP	Off-Network Interworking	On-Network Interworking
Malicious call identification	•		•	
Message waiting notification	•	•		•
Mobile access hunting	•	•	•	•
Multilevel precendence and preemption		•	•	•
Password call acceptance		•		•
Preferred language service	•	•		•
Priority access and channel assignment		•		•
Remote feature call	•	•		•
Reverse charging	•	•	•	
Selective call acceptance	•	•	•	•
Short message service	•	•	•	•
Subscriber PIN access	•	•		
Subscriber PIN intercept		•		•
Three-way calling	•	•		•
Voice mail retrieval	•	•		•
Voice privacy (required service in the United States for PCS)	•	•		•

12.7 ROAMING BETWEEN SIMILAR SYSTEMS

A vital feature for all cellular and PCS systems is the ability to make and receive calls in places other than the home system. From the earliest days of cellular to today, the roaming feature has been a "basic" capability of wireless phones. U.S.-based cellular systems have used a common analog air interface so that any phone could be used in any system. In Europe, different standards were in use in different countries, and it was only the invention of the digital GSM system that brought roaming capabilities to Europe. As PCS is deployed, with its multiplicity of standards, roaming again may not be possible unless efforts are taken to solve many critical issues.

Two aspects of a system are critical when roaming: the air interface protocol and the MAP. When a user roams into a system that supports the same air interface and same MAP, services should work seamlessly, assuming that the designers of the system have done their job correctly. For the purposes of our discussion in this chapter, similar systems are those that support the same MAP but possibly different air interface protocols. Although there are problems, this type of roaming can be solved by assuming that PS manufacturers will develop telephones that support multiple air interface protocols. That such phones exist for dual-mode analog and TDMA or CDMA protocols indicates that the problem can be solved for different digital systems as well. Since the MAP is the same, the user will see services work the same in all systems. With a properly designed PS, the issue of different air interfaces will be completely invisible to the user.

Two key problems must be solved for a dual-mode (or multiple-mode) phone—initial system selection and handoffs. When a phone is first turned on, it goes through a sequence of events that initializes the phone and registers it with a system. A multimode phone must have a set of procedures that will enable it to try frequencies and protocols for two or more air interfaces. It may be necessary for all systems to adopt a common control channel procedure that is used throughout a region, a country, or the world. Alternately, the phone can try multiple sets of frequencies and protocols until it starts decoding information. The start-up sequence may be lengthy since the phone does not know what system it is in when first powered up. The user of the phone may help the phone by giving information about location, but, in general, the phone should do its own analysis. After the phone has registered on a system, the user can place or receive calls as described elsewhere.

After a call is established, it may be necessary to handoff the call to another cell site. All air interfaces support handoff; the issues of handoff between cell sites on the same and different systems are well understood and have been discussed elsewhere in this book. Handoffs between systems essentially require transmission and signaling facilities between the switches of each system. Other than the extra facilities, the handoff between two cell sites on two different systems does not differ from the handoff between cell sites on the same system.

However, when the air interfaces differ between the two cell sites, then the problem is more complicated. All of the digital systems assume that the data receiver can quickly regain synchronization when a call it handed off to another cell. In particular, when CDMA is used, the receiver in the PS receives both cell sites simultaneously and uses signals from both to produce a composite signal. When the handoff occurs between two different air interfaces, new handoff messages will be necessary so that the PS can be informed about the characteristics of the new systems. A second receiver may also be necessary so that the PS can be listening to the new cell before the handoff occurs. Otherwise the handoff delay may be excessive, and calls may be dropped.

12.8 ROAMING BETWEEN DISSIMILAR SYSTEMS

When the MAPs in the two systems are different, then additional problems occur for roaming. The GSM MAP and the IS-41 MAP use different authentication protocols, as described in section 12.5 of this chapter. If the differences between authentication are not solved, then a user of one system will not be able to roam into another system, even if the phone supports both air interfaces and MAPs. For the remainder of this section, we will assume that the security interworking has been solved (although as of this writing it has not been) and examine the other interworking issues.

When dissimilar MAPs are involved, the PS needs software, user interface buttons, and displays that are different for each system. Some may be common (dial pad, SEND, END), but others will be different. The services will work differently in the different systems, and some services may be available in one system and not in the other. Thus, a user may find that a service that is available in the home system may or may not be available in a visited system and may work differently. For example, flexible alerting is supported in the IS-41 MAP but not in the GSM MAP, and a user that frequently roams into multiple cities wants to be alerted in those cities and at home and the office. Unfortunately, if one of the cities supports only the GSM MAP, calls will be missed. Similar disconnects occur from GSM services to IS-41 services.

Once a call is established, handoff becomes even more problematic since the request service may work differently in different systems or not a all.

About the best that can be said for roaming between different MAPs is that, if the security problems are solved, basic call origination and termination services will be available for voice calls. Other capabilities may not be available until well into the next decade.

12.9 EMERGENCY CALLING INTERWORKING

Various countries have established short dialing sequences to reach emergency service personnel (police, fire, first aid). In the United Kingdom, the number 999 was established and quickly spread to other countries. In the United States, 911 is the emergency number, and children are taught to dial the number as soon as they can speak and perform basic reasoning. A key feature of emergency calling is the ability to identify the location (using databases) of the telephone from which the emergency call is placed and to return the call if it is dropped. This enables incapacitated people to place a call, even if they can't speak or in any other way indicate their location. It also enables the emergency service people to confirm (by recalling the number) the emergency if the call is ended before information is received verbally. Wireless phones, with numbers unrelated to their location, have severely complicated emergency call processing.

In the United States, the National Emergency Number Association (NENA) and the Association of Public Safety Communication Officers (APCO) have lobbied for the same capabilities in wireless phones as is currently available in wired phones.

When a multiair interface PS is in use, the additional features of location and call-back are made more complicated. Radio location of a PS may require that all cell sites on the borders of systems support location receivers with multiple air interfaces so that triangulation across multiple cell sites can be done.

When a PCS call ends and the PS locks on a new system, it may take several seconds before all of the HLR and VLR data is updated. If an emergency call ends and the PSAP is trying to reconnect the call, the PS may be paged in the wrong system. A solution to problem must be found that works when the PS is in a new system with the same or differing air interfaces and MAPs. Since the problem of emergency calling from a single air interface phone places demands on the system that exceed current technology capabilities, seamless emergency calling across different systems may not be possible.

12.10 BILLING INTERWORKING

Wireline telephones support billing capabilities by storing the billing data in a switch from the beginning to the end of a call and then, when the call is completed, writing the data to tape (old method) or to a billing data system. When all calls start and end on the same switch (this is the way wireline calls work), then it is possible to keep partial call records in a switch and write the data at the end of a call. When a wireless call starts, remains, and ends on the same PCS switch, then the billing record can be generated by the PCS switch. If the system supports TIA IS-124, then billing records can be sent from one system to the other for eventual delivery of bills to the account holder. If the system does not support IS-124, then arrangements must be made for delivery of the billing data to the home system. The delivery may take weeks or months and can lead to fraud.

When a call starts on one PCS switch and ends on another one because of an intersystem handoff, the location and the format of the billing data becomes more difficult. If a common billing format is not supported across all systems, then billing data may be lost or may be in error.

12.11 SUMMARY

PCS promises the goal of any service available anywhere and anytime throughout the entire world. To place and receive voice and data calls anywhere, anytime, the PCS network must interwork with other networks. In this chapter we discussed some of the issues related to interworking of PCS systems with other PCS systems and with the worldwide wireline network. In some cases solutions were shown. In other cases, solutions have not yet been determined and only alternatives have been shown. It may take until the next millennium before all of the interworking solutions have been determined and implemented.

12.12 PROBLEMS

1. Chapter 12 focuses on adapting various systems to each country of the world. What are the key differences between North American telephone systems and European

telephones systems? Use material from this chapter and from other reference material that you can find.

2. Describe the types of coding and other techniques that you would use on a data link to eliminate the errors when the error rate of the channel is 10-6, 10-2, and 10-1. Use material learned in this chapter and in appendix B to generate your answer.

3. In section 12.5 the interworking between SSD and token-based systems is described to show a protocol failure. Describe the similar set of steps that are necessary when a public key system interworks with a SSD system.

4. List some of the additional messages that are needed in a handoff message when a handoff takes place to a system with a different air interface than currently in use by the PS. How does you answer change if the MAP is different also?

5. Pick two services described in section 12.7 and describe how they might work in two different systems. Then describe how they would need to change so that users roaming from one system to the other would see a common service. HINT: If you do not have access to the detailed operation of GSM- and IS-41-based systems, build your own model of how they might work.

12.13 REFERENCES

1. ATIS, "Air Interface Specification for 1.8 to 2.0 GHz Frequency Hopping Time Division Multiple Access (TDMA) for Personal Communications Services," J-STD-006, 1995.

2. Committee T1, "Stage 2 Service Description for Circuit Mode Switched Bearer Services," Draft T1.704, 1994.

3. "PCS-1900 MHz, IS-136 Based, Mobile Station Minimum Performance Standards," J-STD-009, 1995.

4. "PCS-1900 MHz, IS-136 Based, Base Station Minimum Performance Standards," J-STD-010, 1985.

5. "PCS-1900 MHz, IS-136 Based, Air Interface Compatibility Standards," J-STD-011, 1995.

6. "PCS-1900 Air Interface, Proposed Wideband CDMA PCS Standard," J-STD-007, T1 LB-461, 1995.

7. Personal Access Communications System, Air Interface Standard, Draft J-STD-XXX, LB-477, 1995.

8. Recommendation G162, CCITT Plenary Assembly, Geneva, Blue Book, vol. 111 (May–June 1964): 52.

9. T1P1/94-088, "American National Standard for Telecommunications—Personal Station-Base Station Compatibility Requirements for 1.8 to 2.0 GHz Code Division Multiple Access (CDMA) Personal Communications Systems," Draft, J-STD-008, November 1994.

10. T1P1/94-089, PCS2000, "A Composite CDMA/TDMA Air Interface Compatibility Standard for Personal Communications in 1.8–2.2 GHz for Licensed and Unlicensed Applications," Committee T1 Approved Trial User Standard, T1-LB-459, November, 1994.

CHAPTER 13

Design of a Wireless System:
A Case Study

13.1 INTRODUCTION

In this chapter, we first discuss the planning and engineering of a cellular radio system, including engineering philosophy, engineering considerations, quality-of-service criteria, and types of analyses. We then illustrate the process of growing a wireless system by considering a growth scenario with a frequency reuse factor of 7. We also provide calculations for the frequencies of mobile call originations, mobile call terminations, handoffs, and location updates as a function of a Mobile Switching Center (MSC) serving area for the cellular and the PCS system located in a large metropolitan area.

The process we describe here is an iterative process. An important part of the process is the need to measure information on a system once it is in operation. Although propagation models may indicate that the coverage is adequate in all parts of the system, when the system is installed, holes in the coverage may occur. They will then be filled in with new cells, reoriented antennas, higher power, and so on. Similarly, traffic at one or more cells may be higher than at other cells and may require cell splitting sooner than planned. Measurement data is an important element in correctly planning and growing a cellular system. We discussed the measurement data in chapter 11.

13.2 PLANNING AND ENGINEERING A CELLULAR RADIO SYSTEM

The planning and engineering of a cellular RS requires significant time and judgment. Before planning a system, the engineer should first acquire an understanding of basic cellular technology (as described in chapters 1–9), including the regular hexagonal grid structures, channel assignment, cell splitting, overlaid cells, call processing, radio propagation concepts, and other principles.

In the planning and engineering of cellular systems, a phased approach is often used. The first step is to design the system based on estimates of what will happen in the future. Then, when the system is operational and providing service to customers, measurement data must be collected to determine if the system meets design objectives and to reconfigure the system when it does not. With this approach, cellular engineers need to select sites, determine quantities of equipment to install to meet channel availability objectives, and design the system to provide adequate transmission performance.

The general philosophy in RF engineering for cellular systems is to provide the highest possible performance in the radio portion of the system while minimizing costs. Radio performance includes both the quality of the control transmission path and the quality of the voice transmission path. The measure of transmission performance is the RF signal-to-impairment ratio, $S/(I+N)$, where impairment refers to the power sum of noise (N) and cochannel interference (I).

It is difficult to provide a precise set of rules to achieve performance goals in the complex environment of cellular radio. Engineering judgment is required to achieve compromise solutions and to customize the planning procedure. In some cases it may be necessary for engineers to trade off different types of system impairments. Reducing power at one cell site to solve an interference problem at a nearby cell site, for example, may result in degraded noise or interference performance at the cell site where the power was reduced. There may be other situations when engineers may be forced to improve particularly bad transmission problems in relatively small geographical areas at the expense of slightly degraded performance over a relatively large area and vice versa. The best solutions may not always be obvious. The complexity of these trade-offs is compounded by a lack of reliable data about customer satisfaction related to RF performance. Extensive customer-acceptance data is not available, but several studies have indicated that, in the range of a 15- to 25-dB $S/(I+N)$, customer satisfaction improves significantly with relatively small improvements in $S/(I+N)$. An SNR-plus-interference ratio equal to 17 dB or better in 90% of the geographical area of a system is the recommended performance level. As a general objective, the recommendation for RF engineering of voice channels is to provide a minimum of a 17-dB $S/(I+N)$ in the serving area of each omnidirectional and directional site face and to achieve as high an $S/(I+N)$ performance level above 17 dB as is economically feasible. The $S/(I+N)$ criteria for access channels is also 17 dB or better in 90% of the geographical area.

If the following quality-of-service criteria is used in the design of a cellular system, good coverage of the area will be achieved. The signal strength, in both transmission directions, should result in:

☞ A mean $S/(I+N)$ of $\geq$ 17 dB, over 90% of the service area

☞ A mean $S/(I+N)$ of $\geq$ 5 dB over 99% of the service area

Generally, RF engineers calculate both the average and the worst-case S/(I+N) performance analyses. The average performance calculation provides a measure of overall customer perception of the quality of service and generates the information that will be most useful to system managers. In an average performance analysis, it is possible that poor performance on some channels may be masked by very good performance on other channels. A worst-case S/(I+N) analysis avoids this problem by determining individual channel worst-case performance that is then used to maximize performance on individual channels and thus also maximizes system performance. Even in a system with very good average performance, a small number of poor channels can contribute to a significant lost-call rate.

13.3 OUTLINE OF THE ENGINEERING PROCEDURE

The system engineer starts with specific system configuration and performs a test for adequate performance. If the test fails, then the engineer executes various diagnostic procedures to find the cause of the failure. Once the engineer understands the cause of the failure, he or she makes decisions about which techniques are appropriate and how to apply them to develop a modified system configuration that satisfies the S/(I+N) ratio performance test criteria. If criteria is not met, the engineer repeats the diagnostic and problem-solving procedures until the specified criteria is satisfied (Figure 13.1).

The definition of a "system configuration" is a cellular grid that specifies cell-site locations, transmitter powers, antenna types, and antenna heights. Gen-

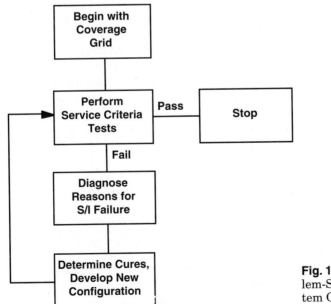

Fig. 13.1 Diagnostic and Problem-Solving Procedures for System Configuration

erally, for designing a new system, all antennas are made omnidirectional, and their heights are determined from the coverage area of the cell; the power outputs of the transmitters are set to the maximum permitted; the voice and setup channel groups are assigned based on the material in the next section. After iterations to support RF coverage into all areas and to support expected traffic, the engineer will determine the final system configuration. This configuration represents the best compromise between the various performance criteria, cost, and practicality.

13.4 FREQUENCY REUSE AND CHANNEL GROUP ASSIGNMENTS

The parameter N is known as the "number of channel groups" or the "number of cells per cluster" and is determined on the basis of cochannel interference considerations. A smaller value of N degrades the S/I ratio performance. A larger value of N improves S/I ratio performance. The significance of N is that the allocated voice channels are divided into N channel "groups," one of which is used at each cell. The value normally used for N is seven in the North American systems and four in the GSM system.

The procedure for assigning channel groups to cells is the same for both omnidirectional and directional cell sites. The assignment patterns that are established at the start-up of a system determine which channel groups are assigned to cell splitting.

At the start-up of a system, the procedure for assigning channel groups begins with overlaying an array of regular hexagons on a map of the service area. Each hexagon should contain exactly one cell site, preferably close to the center. The hexagons may all be of the same size, or, if cell splitting is necessary from the beginning, then there may be two or even three cell-site sizes. Figure 13.2 shows a start-up configuration with two sizes of cell. First, we assign a channel group to each cell in large-cell pattern, thus eliminating all the largest cells from the map of the actual system (see Figure 13.3). Next we assign channel groups to cells of the next smaller size. The process of assigning channel groups to the smaller cells uses the completed pattern of assignments for larger cells.

In Figure 13.4, we add the small cells to produce a system map in which channel-group assignments are shown for:

- ☞ All the actual large cells
- ☞ For any small cell that overlays the center of a hypothetical large cell

Next, we add a few hypothetical smaller cells to the system map to aid in determining the channel-group assignments for the actual small cells. In Figure 13.5 the hypothetical cells are shown by broken lines.

We determine a particular small cell's channel-group assignment by locating a cochannel cell to which a channel group has already been assigned. The

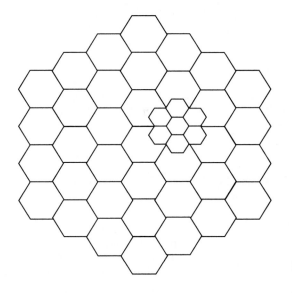

Fig. 13.2 Start-Up System

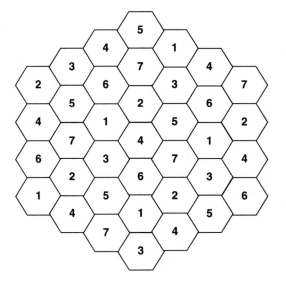

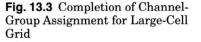

Fig. 13.3 Completion of Channel-Group Assignment for Large-Cell Grid

standard grid of channel assignments (Figure 13.3) holds for any cell size. Then, the same channel group is assigned to the cell in question. For example, in Figure 13.5, the small cell on channel-group 6 is mapped to the hypothetical cell on group 6, and group 4 is mapped to group 4. When we complete this mapping, in Figure 13.6, we have channel-group assignments for all cells.

After the process of regular channel assignment is completed except for deciding how the channel sets within each channel group should be used, traffic engineering determines how many channels are required for an omnidirectional cell site or for each face of a directional cell site.

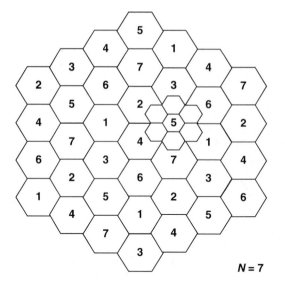

Fig. 13.4 Complete System Map with Channel-Group Assignments for Large Cells

$N = 7$

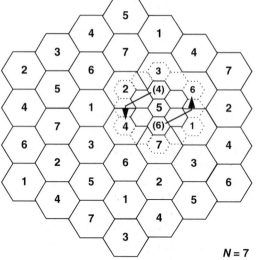

Fig. 13.5 Method of Assigning Channel Groups to Small Cells

$N = 7$

13.5 CONSIDERATIONS FOR A START-UP SYSTEM

In planning a new cellular system, system designers should follow the objective of providing coverage to the intended service area while using a minimum number of cell sites. Providing coverage means that the RF S/(I+N) ratio throughout the service area must have a high probability of being large enough to produce satisfactory transmission quality.

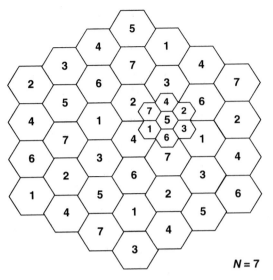

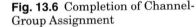

Fig. 13.6 Completion of Channel-Group Assignment

$N = 7$

For geographically small systems ($\approx$ 1,000 square miles or less), the start-up configuration can usually be based on a single cell radius. In other words, all of the idealized hexagonal cells have the same area. The largest systems usually start up with two or even three cell sizes.

When the coverage area is large, RF propagation characteristics may vary so much from one part of the system to another that a single value of cell radius may not be suitable for the entire system. A relatively large radius is appropriate for sections of the system in which path loss increases relatively slowly with distance, but this radius would produce inadequate coverage in the sections of the system with more severe path loss. A relatively small radius is necessary for sections in which path loss is severe, but using this radius in sections that have milder path loss would inflate the system cost by requiring more cell sites than the number necessary for adequate coverage. Cell splitting offers a way to use a larger radius where permissible and a smaller radius where necessary, while maintaining an orderly geographical pattern of cells amenable to regular channel assignment. The radius of the largest cells determines the set of possible values from which all cell radii should be chosen. System designers should choose the largest cell radius that yields a set of cell radii that suit the range of propagation conditions in the system service area.

13.5.1 Cell Splitting

After a system has been in operation, traffic will grow in the system and will require that additional channels be made available. If all channels at a cell have not be installed, then the cell should be grown to it full complement of channels. However, at some point in the growth of a cellular system, the engineer will need to split cells to add additional capacity.

When cell splitting occurs, the designer must minimize changes in the system. The value of the reuse factor, N, must be relatively prime with respect to the value of the type of split. When this relationship occurs, then the voice channel frequency assignments at the existing cell sites remain the same when new cell sites are added. The following splitting patterns can be used for various values of N:

- ☞ For $N = 3$, use 4:1 cell splitting.
- ☞ For $N = 4$, use 3:1 cell splitting.
- ☞ For $N = 7$, use 3:1 or 4:1 cell splitting.
- ☞ For $N = 9$, use 4:1 cell splitting.

This is the way 4:1 splitting works. When the new cell site is located on the border between two existing cell areas, the new cell site will cover an area with a radius of one-half that of the larger cell areas from which it was split. Thus, the area covered by the new cell (the secondary cell) is one-fourth of the area of the larger older cell; hence, this is described as 4:1 splitting. For example, if the old cell had a radius of 10 miles, the secondary cell would have a radius of 5 miles.

Here is the way 3:1 splitting works. When the new cell site is located on the corner between the cell areas covered by three existing cell sites, the radius of the cell area covered by the new cell site would be $1/\sqrt{3}$ of the radius of the cell areas covered by the existing cell sites. Thus, the new coverage area is only one-third of the larger cell's coverage area; hence, this is described to as 3:1 splitting.

13.5.2 Segmentation and Dualization

Sometimes engineers want to add an additional cell at less than the reuse distance without using a complete cell-splitting process. This method might be used to fill in a coverage gap in the system. This can result in cochannel interference. The most straightforward method to avoid an increase in cochannel interference is simply not to reuse them. Segmentation divides a channel group into segments of mutually exclusive voice channel frequencies. Then, by assigning different segments to particular cell sites, cochannel interference between these cell sites is avoided. The disadvantage of segmentation is that the capacity of the segmented cells is lower.

When a cellular system is growing, there may be cells of different radii in the same region of the coverage area. This also can result in cochannel interference. By dividing the radios at the cell site into two separate server groups, one for the larger (overlaid) cell and one for the smaller (underlaid) cell, the interference can be minimized. The radios for the primary server group serve the underlaid cell, and the radios of the secondary server group are used to serve mobiles in the overlaid cell areas. As traffic in the smaller cells grows, more and more channels are removed from the secondary group and assigned to the primary group until the secondary group (and its larger cell) disappears.

Dualization provides for control of cochannel interference and also gives a real increase in engineered traffic capacity.

13.6 THE PROCESS OF GROWING A CELLULAR SYSTEM

In the following example, we illustrate the process of growing a cellular system by considering a growth scenario. The demand for service at start-up is based on measurements of carried traffic load. Coverage and interference are predicted using software tools with actual terrain data and geographical locations for the cell sites. For the sake of clarity, the growth scenario is shown schematically with regular hexagons representing the cell areas (see Figure 13.7). The assignment of voice channel frequencies is done according to a reuse factor of 7.

To meet an increasing demand for service, we increase the engineered traffic capacity of individual cell sites by adding voice channels until the channel group is exhausted. We then employ cell splitting to further increase the capacity of the system. As we add new cell sites, we control cochannel interference by using segmentation, dualization, and sectorization. We have discussed segmentation and dualization in the preceding section and sectorization in the chapter 7.

At start-up (see Table 13.1), we set the radius of each cell at 10 miles. We use a nominal value of 0.026 Erlangs per call (corresponding to an average call duration of 100 seconds) to compute the quantity of voice channels required to meet the demand for service with 2% blocking.

We will now follow a four-step process to grow a system.

1. As demand for service increases, we increase the quantity of voice channels at each cell site to the point where the channel groups are exhausted at both cell sites 3 and 4. To meet any further increase in the demand for service, a new cell site is added between cell sites 3 and 4. Voice channel frequencies from channel group 1 (Figure 13.8) are assigned at cell site 7. Since this channel group is not used at any other cell site, there is still no

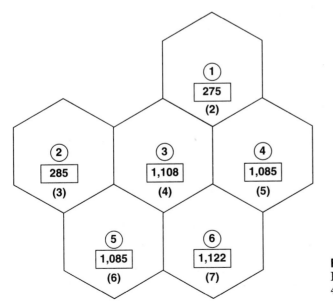

Fig. 13.7 Start-Up (Step 0): Busy-Hour Call Attempts = 4,960

Table 13.1 Growing a Cellular System—Step 1

Cell Site	Channel-Group Number	Traffic Load (Busy-Hour Call Attempts)	Traffic Load (Erlangs)	Number of Voice Channels
1	2	356	9.26	16
2	3	353	9.18	16
3	4	1,307	33.98	44
4	5	1,292	33.59	44
5	6	1,150	29.90	39
6	7	1,196	31.10	41
7	1	554	14.40	22
	Total	6,208	161.41	222

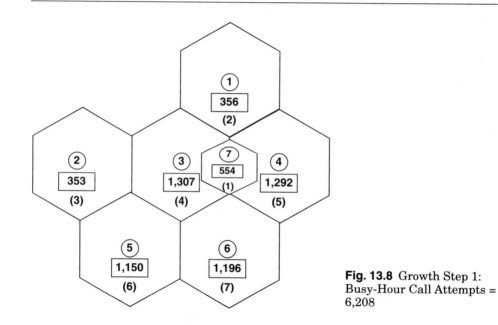

Fig. 13.8 Growth Step 1: Busy-Hour Call Attempts = 6,208

channel reuse in the system. The new configuration is summarized in Table 13.2.

2. As the demand for service continues to increase, the channel groups from which voice channel frequencies are assigned at cell sites 5 and 6 are exhausted. Another new cell is added between them. Voice channel frequencies from channel group 3 are assigned to cell site 8 (Figure 13.9). The demand for service can still be satisfied without channel reuse since channel group 3 can be segmented and different voice channel frequencies assigned at cell sites 3 and 8. The new configuration is summarized in Table 13.3.

Table 13.2 Growing a Cellular System—Step 2

Cell Site	Channel-Group Number	Traffic Load (Busy-Hour Call Attempts)	Traffic Load (Erlangs)	Number of Voice Channels
1	2	386	10.04	17
2	3	400	10.40	17
3	4	1,282	33.33	43
4	5	1,252	32.55	42
5	6	1,265	32.89	42
6	7	1,308	34.00	44
7	1	542	14.09	22
8	3	512	13.31	21
	Total	6,947	180.61	248

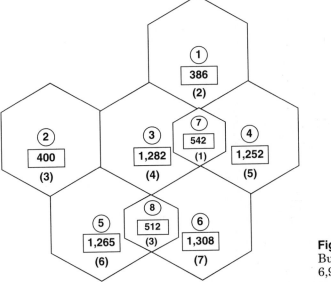

Fig. 13.9 Growth Step 2: Busy-Hour Call Attempts = 6,947

3. As further increases in the demand for service occur, two more new cell sites are added (Figure 13.10). Cell site 9 is added between cell sites 3 and 6, and cell site 10 is added between cell sites 4 and 6. Voice channel frequencies from channel groups 2 and 6 are assigned to cell sites 9 and 10, respectively. Channel group 2 is segmented. However, the demand for service requires channel reuse of the voice channel frequencies of channel group 6. In order to reuse some of the voice channel frequencies of channel group 6, dualization is used at cell site 5 to control cochannel interference. The voice

Table 13.3 Growing a Cellular System—Step 3

Cell Site	Channel-Group Number	Traffic Load (Busy-Hour Call Attempts)	Traffic Load (Erlangs)	Number of Voice Channels
1	2	397	10.32	17
2	3	395	10.27	17
3	4	1,230	31.98	42
4	5	1,232	32.03	42
5 Primary	6	435	11.31	18
5 Secondary	6	639	16.61	24
6	7	742	19.29	27
7	1	541	14.06	21
8	3	409	10.63	17
9	2	469	12.19	19
10	6	459	11.93	19
	Total	6,948	180.65	263

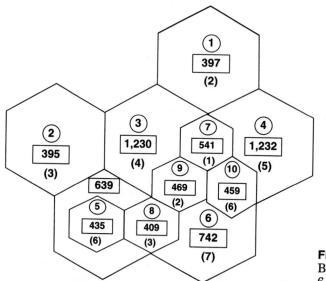

Fig. 13.10 Growth Step 3: Busy-Hour Call Attempts = 6,948

channel frequencies assigned to the primary server group of cell site 5 are reused at cell site 10. The new configuration is summarized in Table 13.4.

4. As the demand for service continues to increase, more and more cell sites are added, and dualization is used at more and more cell sites. Eventually

Table 13.4 Growing a Cellular System—Step 4

Cell Site	Channel-Group Number	Traffic Load (Busy-Hour Call Attempts)	Traffic Load (Erlangs)	Number of Voice Channels[*]
1 Primary α	2	15	0.39	2
1 Primary β	2	106	2.76	4
1 Primary γ	2	17	0.44	2
1 Secondary	2	69	1.79	5
2	3	386	10.04	17
3	4	1,099	28.57	36
4 Primary	5	645	16.77	21
4 Secondary	5	443	11.52	18
5 Primary	6	680	17.68	25
5 Secondary	6	446	11.60	17
6 Primary	7	468	12.17	17
6 Secondary	7	669	17.39	25
7	1	659	17.13	23
8	3	671	17.45	23
9 α	2	229	5.96	8
9 β	2	233	6.06	8
9 γ	2	229	5.96	8
10	6	512	13.31	18
11	1	712	18.51	22
12	5	683	17.76	21
13	7	506	13.16	17
14	1	327	8.50	13
15	3	487	12.66	17
16 α	2	27	0.70	2
16 β	2	114	2.96	5
16 γ	2	182	4.73	7
17 α	2	167	4.34	6
17 β	2	112	2.91	5
17 γ	2	39	1.02	3
Total		10,392	284.24	395

[*]In assigning the voice channel frequencies to the cell sites, an engineering judgment has been used where the quality of service in the lightly loaded sectorized and dualized cell sites is degraded to maintain total number of voice channels equal to 395.

the demand for service requires the same voice channel frequencies from channel group 2 to be assigned at cell sites 1, 9, 16, and 17 (Figure 13.11). This provides the potential for cell site 9 to receive cochannel interference from more than two cell sites. Sectorization is employed using directional antennas at each of the cell sites involved to eliminate the potential for some of the cochannel interface. The new configuration is summarized in Table 13.5.

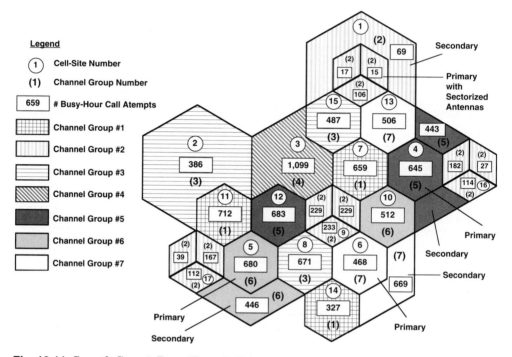

Fig. 13.11 Growth Step 4: Busy-Hour Call Attempts = 10,932

Table 13.5 Distribution of Voice Channels: Step 4

Channel-Group Number	Cell-Site Number	Total Voice Channels in Group
(1)	7, 11, 14	23 + 22 + 13 = 58
(2)	1, 9, 16, 17	13 + 24 + 14 + 14 = 65
(3)	2, 8, 15	17 + 23 + 17 = 57
(4)	3	36
(5)	4, 12	21 + 18 + 21 = 60
(6)	5, 10	17 + 25 + 18 = 60
(7)	6, 13	25 + 17 + 17 = 59
	Total	395

13.7 TRAFFIC CALCULATIONS FOR A CELLULAR AND PCS SYSTEM

In this example we compute mobile originations, terminations, handoffs, and location updates as a function of the MSC serving area for a cellular and PCS system located in a large metropolitan area. Different assumptions are made for cellular and PCS systems. Cellular systems today are primarily used to supplement wireline service. Not all users have cellular telephones, and cellular telephone users will use wired phones when they are available. PCS offers the concept of universal service; therefore, we examine a model where all users in a coverage area have and use wireless PCS telephones.

13.7.1 The Cellular System Model

For the cellular model, we use the following assumptions:

☞ Population density: 45,000 people / km^2

☞ Moving speed of terminal: 6 km/hour

☞ Percentage of subscribers: 10%

☞ Percentage of powered terminals: 50%

☞ Erlangs per terminal: 0.06

☞ Percentage of mobile-origination calls: 60%

☞ Average call holding time: 120 seconds

☞ Hexagonal cell radius: 1.0 km

☞ Cells per location area: 16

☞ VLR area: MSC area

Solution:

$$\text{The area of each cell} = 2.6\ R^2 = 2.6\ \text{km}^2$$

$$\text{MSC serving area} = 16 \times 2.6 = 41.6\ \text{km}^2$$

No. of mobile call originations

$$= \text{area} \times \text{density} \times \%\ \text{subscribers} \times \%\ \text{mobile originations} \times \frac{\text{Erlang}}{\text{terminal}} \times \frac{1}{\text{call holding time}}$$

$$\text{Mobile call originations} = 41.6 \times 45{,}000 \times 0.1 \times 0.6 \times 0.06 \times \frac{1}{\frac{1}{30}} = 202,176\ \text{calls/hour}$$

$$\text{Mobile call terminations} = 134{,}784$$

We need a mobility model to calculate the number of cell boundary, location area, and switch area crossings. We will use a simple flow-based model that assumes people are uniformly distributed in an area A and the direction of travel of each user relative to the border is uniformly distributed on $(0, 2\pi)$. If we define ρ as the density of people per km^2 (with or without terminals), v to be a person's average speed in km/hour, and L to be the length of the perimeter of the area A, then the average number of people leaving the area A per hour is given by:

$$\text{No. of people leaving area } = \frac{\rho v L}{\pi}$$

By conservation of flow, the above equation also provides the number of crossings into the area.

$$\text{Crossings per cell } = \frac{45{,}000 \times 6 \times 6}{\pi} = 515{,}595$$

$$\frac{\text{Crossing}}{\text{area}} = \frac{\rho v L_{area}}{\pi}$$

$$\text{Area } = 41.6 = 2.6 R_{area}^{2}$$

$$\therefore R_{area} = 4.0 \text{ km}$$

$$\therefore \frac{\text{Crossing}}{\text{area}} = \frac{45{,}000 \times 6 \times 24}{\pi} = 2{,}062{,}381$$

Handoffs (area) = Total number of cell departures (interswitch + intra-switch) made by calls in progress in the switch area in crossings/hour.

$$\text{Interswitch handoffs } = \text{No. of cells} \times \frac{\text{crossings}}{\text{area}} \times \% \text{ subscribers} \times \frac{\text{Erlangs}}{\text{terminal}}$$

$$\therefore \text{Total handoffs } = 16 \times 515{,}595 \times 0.1 \times 0.06 = 49{,}497$$

The number of interswitch handoffs generated by leaving the switch area, is also the number of handoffs into the area and is given by:

$$\text{Interswitch handoffs } = \frac{\text{crossings}}{\text{area}} \times \% \text{ subscribers} \times \frac{\text{Erlangs}}{\text{terminal}}$$

$$\therefore \text{Interswitch handoffs } = 2{,}062{,}381 \times 0.1 \times 0.06 = 12{,}374$$

$$\text{Intraswitch handoffs } = 49{,}497 - 12{,}374 = 37{,}123$$

Location update (area) = No. of location area in the switch $\times$ crossings/area $\times$ % subscribers $\times$ (% terminals powered on – Erlang/terminal)

$$\therefore \text{Location update (area) } = 1 \times 2{,}062{,}381 \times 0.1 \times (0.5 - 0.06) = 90{,}745$$

$$\text{Home VLR updates } = 68{,}059 \text{ and}$$

$$\text{Visited VLR update } = 22{,}686.$$

For the PCS system, we use the following assumptions and repeat the preceding calculations:

13.7.2 The PCS System Model

☞ Population density: 45,000 people/km^2
☞ Moving speed of terminal: 6 km/hour
☞ Percentage of powered terminals: 100%
☞ Percentage of subscribers: 80%
☞ Erlangs per terminal: 0.06
☞ Percentage of mobile-origination calls: 60%
☞ Average call holding time: 120 seconds
☞ Hexagonal cell radius: 0.2 km
☞ Cell per location area: 40
☞ VLR area: MSC area

We make the following calculations:

$$\text{The area of each cell} = 2.6\,R^2 = 2.6(0.2)^2 = 0.104 \text{ km}^2$$

$$\text{MSC serving area} = 40 \times 0.104 = 4.16 \text{ km}^2$$

$$\text{Mobile call originations} = 4.16 \times 45,000 \times 0.6 \times 0.6 \times 0.06 \times 0.8 \times \frac{1}{\frac{1}{30}} = 161,741 \text{ calls/hour}$$

$$\text{Mobile call terminations} = 107,827 \text{ calls/hour}$$

$$\text{Crossings per cell} = \frac{45,000 \times 6 \times 1.2}{\pi} = 103,119$$

$$2.6R_{area}^{\,2} = 4.16$$

$$\therefore R_{area} = 1.265 \text{ km}$$

$$\therefore \frac{\text{Crossing}}{\text{area}} = \frac{45,000 \times 6 \times 1.265 \times 6}{\pi} = 652,182$$

$$\therefore \text{Total handoffs} = 40 \times 103,119 \times 0.8 \times 0.06 = 197,988$$

$$\text{No. of intraswitch handoffs} = 166,683$$

$$\text{Interswitch handoffs} = 197,988 - 166,683 = 31,305$$

$$\text{Location update (area)} = 1 \times 652,182 \times 0.8 \times (1 - 0.06) = 490,411 \text{ per hour}$$

$$\text{Home VLR updates} = 412,894 \text{ and}$$

$$\text{Visited VLR update} = 77,547.$$

Table 13.6 provides a summary of the calculations for both systems. From the table, we can identify that the PCS system with its small cell radius has a five-fold increase in the rate of handoffs and a seven-fold increase in VLR updates per call over the cellular system.

Table 13.6 A Comparison of Cellular vs. PCS Systems

	Cellular	PCS
No. of mobile originations/hour	202,176	161,741
No. of mobile terminations/hour	134,784	107,827
(O + T) calls/hour	336,960	269,568
Intraswitch handoffs/hour	37,123 (0.11 per call)	166,683 (0.62 per call)
Interswitch handoffs/hour	12,374 (0.04 per call)	31,305 (0.12 per call)
Total handoffs	0.15 per call	0.74 per call
Home VLR updates	0.20 per call	1.53 per call
Visited VLR updates	0.07 per call	0.29 per call

13.8 SUMMARY

In this chapter, we presented the planning and engineering process that can be used for a cellular radio system. We discussed cell splitting, segmentation, and dualization and illustrated their applications in the process of growing a cellular system using a growth scenario with a frequency reuse factor of 7.

13.9 PROBLEM

Use the call flow diagrams shown in chapters 8, 9, and 10. Estimate the number of signaling messages per mobile call for the registration and authentication process in a PCS system. Compare the results for a GSM and IS-41-based system. The systems have the following characteristics:

- Population density: 60,000 people/km^2
- Speed of PS: 30 km/hour
- Percentage of subscribers in the coverage area: 85%
- Percentage of terminals powered on: 100%
- Erlangs/terminal: 0.06
- Percentage of mobile originations: 60%
- Percentage of mobile terminations: 40%
- Average call holding time: 120 seconds

- $N = 7$, with a cell radius of 0.3 km
- Number of cells/location area: 60
- VLR area equals the MSC area

How does the number of signaling messages vary with respect to cell radius? Plot a curve to show messages/cell vs. cell radius for a cell radius of 1.0, 2.0, 4.0, and 8.0 km.

13.10 REFERENCES

1. May, Adolf D., *Traffic Flow Fundamentals*, Englewood Cliffs, New Jersey: Prentice-Hall, 1990.
2. Meier-Hellstern, K. S., E. Alonso, and D. R. O'Neil, "The Use of SS7 and GSM to Support High Density Personal Communications," *Wireless Communications—Future Direction*, J. M. Holtzman, and D. J. Goodman, eds., Boston: Kluwer Academic Publishers, 1993.

Cellular Digital Packet Data (CDPD) Network

14.1 INTRODUCTION

Some wireless data systems are designed for packet-switched rather than circuit-switched operation. Examples of some wide-area data services are: the Advanced Radio Data Information Service (ARDIS), the RAM Mobile Data System, and the Cellular Digital Packet Data Services. ARDIS and RAM Mobile Data offer wireless packet data messaging service over a dedicated network using the Specialized Mobile Radio (SMR) frequencies near 800/900 MHz. ARDIS offers service in over 400 metropolitan areas in the United States. RAM Mobile Data offers service over its MOBITEX network and provides coverage in over 266 metropolitan areas, with 10 to 30 duplex channels available in each area. The data rate is 8 kbs. MOBITEX networks are operational in other countries. ARDIS and RAM are discussed in more detail in chapter 15.

The Cellular Digital Packet Data network does not require a specialized network but rather uses the existing analog cellular network. CDPD takes advantage of the idle time on the analog AMPS channel to transmit packet data at a rate of 19.2 kbs. It is designed to operate as a transparent overlay on the AMPS system.

In this chapter, we discuss the CDPD network. We will present the theory of packet networks and describe other wireless packet-switched data systems in chapter 15.

14.2 CDPD NETWORK DEFINED

The CDPD network provides wireless packet data connectivity to mobile data communications users. It is designed as an extension to the existing data communications networks. The extension is achieved by defining the CDPD network as a peer-level (multiprotocol connectionless) network to existing data networks. Standard connectionless network protocols are used to access services through the CDPD network. The CDPD network uses an open-system interconnect for network design to allow additional new network protocols when appropriate. It also can be deployed quickly with existing technology whenever possible and appropriate. New technology is used only where necessary to provide services unique to the CDPD network. The primary new technology components used in the CDPD network are a CDPD modem that uses existing cellular channels for air data link services and support for mobility management. The mobility management function focuses on continuous communication to mobile subscribers. The connectionless service allows the CDPD network to provide a simple and reliable cell transfer mechanism.

In summary, the CDPD is a connectionless, multiprotocol network service that provides a peer network extension to existing data communications network.

14.3 THE NETWORK ARCHITECTURE FOR CDPD

Figure 14.1 shows the internal network reference model for the CDPD network architecture. The external subnetwork interface reference points are A, E, and I. This architecture is separate and distinct from the network reference architectures described in chapter 8. In particular, the letter designations of the interfaces are not the same as those in chapter 8.

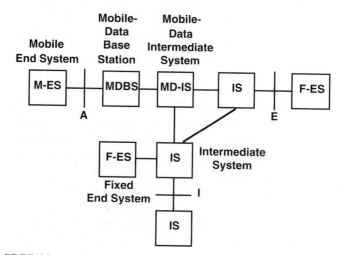

Fig. 14.1 The CDPD Network Architecture

The CDPD network has the following interfaces:

☞ **The Air link interface—reference point A:** The air link interface is the CDPD service provider's interface for providing services over the radio link to subscribers. CDPD subscribers use mobile end systems to access network services through this air link interface.

☞ **The external interface—reference point E:** The external interface is the CDPD service provider's interface to external networks. The external application service providers communicate with CDPD subscribers through this interface.

☞ **The interservice provider interface—reference point I:** The interservice provider interface is a CDPD service provider's interface to cooperating CDPD service provider networks. This interface allows support of CDPD network services across all areas served by CDPD service provider domain. This interface is not visible outside the CDPD network.

There are two basic classes of network entities in the CDPD network. In OSI terminology these are called End Systems (ESs) and Intermediate Systems (ISs). In Internet terminology, end systems are hosts and intermediate systems are routers.

End systems represent the network user host and terminals that communicate with each other. Intermediate systems relay data packets from end systems and route them toward their intended destinations. The network supports mobile-to-mobile, mobile-to-fixed location, and fixed location-to-mobile packet exchange. The network can support fixed location-to-fixed location communication, but it is not the intended mode of service provided to CDPD end systems.

End systems are logical end points of communication and are addressed as source or destination network service access points. In the OSI layered reference model, end systems provide the functions of the layers above the network layer and provide the intersystem communication functions of the lowest three layers. End systems do not provide the network-layer relay function.

The CDPD network makes a distinction between mobile end systems and fixed end systems for the purposes of mobility management.

A detailed description of each network element follows.

14.3.1 Mobile End System (Mobile End Station)

Subscribers gain access to mobile communication services from the subscriber equipment that functions as a mobile end system. Mobile end systems communicate with the CDPD network via the A-Interface. The physical location of mobile end systems may change with time, but continuous network access is maintained. Mobile end systems perform additional functions in the network sublayer and below to allow subscriber equipment to move from cell to cell or network to network in a fashion that is transparent to the protocols above the network-layer. The CDPD network tracks the location of the mobile end systems and routes network-layer datagrams to them and from them accordingly.

Networks traditionally route information based on its destination address. In the CDPD network, the address of a mobile subscriber does not imply location and, therefore, is not used directly for routing. Mobility implies that mobile end systems change their subnetwork point of attachment at will. Therefore, traditional network connectivity and routing functions cannot be used because the location of mobile end systems, and hence the route to reach them, cannot be determined from their network address.

The mobility support functions are separated into two subfunctions:

1. Mobility management functions that support tracking the current subnetwork point of attachment and routing network protocol data units.

2. Radio resources management functions that support maintaining connectivity to a suitable subnetwork point of attachment.

14.3.2 Fixed End System (F-ES)

Fixed end systems are the external data application systems or internal network support and service application systems. A host computer is a typical fixed end system. Their location is fixed, and traditional routing based on a network address implying location is possible. Fixed end systems are not concerned with the mobility issues of the mobile end systems to which they communicate. Existing application systems will not need to be modified to communicate with the mobile end systems. Existing standard protocols at all layers in the fixed end systems will transparently interwork with CDPD mobile subscribers.

For the purposes of security, authentication, and accounting, a distinction can be made between external and internal fixed end systems. External fixed end systems are owned, operated, administered, and maintained outside of the direct control of the CDPD network operator. As a public service provider, the CDPD network operator needs to protect itself from invalid or unauthorized use of the network and needs to put in place controls to limit the privileges and access available to external fixed end systems. Fixed end systems external to the CDPD network exchange network-layer datagrams with the CDPD network over the E-Interface.

Internal fixed end systems are value-added network service application systems or network support services administered and maintained by the CDPD network operator. Not all these end systems are visible or directly addressable by CDPD network subscribers. Examples of internal fixed end systems are:

1. Network support services involved with authentication and authorization, network management, accounting data collection, and so on.

2. Value-added service applications that provide domain name services, location services, or other supplementary services.

14.3.3 Intermediate Systems

An intermediate system performs only functions allocated to lowest three layers of the OSI reference model. The physical equipment performing such functions is called a router. The CDPD network has two classes of intermediate systems based on the requirement for knowledge of mobility in end systems.

Intermediate systems use standard, commercial, off-the-shelf routers that support OSI connectionless network service. This equipment and its associated physical interconnections create the CDPD network backbone. CDPD networks operate as wireless extensions of existing connectionless networks. In the OSI network-layer routing topology, each CDPD network carrier is seen as a separate administrative routing domain.

A second category of intermediate systems is the Mobile Data Intermediate System (MD-IS) that performs mobility routing functions. Mobile data intermediate systems are the only network relay systems that have any knowledge of mobility and operate a CDPD-specific Mobile Network Location Protocol (MNLP) to exchange location information.

The geographic grouping of cells connected to a mobile data intermediate system defines a serving area that may cover multiple Cellular Geographic Service Areas (CGSAs), or it may be only part of a single CGSA. Each cell is controlled by a mobile data base station that acts as a data relay system between a mobile end station and its current serving mobile data intermediate system (Figure 14.2).

The mobile data intermediate system performs two distinct mobility functions: Mobile Home Function (MHF) and Mobile Serving Function (MSF). Every mobile end system logically belongs to a fixed home area. The home area acts as the anchor to provide a mobility-independent routing destination area for intermediate systems and end systems that are not mobile-aware. The mobile home

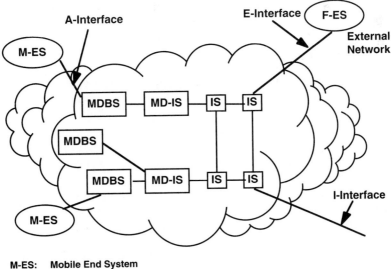

M-ES: Mobile End System
F-ES: Fixed End System
MD-IS: Mobile-Data Intermediate System
MDBS: Mobile-Data Base Station
IS: Intermediate System

Fig. 14.2 The CDPD Network

function in the home area mobile data intermediate system provides a packet forwarding service and maintains a database of the current serving area for each of its mobile end systems. The mobile forwarding function is based on the principle of encapsulating mobile-end-station addressed packets and forwarding them to the mobile serving function in each serving area the mobile end station visits.

The mobile serving function of a mobile data intermediate system handles the routing of packets for all visiting mobile end systems in its serving area. When a mobile end station registers for network access in a mobile data intermediate system serving area, the mobile serving function informs the home mobile data intermediate system of its current location.

The serving mobile data intermediate system layer management functions assist the network support service applications for authentication, authorization, and accounting of the use of the network services by the mobile end station.

On the air link side, a mobile data intermediate system performs the function of a mobile-specific end system intermediate system router (i.e., it routes data for mobile end systems within its area to the current subnetwork point of attachment based on local knowledge of the subscriber's current cell). The mobile data intermediate system utilizes and controls the mobile data base stations, as its agents, that are close to the radio transmitter and receiver. Mobile data intermediate systems interface to the mobile data base station and the intermediate system. Mobile data intermediate systems should need to support only two data link/physical interfaces: the data link to the mobile end systems (via its subordinate mobile data base stations) and a data link to the intermediate systems. In summary, the mobile data intermediate system provides a function analogous to that of the Mobile Switching Center (MSC) in a cellular telephone system.

14.3.4 Mobile Data Base Station (MDBS)

The MDBS provides layer 2 data link or media access relay functions for a set of radio channels serving a cell. The mobile data base station has a responsibility of controlling the radio interface, including radio channel allocation, interoperation with cellular voice channel usage, and radio media access control. The mobile data base station needs to accommodate the physical and Medium Access Control (MAC) layers of the A-Interface and the land interface between it and the mobile data intermediate system.

To facilitate the channel-sharing procedure, the mobile data base stations are expected to be colocated with the cell-site equipment providing cellular telephone service. A cell site may operate multiple CDPD channel streams, with each stream requiring a logical relay function. A channel stream is a bidirectional communications path between a mobile-data base station and a group of mobile end systems, using a single RF channel pair at a time, within a single cell. The RF channel pair used by a channel stream changes over time because of channel hopping to avoid interference or collision with cellular voice users. Because several different RF channels may be used in the same cell at the same time, there may be several channel streams active in a cell concurrently, provided they use distinct RF channels at any one time.

14.4 CDPD PROTOCOLS

The CDPD protocols are in layers 1, 2, 3, and 4 of the OSI model.

14.4.1 Physical Layer

The physical layer accepts a sequence of bits from the MAC layer and transforms them into a modulated waveform for transmission to the remote end. The physical layer also receives a modulated waveform from the remote end, which it converts into a sequence of received bits for delivery to the MAC layer. The RF channel that is used for transmission is one of the 30-kHz forward and reverse RF channel pairs chosen from the range of channels 1 to 1023 (as defined in chapter 8).

Communications between the mobile data base station and the mobile end station take place over a pair of RF channels. RF channels specified as control channels in AMPS or TDMA (chapter 8) are not used for CDPD (i.e., RF channels 313–354, RF channels 688–708, and RF channels 737–757). The data transmission uses Gaussian-filtered Minimum Shift Keying (GMSK) modulation (see chapter 6) at a data rate of 19.2-kbs. The Gaussian pulse-shaping filter is specified to have a bandwidth—time product of $B_w T \approx 0.5$. The specified $B_w T$ product assures a transmitted waveform with a bandwidth narrow enough to meet adjacent-channel interference requirements, while keeping the intersymbol interference small enough to allow a simple demodulation technique. Degraded radio channel conditions may limit the actual information payload throughput rate to lower levels and may introduce additional time delays because of the error detection and retransmission protocols. The power level used for transmissions in CDPD is consistent with power classes defined by TIA's IS-54, IS-55, and IS-56. The received signal strength is measured by both the mobile data base station and the mobile end station to determine whether the mobile end station is transmitting at the appropriate power level.

The physical layer provides the following services:

☞ The ability to tune to a specified pair of RF channels for transmission and reception of bits between the mobile end station and the mobile data base station

☞ The ability to transmit and receive bits between the mobile end station and mobile data base station across a pair of RF channels

☞ The ability to set the power level to be used for transmission of bits between the mobile end station and the mobile data base station

☞ The ability to measure the signal strength of the received bits at the mobile end station and mobile data base station

☞ The ability to suspend and resume monitoring of RF channels in the mobile end station to conserve battery power

14.4.2 Medium Access Control (MAC) Layer

In the CDPD air link interface, the data link layer is divided into two distinct sublayers: the medium access control function and the Logical Link Control (LLC) function. Both MAC and LLC protocols operate within the data link layer of the OSI reference architecture.

The MAC functions arbitrate access to the shared medium between mobile end systems and a mobile data base station. The MAC layer also provides frame recognition, frame delimiting, and error detection/correction.

The medium access method allows two or more mobile end systems to share the common transmission medium. A mobile end station sends the message in bit serial form when it senses that the medium is idle. If, after initiating a transmission, the message collides with that of another mobile end station, the station remains quiet for a random amount of time before attempting to transmit again. This process is described in more detail in chapter 15.

The channel stream comprises a forward channel from the mobile data base station to the mobile end station and a reverse channel from the mobile end station to the mobile data base station (Figure 14.3). The mobile data base station supports full-duplex operation, while a particular implementation of a mobile end station may support either full-duplex or half-duplex operation. The forward channel is a contentionless broadcast channel that carries data packets transmitted by the mobile data base station only. In the forward channel, the mobile data base station forms data frames by adding standard High-Level Data Link Control (HDLC) terminating flags and inserting zero bits and then segments each frame into blocks of 274 bits. Information is received and decoded by all mobile end systems on the channel simultaneously; source and destination addressing is not a function of the CDPD MAC layer.

The reverse channel is shared among all mobile end systems. It carries packets transmitted by the mobile end systems. Arbitration of access and resolution of contention is controlled by each mobile end station and assisted by reverse channel status information transmitted by the mobile data base station on the forward channel.

Information is exchanged between mobile end systems and the mobile data base station in a series of data link layer frames. The MAC layer attempts to reliably deliver frames in sequence without loss or corruption. Because of the inherent noisy nature of the radio medium, or because of channel congestion, some

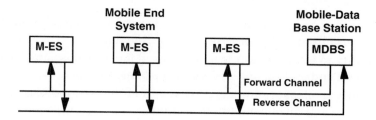

Fig. 14.3 Model of the CDPD Channel Stream

frames may not be delivered error-free within a reasonable period of time. The undelivered frames are discarded and any further recovery action is taken by the higher layers. In the reverse channel direction, the MAC layer at the mobile end station is able to signal the outcome of a transmission attempt to the LLC-layer protocol entity.

The MAC layer performs the following functions:

☞ **Data encapsulation:** LLC layer data is encapsulated in frames by MAC layer functions to preserve the original bit ordering, content, and length. Frame boundary delimitation and data transparency are maintained using flag sequence and zero-bit stuffing techniques. Error detection and correction is performed by using a "forward error correction code." Link layer frames are segmented into Reed-Solomon encoded code word blocks. The coding provides correction of errors in the received blocks with a very low probability of undetected error.

☞ **Medium access management:** Access to the reverse channel is controlled using a slotted nonpersistent Digital Sense Multiple Access with Collision Detection (DSMA/CD) access method. This is similar to Carrier Sense Multiple Access with Collision Detection (CSMA/CD) used for Ethernet (IEEE 802.2) and described in the next chapter.

In DSMA/CD, the reverse channel status is signaled by the transmission of channel status flags at periodic intervals on the forward channel, because the mobile end systems cannot sense the status of the reverse channel directly. Two types of reverse channel status flags are provided:

1. Channel busy/idle status to indicate that the reverse channel is busy or idle

2. Block decode status to indicate that the previous forward error correction block was successfully or unsuccessfully decoded by the mobile data base station

☞ **Channel stream timing and synchronization:** The forward channel transmits synchronization and timing indicators to permit mobile end systems to acquire and discriminate reverse channel status flag, forward error control block boundaries, and synchronize to the DSMA/CD microslot clock.

14.4.2.1 Forward Channel Procedure The forward channel corresponds to the mobile data base station transmit channel and represents data transmission from mobile data base station to the ensemble of mobile end systems. Transmission on the forward channel is considered continuous, although it may be suspended from time to time while the mobile data base station institutes the RF channel hop procedures. The forward channel stream synchronization words allow mobile end systems to synchronize to the forward channel data stream and control flags to control access to the reverse channel by the mobile end systems.

Each mobile end station tuned to the mobile data base station forward channel transmit frequency receives the same forward channel data stream. Within the forward channel data stream are the forward channel synchronization word, the reverse channel busy/idle status flags, the decode failure flags, and the 378-bit Reed-Solomon blocks. Each block contains the 8-bit color code assigned to the channel stream.

The mobile data base station inserts 8-bit color codes prior to every 274 bits of the bit stream; the resultant 282 bits, beginning with 8-bit color codes, are then encoded into a 378-bit coded block using a Reed-Solomon (63, 47) code word over a 64-ary alphabet (6-bit symbols). These bits constitute the block for the forward channel. Each block transmitted on the forward channel is combined with synchronization and other MAC control data prior to transmission (Figure 14.4).

The mobile data base station senses the reverse channel for the presence of a data transmission and sets the busy/idle flag to a busy state if any data transmissions are detected; otherwise an idle state is indicated. The mobile data base station obtains synchronization with blocks in a reverse channel transmission using the 22-bit reverse channel synchronization word.

The mobile data base station decodes the Reed-Solomon code word received on the reverse channel. The status of the Reed-Solomon decoding procedure is indicated on the forward channel with the decode status flag. If the Reed-Solomon decoding procedure is successful, the decode status flag is set to indicate that the block was successfully decoded. If the Reed-Solomon decoding procedure was not successful, the decode status flag is set to indicate that block was not successfully decoded.

14.4.2.2 Reverse Channel Procedure   The mobile data base station has permanent receive access to the reverse channel while the channel is available for CDPD use. It senses the state of the reverse channel and indicates this state to the mobile end station using control flags inserted periodically in the forward channel stream.

Access to the reverse channel is shared among mobile end systems by a random access DSMA/CD mechanism. Any mobile end station wishing to transmit first senses the busy/idle flag in the forward channel stream. If the status of the reverse channel is idle, the mobile end station initiates transmission. If the busy/idle flag is busy, the mobile end station waits for a random entrance delay interval before it senses the channel status again. The random entrance delay corre-

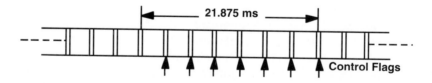

- **One block: 378 bits (63 symbols) of Reed-Solomon encoded data + 42 control bits = 420 bits per 21.875 ms (i.e., 19.2 Kb/s)**
- **6-bit control flags: 5 bits of forward sync work XOR'd with 5-bit busy/idle flags + 1 bit of 7-bit decode status flag**

Fig. 14.4 CDPD Data Service: Forward Channel Block Structure

sponds to the "number of microslots delay" before the channel status is again sensed. This gives the slotted nonpersistent aspect of DSMA (see chapter 15).

The reverse link frames are segmented and encoded into 378-bit blocks using the same Reed-Solomon code as in the forward channel. The mobile end station may form up 64 encoded blocks for transmission in a single reverse channel transmission burst. During the transmission, a 7-bit continuity indicator is interleaved into each coded block, or all zeros are used to indicate that this is the last block of the burst. The reverse channel block structure is shown in Figure 14.5.

Once the mobile end station gains access to the reverse channel, continued access to the reverse channel is controlled by the decode status flag. If the decoding procedure indicated to the mobile end station is unsuccessful, the transmitting mobile end station immediately ceases transmission and attempts to regain access to the channel after an appropriate exponential back-off retransmission delay. If the decoding procedure is successful, the mobile end station continues transmission of any subsequent blocks until all queued blocks are transmitted or the decoding status is indicated to be unsuccessful on one of the subsequent blocks.

14.4.3 Mobile Data Link Protocol (MDLP)

MDLP is a protocol that operates within the data link layer of the OSI architecture to provide logical link control services between mobile end systems and the mobile data intermediate systems. MDLP uses the services of the MAC layer to provide access to the physical channel and transparent transfer of link layer frames between data link entities.

MDLP conveys information between network layer (layer 3) entities across the CDPD air link (A) interface. MDLP supports mobile end systems sharing access to a single channel stream. The channel stream topology is that of point-to-

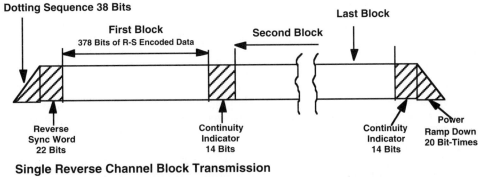

Single Reverse Channel Block Transmission

> = **38 (dotting sequence) + 22 (reverse sync word) + 378 (R-S encoded data) + 14 (continuity indicator) + 20 (power ramp down)**
>
> = **472 bits/24.58 ms (i.e., 19.2 Kb/s)**

Fig. 14.5 CDPD Data Service: Reverse Channel Block Structure

multipoint subnetwork. In this case, direct communication is possible only between the user side and network side of the channel stream. Direct communication between two mobile end systems on the same channel stream is not possible.

All data link protocol data units are transmitted in frames that are delimited by the MAC layer functions. MDLP includes the following functions:

☞ The provision of one or more logical data link connections on a channel stream. Discrimination between data link connections is by means of an address contained in each frame.

☞ Sequence control to maintain the sequential order of frame across a data connection.

☞ Detection of transmission, format, and operational errors on a data link connection.

☞ Recovery from detected transmission, format, and operational errors.

☞ Modification to the management entity of unrecoverable errors.

☞ Flow control.

☞ Sleep function that suspends the operation of a data link connection to allow implementation of mobile end station power conservation mechanism.

Data link layer functions provide the means for information transfer between multiple combinations of data link connection end points. The information transfer may be via point-to-point connections or via broadcast connections.

In point-to-point information transfer, a frame is directed to a single end point. The Temporary Equipment Identifier (TEI) in the frame identifies the user side (i.e., mobile end point) of the transfer. The network side of the transfer is implied by the channel stream topology. Unacknowledged or acknowledged multiple frame operation is possible.

In case of broadcast information, a frame is directed to all user-side end points on the channel stream. Higher-layer protocols may discriminate whether the information is intended for only a subset of all end systems (i.e., layer 3 multicast or broadcast groups). Only unacknowledged operation is possible in broadcast information transfer. Broadcast information transfer is indicated by a predefined value of TEI in the frame.

The concepts and procedures used in MDLP are based on CCITT Q.920 and Q.921 specifications. Many formats and procedures are nearly similar or identical; differences were introduced only where required to reflect the unique aspects of the CDPD environment. Differences between MDLP and LAPD are listed below:

1. Separation of MAC and LLC into separate sublayers. MDLP does not require frame flags, bit stuffing, or frame check sequences.

2. Modified addressing formats

   ✗ The address field is permitted to be variable length, using the optional multi-octet address-encoding technique from ISO-3309.

✗ The Service Access Point Identifier (SAPI) field is removed from the address field format. MDLP requires only a single data link service access point. The Data Link Connection Identifier (DLCI) consists only of the TEI.

✗ Terminology of TEI is changed. In MDLP, TEI refers to Temporary Equipment Identifier, whereas in LAPD it refers to Terminal Endpoint Identifier.

✗ TEI value of zero is reserved for communication between data link layer management entities.

✗ TEI value of 1 is reserved for use as a broadcast data link connection.

3. Modified error recovery procedures for Frame Number [N(S)] sequence errors: the Reject (REJ) supervisory frame is not used. Instead, the Selective Reject (SREJ) supervisory frame is used to request retransmission of a single Information transfer format (I) frame.

4. Additional network management functions: a simple data link loopback test consisting of the TEST command/response frame is available for network management functions.

A management control function to attempt to temporarily disable malfunctioning units is available in MDLP. This consists of the ZAP unnumbered command frame.

5. Modified TEI management formats and procedures: The mobile end station equipment ID is used in place of the Link Access Protocol-D channel (LAPD) random reference number. This guarantees uniqueness and avoids possible collisions between TEI Identity Requests. The Identity Denied and Identity Verify Request messages and procedures are not required and have been removed from MDLP.

6. Sleep mode supervision procedures: MDLP contains procedures to assist in implementing power-conserving strategies in the mobile end station.

7. Modified Exchange Information (XID) procedure: additional CDPD parameters have been identified and may be exchanged via XID procedures. These parameters include MAC and physical layer parameters as well as data link layer parameters. The XID procedure has been modified to allow broadcast by network side of parameters common to all layer management entities. In this case, no response is expected or permitted from the user-side entities.

14.4.4 Subnetwork-Dependent Convergence Protocol (SNDCP)

Connectionless-mode network protocols are intended to be capable of operating over connectionless-mode services derived from a variety of subnetworks and data links. To simplify the specifications of the protocol, their operation is defined with respect to an abstract underlying service. The SNDCP provides this underlying service. The SNDCP provides the following services to the network layer (layer 3):

☞ Connectionless-mode subnetwork service

☞ Transparent transfer of a minimum number of octets of user data

☞ User data confidentially

The SNDCP performs the following functions:

1. Mapping of data primitives.
2. Segmenting and reassembling of Network Protocol Data Units (NPDUs) to make more efficient use of air link resources than comparable segmentation and reassembly provided by the Connectionless Network Layer Service (CLNS).
3. Compression and recovery of redundant protocol control information from NPDUs to increase data link performance and efficiency. The compression protocol used is specific to the particular network layer and transport layer protocols.
4. Encryption/decryption of layer 3 NPDUs and exchange of encryption keys. Encryption is used to provide user data confidentiality over the CDPD A-Interface. This part defines the encoding and structure of Protocol Data Units (PDUs) for exchange of encrypted information.
5. Multiplexing of NPDUs from different layer 3 protocol entities onto a single data link connection. The SNDCP operates over the data link between mobile end station and mobile data intermediate system.

The SNDCP functions allow a single data link connection to be shared by multiple layer 3 protocol entities. Sharing a single data link connection requires that different network layer entities and protocols be identifiable.

Protocol header compression attempts to remove redundant protocol header information during transmission from a sequence of NPDUs transferred between a pair of source and destination addresses. The compression protocol used is specific to the particular network layer and transport layer protocols in use.

Protocol header compression is provided for mobile end systems using the TCP/IP protocol. This compression is specific to TCP/IP datagrams. Figure 14.6 shows the packet formation data flow, and Figure 14.7 indicates the protocol layers and data flow in the CDPD network.

14.4.5 Mobile Network Registration Protocol (MNRP)

This transport protocol is connectionless and is designed to operate in close conjunction with protocols providing connectionless-mode network service, such as Connectionless Network Protocol (CLNP) and Internetwork Protocol (IP). It is also designed to work with MNLP to provide mobility management services in the CDPD network.

This protocol provides configuration information about mobile end systems to mobile data intermediate systems. A mobile data intermediate system is informed of the Network Entity Identifiers (NEIs) supported by each mobile end station and SubNetwork Point of Attachment (SNPA) address of the mobile end

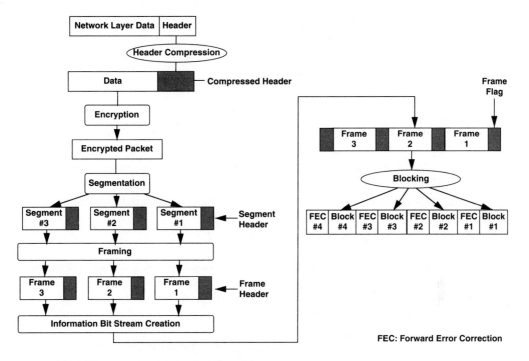

Fig. 14.6 CDPD Packet Formation Data Flow

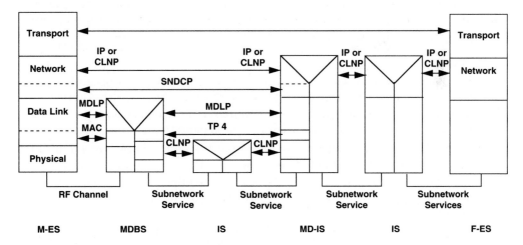

Fig. 14.7 CDPD Protocol Layers and Data Flow

station. That is, a serving mobile data intermediate system obtains information that allows it to associate destination network addresses to specific channel streams and specific data link connections on the channel stream. Once the mobile data intermediate system obtains this information, information and routing metrics about these NEIs may be disseminated to other mobile data intermediate systems to calculate routes to and from each mobile end station on the subnetwork.

The protocol provides the following services:

☞ Registration of NEIs associated with a mobile end station to a serving mobile data intermediate system

☞ Conveyance of data origin authentication information about a mobile end station and its NEIs

☞ Confirmation by mobile data intermediate system of its willingness and ability to provide network routing service to a mobile end station

☞ Deregistration of an NEI with the serving mobile data intermediate system

14.4.6 Radio Resource Management

The cells supporting CDPD services are assumed to be supporting other services, in particular those of AMPS. CDPD transmission may share both frequency and transmission equipment with other types of service. The areas served by a set of cells have some degree of overlap, so that roaming mobile end stations can maintain continuous service as they move from one cell to an adjacent cell. Two cells are considered adjacent if it is possible for a mobile end station to maintain continuous service by switching from one cell to the other. This switching process is called a "cell transfer."

CDPD is a value-added system; other users of RF spectrum are not necessarily aware of the presence of CDPD. This has several important consequences:

☞ The CDPD transmission must not interfere with transmission of other services.

☞ The CDPD transmission must be frequency-agile. It is not possible to dedicate channels for CDPD service. Thus, the only bandwidth available for CDPD use is the idle time between other uses of the same spectrum. Switching of RF frequencies within a cell is called a "channel hop."

☞ The CDPD network control is achieved in-band. A separate dedicated CDPD control channel is not available for the same reasons that a dedicated channel for CDPD data service is not available. To maintain compatibility with AMPS, CDPD control messages are not included in the AMPS control channels.

On the mobile end station side, radio resource management consists of algorithms required to acquire and track CDPD transmissions for initial acquisition, channel hop, and cell transfer. The algorithms are also required to maintain mobile end station transmission power at an appropriate level. These algorithms are designed to operate both with the assistance of information supplied by the network and in the absence of information supplied by the network. Within the mobile end station, the radio resource management entity is responsible for providing the following services:

☞ Acquiring an RF channel pair when the user requires a channel

☞ Ensuring that the channel stream is provided by a service provider acceptable to the mobile end station and, if necessary, that the channel stream lies within the mobile end station's home area

☞ Maintaining the continuity of channel stream when the RF channel changes

☞ Maintaining the continuity of channel stream when cell transfers to other cells occur

☞ Notifying the user of loss of the mobile data link when link loss occurs due to unrecoverable loss of signal

☞ Initiating MNRP procedures when the radio resource management detects a cell transfer to a cell controlled by a different mobile data intermediate system

☞ Monitoring the received signal strength on the forward channel and setting the appropriate transmit power level for the mobile end station

On the network side, radio resource management is implemented in the mobile data base station. It consists of algorithms and procedures concerned with correct configuration and selection of RF channels for the CDPD use. Radio resource management is also responsible for maintaining and distributing network configuration and power control data to mobile end systems to allow the mobile end stations to provide improved performance in tracking channel hops and cell transfers. Facilities are provided to allow the exchange of configuration data between mobile data base stations supporting adjacent cells. Finally, radio resource management provides facilities for signaling that the CDPD channel stream is congested and optionally for implementing load-sharing algorithms through directing individual mobile end systems to other RF channels carrying CDPD signals. Within the mobile data base station, the radio resource management entity provides the following services:

☞ Provisioning an RF channel pair to support a channel stream

☞ Changing the RF channel when the channel is required for non-CDPD activity

☞ Changing the RF channel when it is anticipated that the channel will shortly be required for analog voice activity

☞ Monitoring the received signal strength of the mobile end station's transmission and, if necessary, making adjustments to the mobile end station transmission power level

☞ Providing information so that mobile end systems can determine:

   1. The Local Call Identifier (LCI) of the cell
   2. The Channel Stream Identifier (CSI) of the channel stream within the cell
   3. The service provider operating the channel stream
   4. The home area in which the cell is located for mobile end station

☞ Providing information to the mobile end systems to support rapid reacquisition after channel hopping:

1. The set of RF channels allocated for CDPD use
2. The set of RF channels currently available to carry channel stream after the next channel hop on any channel stream

☞ Notifying the mobile end systems when the load on a channel stream is too high to support new registrations

☞ Notifying the adjacent mobile data base stations of channel hop as they occur

☞ Providing information to the mobile end systems to support rapid reacquisition of an RF channel after cell transfer:

1. The set of RF channel available for CDPD use in adjacent cells
2. The set of RF channels currently in use in adjacent cells

Communication between the mobile end station and mobile data base station radio resource entities take place using the services of the air link layer.

14.4.7 Air Link Security

The air link interface supports the following security functions:

☞ **Data link confidentiality:** All information contained in the information fields of subnetwork-dependent convergence protocols data units, including the network entity identifier of the mobile end systems, is transmitted across the air link in an encrypted form.

☞ **Mobile end station authentication:** Each network entity identifier used by the mobile end station is authenticated by the CDPD network to ensure that only the authorized possessor of the network entity identifier is using that network entity identifier.

☞ **Key management:** The network manages all secret keys required to operate the encryption algorithms used to support the first two functions.

☞ **Upgradability:** The network can support upgrade or replacement of the algorithms used to support the first three functions.

☞ **Access control:** The network can support restrictions on access by or to different NEIs, such as restrictions by location and screening lists.

Both the mobile end station and the mobile data intermediate system perform the following security functions:

☞ Exchange of secret keys to be used for encryption and decryption of data transmitted across the air link

☞ Encryption and decryption of data transmitted across the air link

☞ Exchange of authentication data across the air link

The mobile data intermediate system performs the following additional services:

☞ Exchange of authentication data with the home mobile data intermediate system

☞ Notification to the mobile end station of the results of authentication procedures of the home mobile data intermediate system

14.5 CDPD CAPABILITIES AND SERVICES

From the mobile subscriber's perspective, the CDPD network is simply a wireless mobile extension of traditional networks. By using a CDPD network, a subscriber can access the data applications that reside on traditional data networks (e.g., electronic mail, host access, directory services, transactions services). The user can also access value-added services on the CDPD network. These services are accessed via a variety of capabilities that are supported within the CDPD network.

14.5.1 CDPD Network Capabilities

The following capabilities allow a subscriber to communicate seamlessly with existing data communications services. The unique character of the CDPD network services is that one or both ends of data communications may be a mobile end system.

14.5.1.1 Support for Connectionless Network Services The CDPD network offers data services that are consistent with the Connectionless Network Layer Services (CLNS) defined in ISO-8348. In a connectionless service, the network routes each data packet individually using the destination address carried in the packet and having knowledge of current network topology. Connectionless network service is also called a datagram service.

The CDPD network is capable of offering network services in any of several different network protocols. Initially, only two network layer protocols are used: Connectionless Network Protocol—the standard OSI connectionless network protocol defined in ISO-8473; Internet Protocol (IP)—the network layer protocol of the Internet TCP/IP protocol.

Other network layer protocols can be supported on demand provided that the two communication end systems use the same network layer protocol since the CDPD network does not perform protocol translation.

Connection-oriented service may be provided by end-to-end protocol operating above the network layer (e.g., OSI TP4 or Internet TCP). The CDPD network service provider is not directly involved in the provisioning or the operation of transport layer and above services; this is the responsibility of the communicating end systems.

14.5.1.2 Support for Existing Network Interoperability The CDPD Network is designed to provide effective interconnection service to existing networks (see Figure 14.8). Each CDPD service provider network must appear to existing

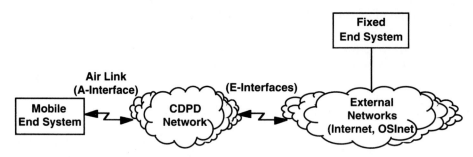

Fig. 14.8 CDPD Network as an Extension of Other Networks

networks as a peer, providing transparent data networking service to the mobile subscriber. To address the variety of existing networks, the CDPD design provides concurrent support of multiple connectionless network layer protocols.

Existing data network services can easily be integrated with the CDPD network. Since external interfaces to the CDPD network use existing network layer protocols, no special technical requirements are imposed on network service providers who wish to offer their services to the CDPD subscribers.

CDPD subscribers access the CDPD network services through the mobile end system, a mobile computing device that conforms to the air link interfaces. With Internet IP and CLNP, the following goals are met:

☞ No changes are needed to networking software of mobile end systems at layers above the network layer.

☞ Commercially available routers may be used without change.

14.5.1.3 Support for Mobility Management   The CDPD network is designed to provide services to manage data communications to the subscriber over a wide geographic area. The data connectivity is maintained as long as the mobile user remains within communications range of the CDPD service provider network. If the mobile subscriber roams into a geographic area serviced by a different CDPD service provider network, interworking procedures must be available between the separate CDPD service provider networks to continue the data connectivity. This mobility management service is, therefore, a distinguishing feature of CDPD network service.

14.5.2 CDPD Network Application Services

A CDPD service provider may provide specific value-added network application services through the CDPD network. The CDPD network permits access by its subscribers to a variety of network application services offered by other service providers. Since CDPD extends only to layer 3 (the network layer), many application services are available through the CDPD network. These services may be offered by the CDPD service providers and by application providers external to the CDPD network. Figure 14.9 shows the CDPD network as a collection of interconnected CDPD service provider networks. Each CDPD service provider serves a coverage area through its own network. A mobile end station in a CDPD service provider's serving area can have access to application services provided by many types of network application service providers. Three categories of network application service providers are identified in Figure 14.9.

1. CDPD service provider network applications: the X system in Figure 14.9.
2. CDPD-specific network applications provided by external sources: the Y system in Figure 14.9.
3. Generic network application made available to CDPD subscribers: the Z system in Figure 14.9.

14.5.2.1 CDPD Service Provider Network Applications A variety of network application services specifically designed for the CDPD subscribers may be offered by CDPD service providers. They are:

☞ **Broadcast services** that are sent to any or all CDPD subscribers in a fixed geographic coverage area (examples are local traffic information, weather information, and flight schedule information).

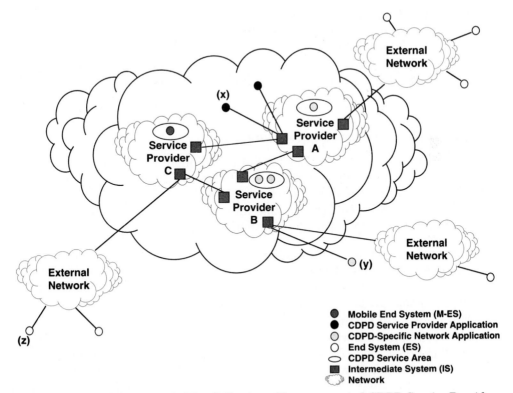

Fig. 14.9 The CDPD Network: The Collection of Interconnected CDPD Service Provider Networks

☞ **Multicast services** that deliver the same message to a specific group of subscribers. The CDPD multicast service is point-to-multipoint, or one-way, unacknowledged message transfer service. Multipoint service sends the message to all members of a multicast group regardless of their current location (examples are fleet management services and financial news subscription services).

☞ **Domain name services** provide the translation between host names and addresses in the CDPD network. This service allows humans and applications to use names formed from text rather than addresses formed from numbers.

☞ **Subscriber location services** that report the location of mobile end stations. This service may be of value to delivery services and fleet vehicle owners to track the location of their vehicles.

☞ **Message handling services** provide store-and-forward messaging capability. This service may include either e-mail services or automated storage of messages on failure of normal delivery mechanisms.

☞ **Directory services** provide access to a collection of information about objects of interest to CDPD network users that is typically found in a directory information base.

☞ **Individual value-added host services** may also be provided by service providers external to the network. These services are specially designed for use by CDPD subscribers and may target provision of timely information to the mobile user. Examples of such services include customized access to dynamic information bases and periodic reporting of a subscriber's stock portfolio.

14.5.3 CDPD Network Support Services

The CDPD network is a commercial public mobile data communications network that provides additional services to allow successful operation of the network. These services perform functions to assist the CDPD service provider in managing the network, collecting usage information to allow billing, protecting the service provider network from unauthorized usage, and so on.

☞ **Network entity identifier (NEI) authentication service** provides identification and verification. In the CDPD connectionless network service, data origin authentication is provided. This service provides the corroboration that the source of data received is as claimed.

☞ **Data confidentiality** ensures that information is not made available or disclosed to unauthorized individuals, entities, or processes. The CDPD network provides basic data confidentiality over the air link that protects the subscriber's data from casual eavesdropping. A CDPD service provider may provide additional value-added services that ensure higher levels of data content protection.

☞ **Access control service** provides the protection of unauthorized use of a resource, including the prevention of use of a resource in an unauthorized manner.

☞ **Accounting services** allow the CDPD service provider to collect network usage data on each subscriber for network billing. The accounting service collects all necessary data to provide accurate computation of charges. The data collected may include packet count, packet size, source and destination addresses, geographic location of end systems, time of transmission, and so on. This service also provides near-real-time information of network usage and facilitates CDPD service providers in providing up-to-date billing information on demand. Usage accounting data also allows service providers to plan system configuration on the basis of actual usage.

☞ **Network management services** ensure a high level of network availability. The CDPD network is designed to incorporate network management services to allow CDPD service providers to operate the network. The network management services provide timely information to the network operators to detect network faults in order to exercise controls to correct faults and to configure the network for optimal operation.

☞ **Network administrative services** provide functions and procedures necessary to maintain the CDPD network service. The CDPD network is designed as an interconnected set of CDPD service provider networks. For successful operation in such an environment, effective procedures should be established to coordinate shared resources (such as address assignment and interarea cell handoff) and shared information (such as accounting data for roaming subscribers).

14.6 SUMMARY

In this chapter, we discussed the Cellular Digital Packet Data (CDPD) network. We presented the network layer reference model for the CDPD network, including interfaces and entities. We provided details of the physical, medium access control, and logical link control layer and discussed the procedure for the forward and reverse channel between the mobile end systems and mobile data base station. We also presented the mobile data link protocol, the subnetwork-dependent and convergence protocol and the mobile network registration protocol. Details of radio resource management and air link security were also given. Finally, we concluded the chapter by providing a brief description of the CDPD services.

14.7 REFERENCES

1. CDPD Industry Coordinator, "Cellular Digital Packet Data Specification, Release 1.0," Kirkland, Washington, 1994.

2. International Standards Organization, "Data Communications—Network Service Definition," ISO 8348, April 1987.

3. International Standards Organization, "Protocol for Providing the Connectionless-Mode Network Service," ISO 8473, May 1987.

4. Pahlavan, K., and A. H. Levesque, "Wireless Data Communications," *Proceedings of the IEEE*, 82, no. 9 (September 1994): 1398–430.

5. Quick, R. R., and K. Balachandran, "Overview of the Cellular Digital Packet Data (CDPD) System," *Proceedings of the PIMRC '93*, Yokohama, Japan, (1993): 338–43.

6. TIA TR45.3 Subcommittee, "Stage 2 Description for Asynchronous Data," draft, September 1993.

7. TIA/EIA IS-54, "Cellular System Dual-Mode Mobile Station-Base Station Compatibility Standard," Telecommunications Industry Association, April 1992.

CHAPTER 15

Packet Radio Systems

15.1 INTRODUCTION

As the wireless world has moved from wireless voice to wireless data, a variety of solutions have been proposed and have resulted in several different approaches to providing data services.

In the previous chapter we discussed Cellular Digital Packet Data (CDPD) that is an overlay to cellular systems. In chapters 8, 9, and 12 we discussed, briefly, the support for data by PCS. In this chapter we will discuss systems that operate independently of cellular and PCS networks and currently provide data services. An entire book could be devoted to wireless data; we provide only an overview in this chapter. We encourage you to consult the references at the end of the chapter for more information.

All of these systems can trace their origins to work that was done at the University of Hawaii in the 1970s. Therefore we first examine a basic packet system, the ALOHA system, and calculate its throughput and delay characteristics. After we explain the basic operations, we examine the system architecture and operation of two commercial packet radio providers—ARDIS, a subsidiary of Motorola, and RAM Mobile Data. Finally we examine low-cost packet radio systems using the AX.25 protocol and the Network Operation System (NOS) with TCP/IP. Although these low-cost systems were designed by and are used by amateur (ham) radio operators, they have been deployed by commercial users throughout the world.

The materials presented in section 15.2 are also applicable to personal station operation on a control channel. When the original AMPS system was designed, the designers modeled the operation of the control channels after the ALOHA system. Subsequent system designs for other protocols used the same concepts. Thus, the operation of the ALOHA channel is applicable to a personal station accessing the network on the reverse control channel. We can understand the issues of access channel performance, overload control, multilevel precedence and preemption, and related access channels by studying the material in this chapter.

15.2 PACKET RADIO BASICS

Packet radio systems are used to provide a data network operation over a wide area using wireless instead of wired connections. The motivation for the use of wireless may be economic—where the costs of the wired connections are prohibitive—or may be physical—where the wired connections do not exist or the need for mobile communications may exist. Usually, one or more motivations exist at the same time. For example, in many developing countries, the communications infrastructure is either nonexistent in certain regions or the terrain prevents its easy deployment. Even developed countries may find regions within the country where communications are needed and the costs of the facilities are prohibitive.

In the 1970s, the University of Hawaii found itself in this position. It needed to connect its data center with terminals throughout the Hawaiian Islands. Researchers at the university applied packet data principles to a wireless environment. However, a wireless packet network has additional protocol issues that do not exist in a wired packet network. In a wired packet network, each receiver can hear the packet transmissions of all transmitters in the network and inhibit their companion transmitter until the channel is idle. In a wireless network, only the receiver at the network controller can receive all transmissions. Figure 15.1 shows an example of a packet data network. Stations S1, S3, and S6 are behind hills, and only the data center can receive their transmissions. Stations S2 and S4 are close to the data center and can receive each other's transmissions. Stations S5 and S7 are at a significant distance from the data center and therefore must use directional gain antennas to provide good link performance. Neither terminal can then hear any other terminal. Thus station S2 can transmit even though another station is transmitting and thus prevent either transmission from being received.

A packet station (Figure 15.2) consists of a data terminal, a Packet Assembler and Disassembler (PAD), a transceiver (transmitter and receiver), and an antenna. The terminal can be a data terminal, a personal computer, a Personal Digital Assistant (PDA), another data center, or any device that generates data. The PAD receives data messages from the terminal, disassembles the messages into short packets, and delivers them to the transmitter. The PAD also receives packets from receiver, assembles them into longer data messages, and delivers

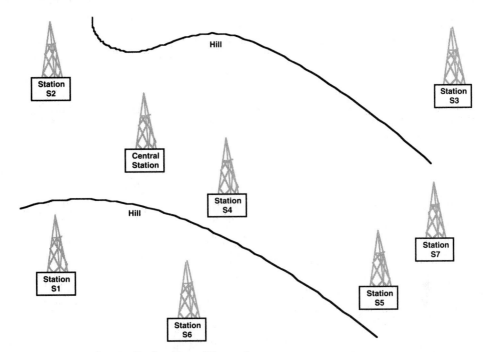

Fig. 15.1 Example of a Packet Data Network

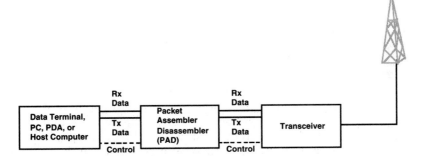

Fig. 15.2 Block Diagram of Packet Station

them to the terminal. In the simplest system, whenever a PAD has a packet to send, it delivers it to the transmitter for immediate transmission, independent of the transmission of data by other transmitters. This may result in a collision of packets from two or more stations with the result that none of the data is received. All stations must then retransmit their data. In the next section, as we study the throughput characteristics of the network, we will describe methods to minimize or avoid packet collisions. The data center will confirm the receipt of a packet by sending an acknowledgment. If the station does not receive an acknowledgment, then it retransmits the packet. Eventually all packets will be acknowledged.

15.2.1 Throughput Characteristics of a Packet Radio Network

In this section we examine the throughput characteristics of a packet radio network by making simplifying assumptions so that we can do the calculations. Designers of new packet networks must make computer models of the systems and do extensive simulations to determine actual performance. The simplified model demonstrates the key issues in packet network performance and permits us to examine methods of improving performance. Researchers have published extensively [1, 2, 6, 9, 10] on the performance modeling of packet networks.

Consider a system of packet radio stations where the network design forces packets to be of fixed length, τ seconds, and packets enter the system at a rate of λ per second. If two or more packets entering the system collide, then we assume that all packets interfere with each other and none are received. Whenever a packet collision occurs, the station does not receive an acknowledgment and retransmits the packet. If it retransmitted immediately, a collision would occur again and no data would ever get through the link. Therefore, whenever a collision occurs, the station will delay retransmission for a random time, X, with average value X_{avg}. Thus no packets ever leave the system; they only are delayed. For simplicity, we assume that X_{avg} is much greater than τ. We can then say that λ is the combined arrival rate for new packets and retransmitted packets.

The capacity of the channel, C, is the number of packets per second that are successfully transmitted and is lower than λ, the arrival rate. If we could stack the packets perfectly, with no overlap, then collisions would not occur and the channel would have capacity, C_{max}, equal to $1/\tau$ packets per second since the channel uses all available transmission time (see Figure 15.3).

$$C_{max} = \frac{1}{\tau} \tag{15.1}$$

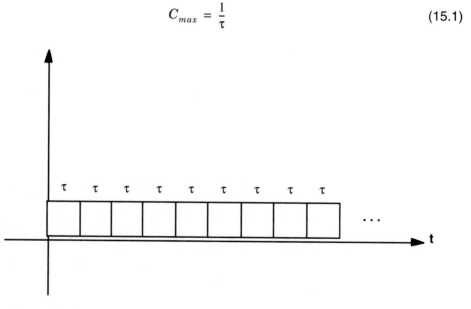

Fig. 15.3 Fixed-Length Packets Perfectly Spaced

In real channels, the packets arrive randomly (Figure 15.4) and collide. Thus the capacity is less than the arrival rate. We make an assumption that all packets arrive independently of all other packets and follow a Poisson distribution. This results in a probability density function of the interarrival times that has an exponential distribution.

$$P_s((t_{n+1} - t_n) < t) = \lambda e^{-\lambda t} \tag{15.2}$$

We consider a packet (Figure 15.5) that starts transmission at time t_n; its transmission will be successful if no new packet arrives during the interval from t_n to $t_n + \tau$, and no packet started transmission from $t_n - \tau$ to t_n. The probability of success will be the product of these two probabilities:

$$P_s = P[(t_{n+1} - t_n) > \tau] P[(t_n - t_{n-1}) > \tau] \tag{15.3a}$$

$$= \left(\int_\tau^\infty \lambda e^{-\lambda t} dt \right) \left(\int_\tau^\infty \lambda e^{-\lambda t} dt \right) \tag{15.3b}$$

$$= e^{-2\lambda \tau} \tag{15.3c}$$

The total number of successful packets, C, is the packet rate times the probability of success for each packet.

$$C = \lambda e^{-2\lambda \tau} \tag{15.4a}$$

$$C = C_{max} \lambda \tau e^{-2\lambda \tau} \tag{15.4b}$$

If we normalize with respect to λ by replacing $\lambda \tau$ with λ (thus letting $\tau = 1$, and $C_{max} = 1$), we get the standard normalized capacity equation for a packet radio system, called the ALOHA System:

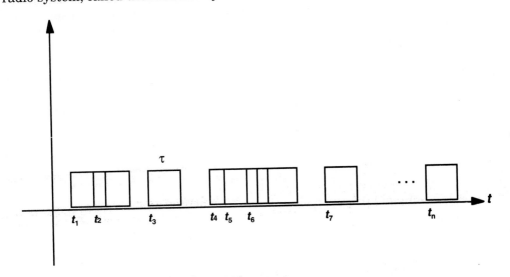

Fig. 15.4 Fixed-Length Packets Randomly Spaced

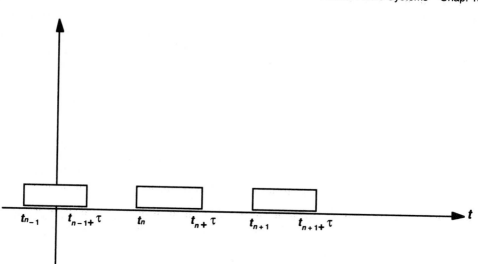

Fig. 15.5 Nonoverlapping Packets

$$C = \lambda e^{-2\lambda} \tag{15.5}$$

Eq. (15.5) is plotted in Figure 15.6 with the carried load, C, as a function of the offered load, λ. From this plot, we can see that the maximum capacity of the channel is 1/2e, at $\lambda = 0.5$; at high offered loads, the capacity of the channel is near zero. At very low traffic, most packets get through, and the carried load is only slightly smaller than the offered load. As traffic increases, collisions consume some of the capacity of the channel. At very high offered loads, almost every packet transmission has a collision, and very few packets are successful. As the system load increases, packet delays increase since packets must be retransmitted many times before success.

We calculate the channel delay for a simple ALOHA channel this way. Let the one-way propagation delay of the channel be $\alpha\tau$ and the acknowledgment packet length be $a\tau$, and $X_{avg} = \delta\tau$ is the average retransmission delay. Let the average time between retransmitted packets be $R\tau$, where R is the average delay measured in packet lengths. Then

$$R\tau = \tau + \alpha\tau + a\tau + \alpha\tau + \delta\tau \tag{15.6a}$$

$$R = 1 + a + 2\alpha + \delta \tag{15.6b}$$

For an offered load of λ and a carried load of C, the probability of success of a packet attempt is

$$P_s = \frac{C}{\lambda} \tag{15.7}$$

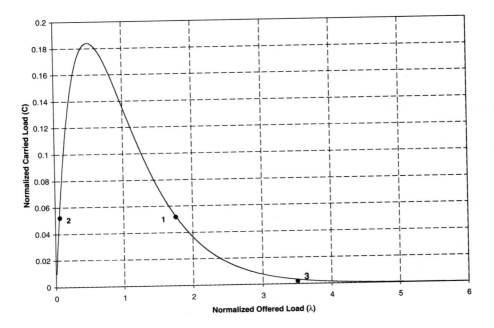

Fig. 15.6 Capacity of ALOHA Channel

A packet will be successful on the nth attempt if it failed on all $n-1$ attempts and succeeded on the nth attempt.

$$(P_s)_n = \left(1 - \frac{C}{\lambda}\right)^{n-1}\left(\frac{C}{\lambda}\right) \tag{15.8}$$

The average number of transmissions is therefore

$$N_{avg} = \sum_1^\infty \left(1 - \frac{C}{\lambda}\right)^{n-1}\left(\frac{C}{\lambda}\right) \tag{15.9}$$

The average number of retransmissions is

$$(N_{avg})_r = N_{avg} - 1 = \frac{\lambda}{C} - 1 \tag{15.10}$$

The average delay, $D\tau$, is the time to transmit one packet plus the time between transmissions multiplied by the average number of retransmissions plus the propagation delay.

$$D\tau = \tau + \left(\frac{\lambda}{C} - 1\right)R\tau + \alpha\tau \tag{15.11a}$$

$$D = 1 + \left(\frac{\lambda}{C} - 1\right)R + \alpha \tag{15.11b}$$

By examining Figures 15.6 and 15.7 we can see that the delays are long for capacities approaching 0.2 of maximum.

E X A M P L E   1 5 – 1

Problem Statement

For a simple ALOHA channel, we have used the following design parameters. What is the maximum capacity of the channel in packets per second? Calculate the carried load for the specified offered load. What is the probability that a packet will be successful? What is the average delay?

(a) The offered load, λ, is 20 packets per second.
(b) The packet length is 7.2 ms.
(c) The acknowledgment packet length is 0.5 ms.
(d) The average packet retransmission delay is 15 packet lengths.
(e) The propagation delay is 1 ms.

Solution

Since the average retransmission delay is much higher than the propagation delay, the Poisson arrivals assumption holds.

The maximum capacity, $C_{max} = 1/\tau = 1/0.0072 = 138.89 \approx 139$ packets per second.

The carried load, C, is

$$C = C_{max}\lambda\tau e^{-2\lambda\tau} = 139 \times 20 \times 0.0072 \times e^{-2 \times 20 \times 0.0072} = 15.0 \text{ packets per second}$$

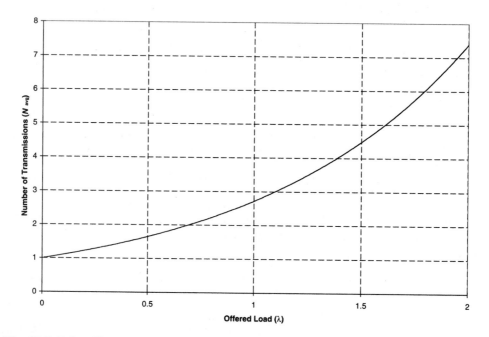

Fig. 15.7 Delay Characteristics of ALOHA Channel

$$P_s = \frac{15.0}{20} = 0.75 = 75\%$$

The propagation delay is $1/7.2 = 0.14$ packet lengths.

The packet acknowledgment is $0.5/7.2 = 0.07$ packet lengths.

$$R = 1 + 2 \times 0.14 + 0.07 + 15 = 16.35$$

$$D = 1 + \left(\frac{20}{15.0} - 1 \right) 16.35 + 0.14$$

$$D = 6.6 \text{ packet lengths}$$

Notice that although the average delay is less than the retransmission delay, some packets see a short delay and some see a long delay of more than 15 packet lengths. It is important to choose the proper values for the system design to assure optimum system operation.

A second effect that occurs because of the characteristic of the carried load versus offered load equation is that the channel becomes unstable. There are two stable states (states 1 and 2 in Figure 15.6) for a given carried load. State 1 has high offered load and long delays, and state 2 has low offered load and short delays. Since the offered load is the combination of newly arriving packets and older packets that were unsuccessful, even channels that have a low rate of newly arriving traffic can be in either state. When traffic smoothly and slowly increases and then smoothly and slowly decreases, the channel will remain stable, and lowering of the offered load will result in operation at point 1 on the curve. However, if the traffic suddenly increases (to point 3) for a period of time and then decreases suddenly, the channel may remain in state 2 for a long time before finally returning to state 1. The dynamic operation of the channel requires computer simulation. Any event that causes the system to suddenly jump from state 1 to state 3 can cause problems with a simple ALOHA channel. Examples of such events are: high calling volumes on an access channel during emergencies (e.g., snowstorms, hurricanes) and many users attempting to access the same data at the same time (news reports, weather information, stock transaction data, and so on).

15.2.2 Methods to Improve the Capacity of a Packet Radio Network

In the preceding section we discussed a simple ALOHA channel. In this section we will discuss system characteristics to improve the channel capacity. The first step to improve capacity is to segment the transmission times into slots of length τ (Figure 15.8). The transmitter can then send a packet only at the beginning of a slot. Thus, any transmitters that are ready during a slot must wait for the next slot to transmit.

Kleinrock and Tobagi [6] show that the normalized channel capacity for a slotted ALOHA channel is

$$C = \lambda e^{-\lambda} \tag{15.12}$$

Figure 15.9 shows the capacity of the slotted ALOHA channel as a function of the offered load. Notice that the capacity peaks at $1/e$, at $\lambda = 1.0$.

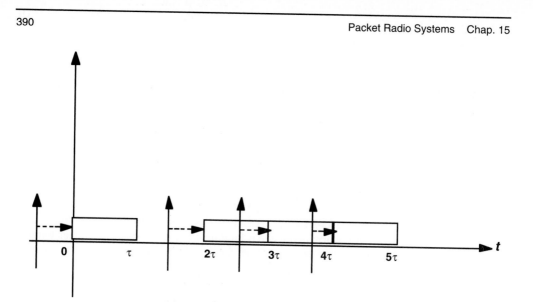

Fig. 15.8 Slotted ALOHA Channel Operation

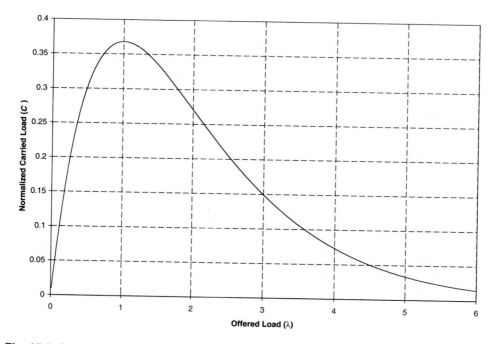

Fig. 15.9 Capacity of Slotted ALOHA Channel

E X A M P L E   1 5 – 2

Problem Statement

For the same conditions as Example 1, calculate the capacity of the channel.

Solution

Since the average retransmission delay is much higher than the propagation delay, the Poisson arrivals assumption holds.

The maximum capacity, $C_{max} = 1/\tau = 1/0.0072 = 138.89 \approx 139$ packets per second.

The carried load, C, is

$$C = C_{max}\lambda\tau e^{-\lambda\tau} = 139 \times 20 \times 0.0072 \times e^{-20 \times 0.0072} = 17.32 \text{ packets per second}$$

$$P_s = \frac{17.32}{20} = 0.87 = 87\%$$

The propagation delay is $1/7.2 = 0.14$ packet lengths.

The packet acknowledgment is $0.5/7.2 = 0.07$ packet lengths.

$$R = 1 + 2 \times 0.14 + 0.07 + 15 = 16.35$$

$$D = 1 + \left(\frac{20}{17.32} - 1\right)16.35 + 0.14$$

$$D = 3.67 \text{ packet lengths}$$

Notice that we have reduced the delay and simultaneously improved the capacity of the channel by using a slotted channel.

In both the ALOHA and the slotted ALOHA channels, the transmitter sends a packet without checking channel status (busy or idle). In many ALOHA systems, the transmitter cannot determine if the channel is being used. In Carrier Sense Multiple Access (CSMA) systems, the transmitter senses the state of the channel before transmitting. If the channel is busy, then the transmitter waits until the next slot. Thus, collisions during a transmission are avoided but not at the start of a slot, and the capacity of the channel improves somewhat. When a CSMA system does not work because some of the transmitters are hidden, then some of the capacity of the channel must be used to send the status of the reverse channel. Tobagi and Kleinrock [10] called this a "busy tone" solution. We see this in cellular and PCS systems that send busy-idle bits on the forward control channel to indicate reverse channel status.

The systems we have described so far transmit a packet immediately. Thus, any transmitters receiving a packet for transmission during a slot will transmit the packet in the next slot. If all transmitters delay by a random delay before transmitting, the traffic spreads out and the capacity of the channel improves. Kleinrock and Tobagi call this channel a "nonslotted, nonpersistent channel" and calculate the capacity of the channel as:

$$C = \frac{\lambda\tau e^{-\alpha\lambda\tau}}{\lambda\tau(1 + 2\alpha) + e^{-\alpha\lambda\tau}} \tag{15.13}$$

For the slotted, nonpersistent channel, they assert that the capacity (Figure 15.10) is

$$C = \frac{\alpha\lambda\tau e^{-\alpha\lambda\tau}}{\alpha + 1 - e^{-\alpha\lambda\tau}} \tag{15.14}$$

For both channels when the propagation delay is 0, (i.e., limit $\alpha \to 0$), then the capacity of the channel is

$$C = \frac{\lambda\tau}{1 + \lambda\tau} \tag{15.15}$$

The nonpersistent channel can therefore approach a capacity of 1 as the offered load increases. This is the ideal approach. The optimum values of the initial delay and the retransmission delay are functions of the offered load. Therefore, at high offered load, the central control of the system must send information to all transmitters to modify its parameters. We see this control capability on the control channels in cellular and PCS systems.

E X A M P L E   1 5 – 3

Problem Statement

For a nonpersistent slotted ALOHA channel, we use the following design parameters. What is the maximum capacity of the channel in packets per second? Calculate the carried load for the specified offered load. What is the probability that a packet will be successful? What is the average delay?

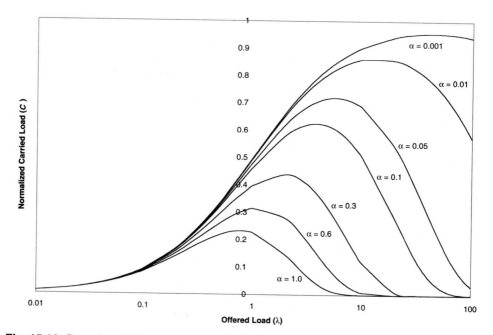

Fig. 15.10 Capacity of Slotted, Nonpersistent ALOHA Channel

(a) The offered load, λ, is 20 packets per second.

(b) The packet length, τ, is 7.2 ms.

(c) The acknowledgment packet length is 0.5 ms.

(d) The average packet transmission delay is 10 packet lengths.

(e) The average packet retransmission delay is 15 packet lengths.

(f) The propagation delay is 1 ms.

Solution

Since the average retransmission delay is much higher than the propagation delay, the Poisson arrivals assumption holds.

The maximum capacity, $C_{max} = 1/\tau = 1/0.0072 = 138.89 \approx 139$ packets per second.

The propagation delay is $1/7.2 = 0.14$ packet lengths.

The carried load, C, is

$$C = \frac{C_{max}\alpha\lambda\tau e^{-\alpha\lambda\tau}}{\alpha + 1 - e^{-\alpha\lambda\tau}} = \frac{139 \times 20 \times 0.0072 \times 0.14 \times e^{-0.14 \times 20 \times 0.0072}}{(0.14 + 1 - e^{-0.14 \times 20 \times 0.0072})}$$

$$C = 17.16 \text{ packets per second}$$

$$P_s = \frac{17.16}{20} = 0.86 = 86\%$$

The packet acknowledgment is $0.5/7.2 = 0.07$ packet lengths.

$$D = \left[\frac{20}{17.16} - 1\right][1 + 2 \times 0.14 + 0.07 + 15] + 1 + 10 + 0.14$$

$$D = 13.85 \text{ packet lengths.}$$

Notice that we have increased the delay and lowered the capacity at low offered loads. In the next example we will see how the channel performs at high loads.

E X A M P L E   1 5 – 4

Problem Statement

In Example 3, for a packet load of 100 packets per second, compare the carried load of the nonpersistent slotted ALOHA channel with the persistent slotted ALOHA channel.

Solution

Since the average retransmission delay is much higher than the propagation delay, the Poisson arrivals assumption holds.

The maximum capacity, $C_{max} = 1/\tau = 1/0.0072 = 138.89 \approx 139$ packets per second.

The propagation delay is $1/7.2 = 0.14$ packet lengths.

The carried load, C, for the nonpersistent channel is

$$C = \frac{C_{max}\alpha\lambda\tau e^{-\alpha\lambda\tau}}{\alpha + 1 - e^{-\alpha\lambda\tau}} = \frac{139 \times 100 \times 0.0072 \times 0.14 \times e^{-0.14 \times 100 \times 0.0072}}{(0.14 + 1 - e^{-0.14 \times 100 \times 0.0072})}$$

$$C = 53.66 \text{ packets per second}$$

The carried load, C, for the persistent slotted ALOHA channel is

$$C = C_{max}\lambda\tau e^{-\lambda\tau} = 139 \times 100 \times 0.0072 \times e^{-100 \times 0.0072} = 48.68 \text{ packets per second}$$

Notice that the capacity has improved but the round-trip propagation delay limits the results.

15.3 THE ARDIS PACKET RADIO NETWORK

ARDIS [7] was originally a joint venture by IBM and Motorola to provide a wireless data network. It is currently a wholly owned subsidiary of Motorola. ARDIS uses frequencies in the 800-MHz band to provide data transmission and reception to mobile terminals on the street and in buildings. ARDIS uses a single frequency in a city with multiple base stations. It is not a true cellular system in that there are no handoffs or frequency reuse in the system. The goal of ARDIS is deep coverage into buildings of a city. Thus, the coverage area of each base station significantly overlaps with the coverage area of neighboring base stations (Figure 15.11). A terminal in any given building in a city may receive signals from multiple base stations, thus guaranteeing coverage inside the building.

ARDIS terminals often transmit to multiple base stations simultaneously, and the network manages the signals received at the multiple base stations to select the signal with the lowest error rate. Each base station connects to a radio network controller, and the network controllers connect to one of four network nodes (Figure 15.12).

ARDIS uses 45-MHz separation between transmit and receive frequencies and uses the standard 25-kHz channel spacing for commercial two-way radio services. It does not use the same frequency band or channel spacing as the AMPS cellular system. The modulation method is FSK, with two or four frequen-

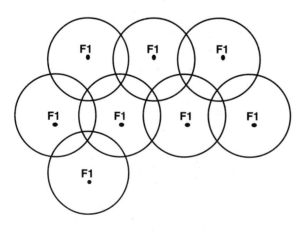

Note that only one frequency (F1) is used.

Fig. 15.11 Example of ARDIS Coverage of an Area

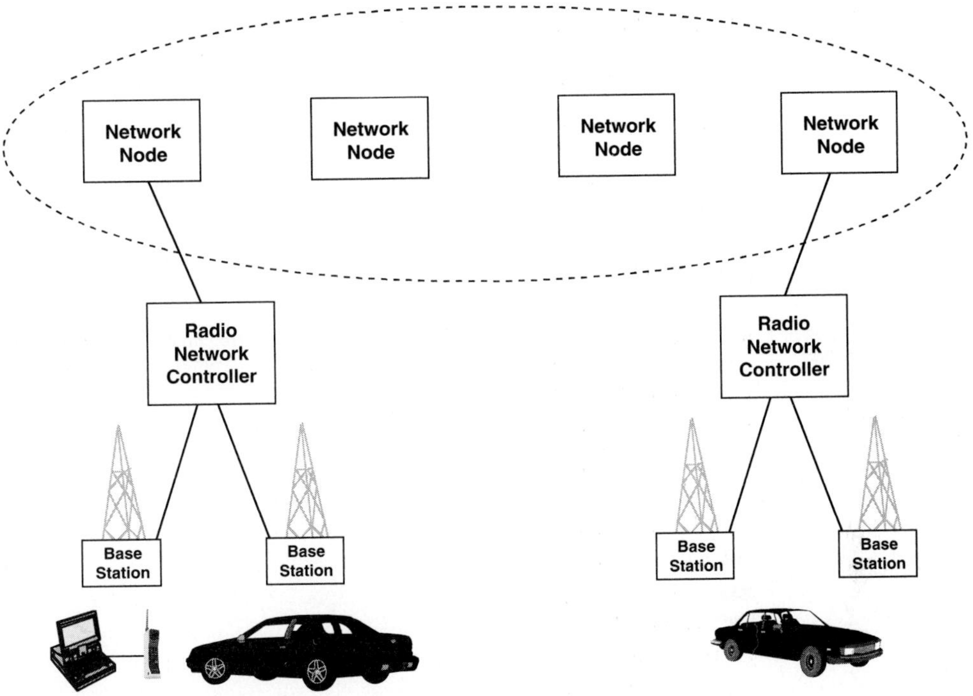

Fig. 15.12 ARDIS Network Architecture

cies, at a data rate of 4.8 or 19.2 kbs. The architecture of ARDIS is closed and minimum details are available [5].

15.4 THE RAM MOBILE DATA PACKET RADIO NETWORK[*]

The RAM Mobile Data Network [8] is a joint venture between BellSouth and RAM broadcasting to provide packet data throughout the United Kingdom and the United States using the Ericsson MOBITEX standard. The system provides a cellular packet system that uses multiple channels in a region with frequency reuse. The system provides data access for computerized dispatch services, wireless messaging, remote data collection, remote database access, wireless access to credit card verification, and automatic vehicle location. The system supports mobile-to-mobile communications and mobile-to-host-computer applications.

   Figure 15.13 shows the general architecture of the RAM system and Figure 15.14 shows the U.S.-specific architecture. Base stations, in a cellular pattern, connect to local switches for control and data transmission. Local switches also provide connections to host computers. Several local switches are connected to a

---

[*]The material in this section and Figures 15.13, 15.14, 15.15, and 15.16 are copyright RAM Mobile Data and are used with permission.

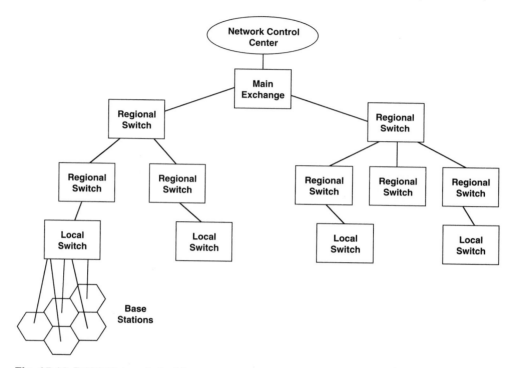

Fig. 15.13 RAM Network Architecture

regional switch, and the regional switches are connected to second-tier regional switches that are interconnected via a main exchange. Managing the network is a Network Control Center (NCC). All network elements have backup power and function uninterrupted during power outages.

In the United States, local and long distance telephone companies supply telephone service. The U.S. government has segmented the market into local areas (LATAs). The long-distance provider handles calls between LATAs. The overview of the RAM system in the United States (Figure 15.15) shows that the host computer and the mobile terminal can be in the same region (LATA) or in different regions of the country. When they are in different regions, the long-distance company provides the data communications service between the regions. In the U.S. network, host computers are connected to the local switches through a Front-End Processor (FEP).

Each of the network elements provides the following capabilities:

☞ The **Network Control Center** does not take part in traffic routing but supervises the entire RAM system to perform: operations and maintenance, provisioning of subscribers and the network, alarm handling and statistics, and traffic calculations and statistics. For a general description of these functions (but not specific to RAM), see chapter 11.

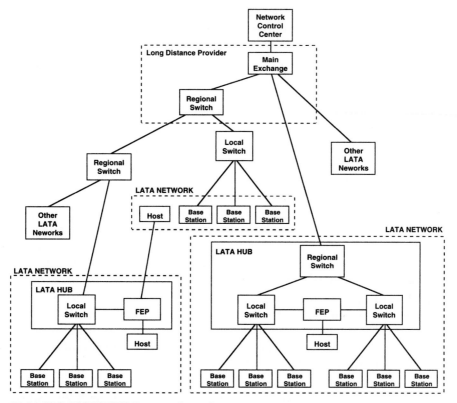

Fig. 15.14 U.S. MOBITEX System Overview

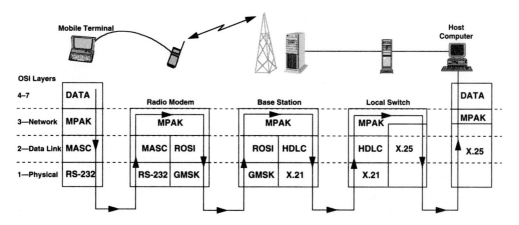

Fig. 15.15 Mobile-to-Host Communications Protocols

☞ The **main exchange** is the uppermost switch in the hierarchy and is the traffic node between regional switches. The functionality of the main exchange and the regional switches are virtually identical.

☞ The **regional switch** is a software-controlled, packet-switching data exchange. It handles traffic between local switches (subnets) and can function autonomously of the main exchange if links are lost. The regional switch provides protocol handling (HDLC and X.25), subscriber data for nodes below it, multiple connections to other switches, and alternate routing to other switches.

☞ The **local switch** is similar to the regional switch and handles communications with base stations. It is the only node that provides connections to host computers.

☞ The **base station** provides the radio interface to the mobile terminals. It also includes a software-controlled packet switch. The base station provides traffic routing within the radio coverage area, support for roaming terminals, subscriber database, protocol-handling operations, and security. It can function autonomously of the local switch if links are lost.

☞ The **Front-End Processor (FEP)**, running Mobigate software, provides protocol conversion to hosts supporting X.25, TCP/IP, and SNA protocols. It also provides the conversion from the connectionless protocol used in RAM to the connection-oriented protocol used by many host computers.

☞ The **backbone network** provides primary and alternate communication facilities between all subnets, regional switches, and the main exchange. The backbone network protocol may be either HDLC or X.25 and connections may consist of one or more 9.6- to 256-kbs links, depending on traffic requirements.

☞ The **host computers** are not part of the RAM network but are part of the customer's network. Applications on the host will depend on the user's needs. The applications can support X.25, TCP/IP, and SNA via the front-end processor.

Figure 15.15 shows the protocols used by the system. MOBITEX supports the seven-layer OSI protocol model. User data from a mobile terminal (either ASCII or specialized data) is segmented into MOBITEX PAcKets (MPAKs) of a maximum length of 512 user bytes plus addressing and network data; then it is delivered to the MOBITEX ASynChronous Communications (MASC) protocol at layer 2 and sent over an RS-232 link to the radio modem. The radio modem recovers the MPAK packets and sends them to the RAdio SIgnaling (ROSI) link layer where they are modulated onto a carrier (80 to 900 MHz, depending on frequency availability in a country) using Gaussian Minimum Shift Keying (GMSK). (See chapter 6 for a description of GMSK.) The base station recovers the MPAKs and packages them into an HDLC link layer with an X.21 physical layer for delivery to the local switch. The switch and the host computer communicate using the X.25 protocol (MPAK, LAP-B, and X.21).

In most cases, users of the system need to connect to one or more host computers or need to communicate with a group of terminals. Users of the RAM MOBITEX system require a subscription. Subscriptions are offered for host computers, individual mobile terminals, individuals (at terminals supporting individual accounts), groups of terminals, and groups of hosts. Access security is based on password protection, electronic serial numbers, and closed user groups. Mobile subscriptions are linked to a particular radio modem. The interface between the radio modem and the attached terminal (e.g., PC, printer, simple terminal, PDA) is via an RS-232 interface or other specialized interface. On the RS-232 interface, the radio modem supports the MASC protocol or an enhanced version of the industry standard "AT" dialing command set for modems.

Mobile radio modems can stay in constant touch with the system through the roaming capabilities of the system. Mobiles monitor and evaluate signals from surrounding base stations, and an algorithm in the mobile determines if and when a transfer to another base station is necessary. When a radio modem powers up, it starts monitoring the last base station's system channel stored in its memory. It will register at that base station if the received signal is stronger than the value good_base and will stay on that channel as long as the signal is stronger than bad_base and no neighboring base station's signal is stronger than the current better_base. If the signal drops below bad_base or a packet transmission fails, the mobile will start searching for a new base station. The mobile first searches all base stations on a neighbor list and if that fails it uses a default set of base stations. The search continues until a suitable base station is found. The base station periodically provides the information needed by the mobile station (good_base, bad_base, neighbor list, and so on) via "sweep and roam" signals (Figure 15.16).

The mobile transceiver operates on 896–901 MHz in the United States and 425–460 MHz in the United Kingdom at 8 kbs using GMSK modulation with a Bandwidth Time (BT) product of 0.3. The power output is 10 W for mobiles and

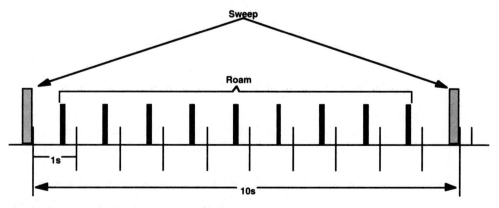

Fig. 15.16 Base Station Sweep and Roam Signals

2–4 W for portables, with power control at the levels of 0, –6, –12, and –18 dB below maximum power. The transmit-receive spacing is 39 MHz in the United States and either 14.5 or 6.5 MHz in the United Kingdom.

RAM states that their system has a radio channel capacity of 2,800 to 13,000 packets per hour, depending on the average message length. Additional radio link capacity can be added by adding more base stations. Each node has a capacity of 108,000 to 252,000 packets per hour, and new switches are being added to increase throughput to 750,000 packets per hour.

Additional details on the RAM system can be obtained from reference [8] or directly from RAM.

15.5 SIMPLE PACKET NETWORKS USING AX.25

In the early 1980s, a group of amateur radio operators formed the American Radio Relay League (ARRL) Ad Hoc Committee on Digital Communications. Their goal was to formulate a protocol that amateur radio operators could use to communicate digitally. Since one of the committee members was an author of the CCITT X.25 protocol, the committee adapted that protocol for use by amateurs and called it Amateur X.25, or AX.25. AX.25 is the layer 2 (link layer) protocol in the OSI reference model. Layers 3–7 are empty and the applications software is either the user typing or some application program (file transfer, for example). The physical layer is typically an FSK modem.

Since the specification of the protocol in 1984, amateurs have constructed an extensive international network of packet radio stations, bulletin boards, and relay nodes enabling amateurs in any portion of the world to communicate with amateurs in any other portion of the world. This network is independent of the Internet. Messages from the amateur network can connect to the Internet, but connections from the Internet to the amateur radio world are limited by the non-commercial nature of amateur radio. The success and low cost of the amateurs' solution has resulted in commercial networks using the same equipment, and that is our motivation for studying this packet network.

The AX.25 protocol adds the following capabilities to the X.25 protocol:

- The addition of an Unnumbered Information (UI) frame, thus permitting connectionless service for broadcast messages.
- The operation of the PAD in half-duplex and full-duplex radio environments.
- The support for a peer-to-peer connection rather than a master-slave connection that is typical of a terminal connected to a host computer.
- The ability to connect to other stations (for a two-way conversation) or to connect to a host computer using a multiport controller.
- The replacement of the X.25 address with an extended ASCII field to support amateur call signs (i.e., AD7I, WA2SFF, and W2CQH, to name a few). The commercial sector has extended this field further to support commercial call signs that are longer than amateur call signs.

The network architecture is similar to that shown in Figure 15.1 except that stations can connect to each other instead of only with the central data controller. The protocol does not support busy-tone signals; therefore it has limited channel capacity and hidden transmitter problems. While the system designers recommend that all stations use omnidirectional antennas to enable everyone to hear everyone else, many stations use directional antennas to improve range at the expense of collisions on the channel. During times of high channel occupancy, low-power stations often have trouble communicating since other stations cannot hear them and fail to inhibit their transmitters.

An amateur (or commercial) station consists of a data terminal or personal computer connected to a Terminal Network Controller (TNC), with the TNC connected to an FM radio transceiver. The current data rate is 1,200 baud and is limited by existing FM radio design. As manufacturers make new FM transceivers available with direct inputs to the modulators and demodulators, amateurs are moving to higher data rates (up to 9,600 baud). A typical packet node runs bulletin board software that supports e-mail and public bulletins. Often, public domain software is available for downloading. The packet node connects with other nodes via leased lines and high-speed packet modems (9.6 to 56 kbs) on other frequencies, thus forming a worldwide network.

The TNC (Figure 15.17) supports a dumb terminal via a Universal Asynchronous Receiver Transmitter (UART) and therefore has a native command-mode interpreter. The TNC sends messages from the terminal to the PAD and on to the AX.25 and HDLC protocol levels. The modem uses audio FSK (1,200 and 2,200 Hz) and connects to the FM transceiver. Typical TNCs cost from $120 to over $1,000, with a typical cost of $300. When a PC is used in place of a dumb terminal, the PC can either run terminal emulation software or host-mode software (Figure 15.18). When the PC runs host-mode software, the TNC must also operate in the host mode.

When the amateurs use a dumb terminal, they communicate with each other and with nodes via ASCII messages. When they use PCs, then communications can be ASCII or binary. Most PC software and most nodes support binary

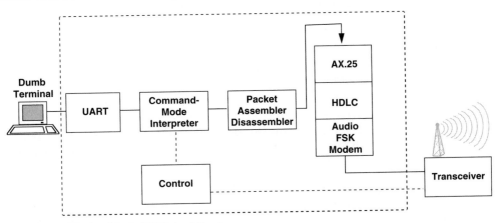

Fig. 15.17 Native-Mode TNC Architecture

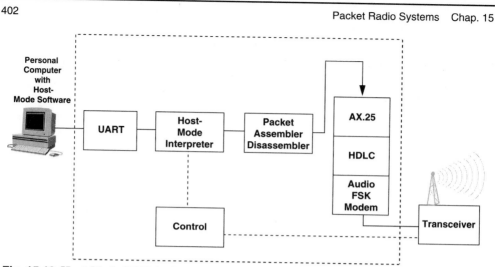

Fig. 15.18 Host-Mode TNC Architecture

file transfer via common protocols (Kermit, XMODEM, among others). Thus, users of the system can exchange software and messages.

15.6 THE NETWORK OPERATING SYSTEM AND TCP/IP

As the rest of the world has moved to TCP/IP, so have amateurs, using the KA9Q Network Operating System (NOS) [11]. NOS runs on the most popular microcomputer systems under MS-DOS, OS/2, VMS, and UNIX. The software is available for "free" via the Internet.* When the microcomputer runs NOS, the TNC must be placed in a transparent mode (KISS mode) that bypasses the command interpreter and the AX.25 layer firmware (see Figure 15.19). NOS then provides the services, the protocols, and the PAD functions.

The TCP/IP protocol is really a collection of programs and their associated protocols for sending data across computer networks. With NOS, the user has the full range of TCP/IP services available. Examples of these services are:

☞ **Address resolution** using the Address Resolution Protocol (ARP) resolves IP addresses with their correct hardware address (in this case, their ham-radio call sign or commercial call sign) to allow packets to be sent to the correct destination. This is done by sending a broadcast message to the host. If the host is available, it will return its address.

☞ **Directory services** using the Domain Name Service (DNS). Most people communicate with other people or host computers via their names, not their

---

*Use anonymous ftp to ftp.ucsd.edu. The files are found in the directory hamradio/packet/tcpip. Some of the documentation is copyrighted, but the software is available for free. Other sites on the Internet may also have the software, and periodically compact disks are published that include the software.

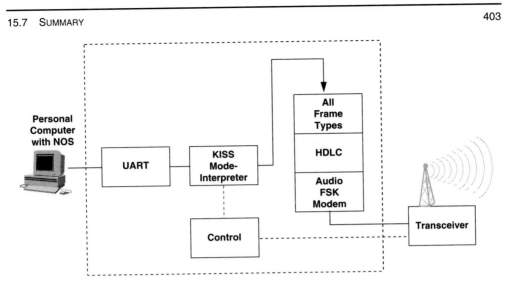

Fig. 15.19 KISS-Mode TNC Architecture

IP addresses. While the local computer may have a list of some IP addresses, the DNS can access a domain server (if one is available) to resolve IP addresses.

☞ **Electronic mail** using either Simple Mail Transfer Protocol (SMTP) or the Post Office Protocol (POP). POP supports hosts that may not always be up by storing them in a mail server until the destination host is available.

☞ **File transfers** using the file transfer protocol (ftp) provides a means of sending ASCII (text) or binary files between hosts.

☞ **Finger** provides a method of finding out information about another user on the network. The returned information is limited by the data stored by the user at his or her own host. Typical information is name, address, and phone number. Some people provide their schedule of meetings or alternate addresses at which they can be found.

☞ **Packet Internet Groper (PING)** provides a method to determine if a device at a given IP address is available.

☞ **Terminal emulation** using the Telnet program permits a user at one host to remotely access another remote host.

More information about NOS and TCP/IP can be obtained from the references or from the files on various ftp sites in the world.

15.7 SUMMARY

In this chapter we have examined the operation of a packet radio network. We showed that packet collisions limit the capacity of the system, and, without controls, high offered loads on the system can cause the channel capacity to decrease toward zero. We then examined two different packet radio providers—ARDIS and RAM mobile data. These two providers have taken different approaches to solving the problem of data transmission and reception. ARDIS uses extensive frequency reuse to cover the interior of buildings, whereas RAM uses a cellular concept. We then studied packet radio systems that use the AX.25 protocol and the Network Operating System with TCP/IP. While amateur radio operators designed the latter two systems, commercial users have adopted their low-cost solutions. We encourage you to consult the references for additional information about this topic.

15.8 PROBLEMS

1. Find at least three regions of the world where packet data communications are preferred over constructing a wired packet network.

2. Show that in Eq. 15.9 $\sum_{1}^{\infty} n \left(1 - \dfrac{C}{\lambda} \right)^{n-1} \dfrac{C}{\lambda} = \dfrac{\lambda}{C}$.

3. Calculate the maximum traffic-handling capacity for a simple ALOHA channel and for the slotted ALOHA channel.

4. For a simple ALOHA channel, the following design parameters have been identified. What is the maximum capacity of the channel in packets per second? Calculate the carried load for the specified offered load. What is the probability that a packet will be successful? What is the average delay? The offered load is 10 packets per second. The packet length is 20 ms. The acknowledgment packet length is 1.0 ms. The average packet retransmission delay is 10 packet lengths. The propagation delay is 1 ms. In what mode (high or low carried load) is the system operating?

5. Show that for a slotted ALOHA channel the normalized capacity is given by $C = \lambda e^{-\lambda}$.

6. Derive Eq. 15-13.

7. Derive Eq. 15-14.

8. Repeat Example 3 for a propagation delay of 0.1 ms.

9. For example 4 compute the packet delay.

15.9 REFERENCES

1. Abramson, N., "Multiple Access in Wireless Digital Networks," *Proceedings of the IEEE* (September 1994).

2. Abramson, Norm, "The Throughput of Packet Broadcasting Channels," in *Multiple Access Communications—Foundations for Emerging Technologies*, edited by Norm Abramson, 233–44.

3. Ackermann, John, "Getting Started with TCP/IP on Packet Radio," available through anonymous ftp at ftp.ucsd.edu as file intronos.zip.

4. AX.25 Amateur Packet-Radio Link-Layer Protocol, Version 2.0, *American Radio Relay League,* October 1984.

5. Gerhards, R., and P. Dupont, "The RD-LAP Air Interface Protocol," *INTERCOMM 93.*

6. Kleinrock, L., and F. A. Tobagi, "Packet Switching in Radio Channels: Part I Carrier Sense Multiple Access Modes and Their Throughput-Delay Characteristics," *IEEE Transactions on Communications,* COM 23, no. 12 (December 1975).

7. Pahlavan, K., and A. H. Levesque, "Wireless Data Communications," *Proceedings of the IEEE* (September 1994).

8. RAM Mobile Data—System Overview, Release 5.3, RAM Mobile Data, Woodbridge, New Jersey, November 1994.

9. Special Issue on Packet Radio Network, *Proceedings of the IEEE* (January 1987).

10. Tobagi, F. A., and L. Kleinrock, "Packet Switching in Radio Channels: Part II The Hidden Terminal Problem in Carrier Sense Multiple-Access and the Busy-Tone Solution," *IEEE Transactions on Communications,* COM 23, no. 12 (December 1975).

11. Wade, Ian, *NOSintro, TCP/IP over Packet Radio, An Introduction to the KA9Q Network Operating System,* Dowermain LTD, Luton, Bedfordshire: United Kingdom, 1992.

Channel Coding

A.1 INTRODUCTION

Data is typically segmented into information blocks, with each information block containing unique added redundancy. The resultant output is a data stream of n-bit blocks where each block consists of k information bits plus $(n-k)$ redundant bits. The addition of redundant bits is a trade-off between bit rate and the probability of bit error, i.e., bit error rate.

The channel coding mechanism adds redundancy to the source code by adding extra code digits in a controlled manner so that the receiver can detect and correct errors caused by channel. The resulting channel codes are usually binary, and they can be classified as block codes and convolutional codes.

Block codes divide the sequence of source digits into sequential blocks of k digits. Each k-digit block is mapped into an n-digit block of output digits, where $n > k$. Coded and uncoded blocks are compared on the basis that both systems use the same total time duration for a transmitted work. A block code can be a linear block code or nonlinear block code. A linear block can be a cyclic code or other. A cyclic code can have correction capability. The ratio of total number of information bits to the total number of bits in the code word k/n is referred to the **code rate** or **code efficiency**. The difference $(1 - k/n)$ is called the **redundancy**. The encoder is said to produce an (n, k) code. Block codes are memoryless codes because each output code word depends on only one source k-bit block and not any preceding blocks of digits. The higher the rate, the more efficient the code is.

Convolutional codes involve memory implemented in the form of a binary shift register having m storage elements or stages. The sequence of source digits, the message, is shifted into and along the register, a bit at a time. The output code becomes the sequence of v output bits for each message bit. A convolutional code (n, k, m) implies that it has n code word length, k message block length, and m memory blocks.

Linear block codes either may be systematic codes in which information bits are identifiable from the codeword or they may be scrambled codes in which information bits are not identifiable from the code word. If the information bit stream is

$$0\,0\,0\,0\,0\,0 \quad 0\,0\,0\,0\,0\,1 \quad 0\,0\,0\,0\,1\,1...$$

then

$$0\,0\,0\,0\,0\,0\,x\,x\,x \quad 0\,0\,0\,0\,0\,1\,x\,x\,x \quad 0\,0\,0\,0\,1\,1\,x\,x\,x...$$

or

$$x\,x\,x\,0\,0\,0\,0\,0\,0 \quad x\,x\,x\,0\,0\,0\,0\,0\,1 \quad x\,x\,x\,0\,0\,0\,0\,1\,1$$

are examples of systematic codes, and

$$0\,0\,x\,0\,0\,0\,x\,x\,0 \quad 0\,0\,x\,x\,0\,x \quad 0\,0\,1 \quad 0\,x\,x\,0\,0\,1\,x\,1...$$

is an example of scrambled code, where x is the added redundant bits.

In a coding system where $m(x)$ or m is the information code word block length, $g(x)$ or G is the linear block code encoder, and $v(x)$ or V is the linear block code block length; then we can write

$$v\,(x) \;=\; m\,(x)\,g\,(x) \tag{A.1a}$$

$$V \;=\; mG \tag{A.1b}$$

E X A M P L E   A – 1

Find G if $g(x) = 1 + x + x^3$ for a (7, 4) code.

$$n - k \;=\; 7 - 4 \;=\; 3$$

$$r_j \;=\; \text{Remainder of} \left[\frac{x^{n-k+j-1}}{g\,(x)} \right] j = 1,\dots,k$$

$$G \;=\; \begin{bmatrix} r_1 & 1 & 0 & . & . & 0 \\ r_2 & 0 & 1 & . & . & 1 \\ . & . & . & . & . & . \\ . & . & . & . & . & . \\ . & . & . & . & . & . \\ r_k & 0 & 0 & . & . & 1 \end{bmatrix} \;=\; [r|I]$$

$$r_j = Re\left[\frac{x^{3+j-1}}{1+x+x^3}\right] \quad j = 1, 2, 3, 4$$

$$r_1 = Re\left[\frac{x^3}{1+x+x^3}\right] = 1+x \rightarrow r_1\begin{bmatrix}1 & 1 & 0\end{bmatrix}$$

$$r_2 = Re\left[\frac{x^4}{1+x+x^3}\right] = x+x^2 \rightarrow r_2\begin{bmatrix}0 & 1 & 1\end{bmatrix}$$

$$r_3 = Re\left[\frac{x^5}{1+x+x^3}\right] = 1+x+x^2 \rightarrow r_3\begin{bmatrix}1 & 1 & 1\end{bmatrix}$$

$$r_4 = Re\left[\frac{x^6}{1+x+x^3}\right] = 1+x^2 \rightarrow r_4\begin{bmatrix}1 & 0 & 1\end{bmatrix}$$

$$G = \begin{bmatrix} 1 & 1 & 0 & 1 & 0 & 0 & 0 \\ 0 & 1 & 1 & 0 & 1 & 0 & 0 \\ 1 & 1 & 1 & 0 & 0 & 1 & 0 \\ 1 & 0 & 1 & 0 & 0 & 0 & 1 \end{bmatrix} = [r|I]$$

E X A M P L E   A – 2

$m = [1\ 1\ 0\ 1]$ and

$$G = \begin{bmatrix} 1 & 1 & 0 & 1 & 0 & 0 & 0 \\ 0 & 1 & 1 & 0 & 1 & 0 & 0 \\ 1 & 1 & 1 & 0 & 0 & 1 & 0 \\ 0 & 0 & 1 & 0 & 0 & 0 & 1 \end{bmatrix}$$

$$V = mG$$

$$V = \begin{bmatrix}1 & 1 & 0 & 1\end{bmatrix}\begin{bmatrix} 1 & 1 & 0 & 1 & 0 & 0 & 0 \\ 0 & 1 & 1 & 0 & 1 & 0 & 0 \\ 1 & 1 & 1 & 0 & 0 & 1 & 0 \\ 0 & 0 & 1 & 0 & 0 & 0 & 1 \end{bmatrix} = \begin{bmatrix}0 & 0 & 0 & 1 & 1 & 0 & 1\end{bmatrix}$$

E X A M P L E   A – 3

Determine the Cyclic Redundancy Check (CRC) for the character $a = [0\ 1\ 1\ 0\ 0\ 0\ 0\ \underline{1}]$, where $\underline{1}$ is the parity bit, if $g(x) = x^5 + x^2 + 1$.

$$k = 8; n - k = 5, \text{ therefore } n = 13$$

$$m(x) = \begin{bmatrix}0 & 1 & 1 & 0 & 0 & 0 & 0 & 1\end{bmatrix} \text{ or } x^6 + x^5 + 1$$

$$v(x) = f(x) + R(x)$$

where $f(x) = x^{n-k}m(x) = x^5(x^6 + x^5 + 1) = x^{11} + x^{10} + x^5$

and

$$R(x) = Re\left[\frac{f(x)}{g(x)}\right] = Re\left[\frac{x^{11} + x^{10} + x^5}{x^5 + x^2 + 1}\right] = x^4 + x + 1$$

$$\therefore v(x) = x^{11} + x^{10} + x^5 + x^4 + x + 1$$

$$v(x) = \begin{bmatrix} 0 & 1 & 1 & 0 & 0 & 0 & 0 & 1 & 1 & 0 & 0 & 1 & 1 \end{bmatrix}$$

A.2 HAMMING CODE

Hamming codes are linear cyclic systematic codes.

Hamming codes with $n = 2^j - 1$ and $k = n - j$ exist for any integer $j \geq 3$. The rate of this code is $R = k/n = (2^j - j - 1)/(2^j - 1)$ which approaches 1.0 as $j \rightarrow \infty$. Table A.1 lists n, k, and R for the first seven Hamming codes.

The Hamming codes are defined by parity-check matrix containing $n - k = j$ rows and n columns. The columns consist of all possible nonzero j-component vectors.

$$[H] = [r^T | I] \tag{A.2}$$

where I = unit matrix and r_i = Remainder of $[(x^{n-k+i-1})/g(x)]$ $i = 1, 2, ..., k$ and $g(x)$ is the generator polynomial. Table A.2 gives Hamming code generator polynomials up to block length $n = 2^{10} - 1$.

Table A.1 Hamming Codes

j	n	k	R
3	7	4	0.57
4	15	11	0.73
5	31	26	0.84
6	63	57	0.90
7	127	120	0.94
8	255	247	0.97
9	511	502	0.98
10	1,023	1,013	0.99

Table A.2 Hamming Code Generators

j	$g(x)$
3	$1 + x + x^3$
4	$1 + x + x^4$
5	$1 + x^2 + x^5$
6	$1 + x + x^6$
7	$1 + x^3 + x^7$
8	$1 + x^2 + x^3 + x^4 + x^8$
9	$1 + x^4 + x^9$
10	$1 + x^3 + x^{10}$

The Hamming weight of code word V [of $v(x)$] is

$$W(V) = \sum_1^n 1\text{'s in } V \text{ and}$$

$$d_{min} = min\,[W\,(v \subset V)]$$

for any nonzero V. If p denotes the number of errors detected and q is the number of errors corrected, then the detection capability of (n, k) codes will be $p = d_{min} - 1$, and correction capability will be $q = (d_{min} - 1)/2$. Table A.3 provides a summary of the Hamming codes.

Table A.3 Summary of Hamming Codes[*]

	Hamming Single-Error Correction Codes	Hamming Double-Error Detection Codes
Code length (n)	$n = 2^j - 1$	$n = 2^j - 1$
Message length (k)	$k = 2^j - j - 1$	$k = 2^j - j - 2$
Parity-check length $(n - k)$	j	$j + 1$
Message	$m(x)$	$m(x)$
Generator polynomial	$g(x)$	$p(x) = (1 + x)\,g(x)$
Code word	$v(x) = m(x)\,g(x)$	$v(x) = m(x)\,p(x)$

[*]j is an integer.

E X A M P L E   A – 4

Problem Statement

Choose a Hamming single-error correction code. Find the code word for $m = [1\ 0\ 0\ 1]$

Solution

$$m\ (x)\ =\ 1 + x^3$$

$$k\ =\ 2^j - j - 1$$

$$4 = 2^j - j - 1,\ \ j\ =\ 3,\ \ g\ (x)\ =\ 1 + x + x^3$$

$$n\ =\ 2^3 - 1\ =\ 7$$

$$v\ (x)\ =\ (1 + x^3)\ (1 + x + x^3)\ =\ 1 + x + x^4 + x^6\ =\ \begin{bmatrix} 1\ 0\ 1\ 0\ 0\ 1\ 1 \end{bmatrix}$$

A.3 BCH Codes

A BCH code is defined as (n, k, q) where n = code word length = $2^j - 1$, k = information length, and q = error correction capability. The following constraints are imposed:

$$(n - k) \leq jq$$

$$d_{min} \geq jq + 1$$

Table A.4 specifies some of the BCH codes.

E X A M P L E   A – 5

Problem Statement

Choose a BCH double-error correction code to transmit a 7-bit block signal. Determine the code word if $m = [1\ 0\ 1\ 0\ 1\ 0\ 1]$

$$(n - k) \leq 4 \times 2$$

Table A.4 BCH Codes

(n, k, q)	j	$g(x)$
$(7, 4, 1)$	3	$1 + x + x^3$
$(15, 11, 1)$	4	$1 + x + x^4$
$(15, 7, 2)$	4	$1 + x^4 + x^6 + x^7 + x^8$
$(15, 3, 3)$	4	$1 + x + x^2 + x^4 + x^5 + x^8 + x^{10}$

$$n = 8 + k = 15$$

$$\text{BCH} = \begin{bmatrix} 15, 7, 2 \end{bmatrix}$$

$$g(x) = 1 + x^4 + x^6 + x^7 + x^8$$

$$v(x) = m(x) g(x)$$

$$v(x) = (1 + x^2 + x^4 + x^6)(1 + x^4 + x^6 + x^7 + x^8)$$

$$v(x) = 1 + x^2 + x^6 + x^7 + x^8 + x^9 + x^{10} + x^{11} + x^{13} + x^{14}$$

$$v = \begin{bmatrix} 1\ 1\ 0\ 1\ 1\ 1\ 1\ 1\ 1\ 0\ 0\ 0\ 1\ 0\ 1 \end{bmatrix}$$

A.3.1 Nonbinary BCH Codes

These codes are defined as

☞ Block length = $n = q^j - 1$
☞ Parity check digits = $n - k \le 2jt$
☞ Maximum t-error correction if $d_{min} \ge 2t + 1$
☞ $q = p^m$; p = prime number

where m, j, and t are integers.

A.4 REED-SOLOMON CODES

The Reed-Solomon codes are defined as:

☞ Block length = $n = q - 1$ where $q = 2^m$
☞ Parity check digits = $n - k = 2t$
☞ $d_{min} = 2t + 1$

Orthogonal Functions

The Code Division Multiple Access (CDMA) system requires orthogonal functions for channel selection. Since CDMA systems use the same frequency bands for all transmissions, only a different code can be used to select a channel. This appendix examines how CDMA forms those orthogonal functions.

Orthogonal functions have the characteristic that:

$$\sum_{k=0}^{M-1} W_i(k\tau)\, W_j(k\tau - n\tau) \ = \ 0, \quad i \neq j, \ \mathrm{n} = 0, ..., M-1 \tag{B.1}$$

$$\sum_{k=0}^{M-1} W_i(k\tau)\, W_i(k\tau - n\tau) \ = \ 0, \ \mathrm{n} = 1, ..., M-1 \tag{B.2}$$

where:
$W_i(k\tau)$, and $W_j(k\tau)$ are the i-th and j-th orthogonal members of an orthogonal set,
M is the length of the set, and
τ is the symbol duration.

One orthogonal function is the Walsh function. The basic set of Walsh functions is the set of four patterns—0000, 0101, 0011, 0110. When the Walsh functions modulate the transmitter, biphase shift keying is used. Thus, 0 represents 0° phase shift and 1 represents 180° phase shift. The 64 orthogonal Walsh functions provide the basis for the CDMA modulation method used in PCS. Table B.1 describes the bit patterns for the 64 orthogonal Walsh codes. For example,

W_{27} = 0110 0110 1001 1001 1001 1001 0110 0110 0110 0110 1001 1001 1001 1001 0110 0110

It is left as an exercise for the reader to show that all 64 codes are orthogonal to each other.

The W-CDMA system requires additional orthogonal codes and uses combinations of the basic Walsh 64 set to build code sets for 5-MHz and 10-MHz CDMA systems. The code set for the 5-MHz bandwidth system is constructed by using sets of the $Walsh_{64}$ and the inverted $Walsh_{64}$ sets (shown with a bar over the set) as described in Tables B.3 to B-5. These combinations are chosen to maintain the orthogonality of the codes sets.

When the W-CDMA system uses 15-MHz bandwidth, then a Hadamard code set is used. Tables B.6 to B.10 describe the characteristics of the 768-code Hadamard code set.

Table B.1 Walsh Chip within a Walsh Function[*]

	0123	4567	11 8901	1111 2345	1111 6789	2222 0123	2222 4567	2233 8901	3333 2345	3333 6789	4444 0123	4444 4567	4455 8901	5555 2345	5555 6789	6666 0123
0	0000	0000	0000	0000	0000	0000	0000	0000	0000	0000	0000	0000	0000	0000	0000	0000
1	0101	0101	0101	0101	0101	0101	0101	0101	0101	0101	0101	0101	0101	0101	0101	0101
2	0011	0011	0011	0011	0011	0011	0011	0011	0011	0011	0011	0011	0011	0011	0011	0011
3	0110	0110	0110	0110	0110	0110	0110	0110	0110	0110	0110	0110	0110	0110	0110	0110
4	0000	1111	0000	1111	0000	1111	0000	1111	0000	1111	0000	1111	0000	1111	0000	1111
5	0101	1010	0101	1010	0101	1010	0101	1010	0101	1010	0101	1010	0101	1010	0101	1010
6	0011	1100	0011	1100	0011	1100	0011	1100	0011	1100	0011	1100	0011	1100	0011	1100
7	0110	1001	0110	1001	0110	1001	0110	1001	0110	1001	0110	1001	0110	1001	0110	1001
8	0000	0000	1111	1111	0000	0000	1111	1111	0000	0000	1111	1111	0000	0000	1111	1111
9	0101	0101	1010	1010	0101	0101	1010	1010	0101	0101	1010	1010	0101	0101	1010	1010
1	0011	0011	1100	1100	0011	0011	1100	1100	0011	0011	1100	1100	0011	0011	1100	1100
11	0110	0110	1001	1001	0110	0110	1001	1001	0110	0110	1001	1001	0110	0110	1001	1001
12	0000	1111	1111	0000	0000	1111	1111	0000	0000	1111	1111	0000	0000	1111	1111	0000
13	0101	1010	1010	0101	0101	1010	1010	0101	0101	1010	1010	0101	0101	1010	1010	0101
14	0011	1100	1100	0011	0011	1100	1100	0011	0011	1100	1100	0011	0011	1100	1100	0011
15	0110	1001	1001	0110	0110	1001	1001	0110	0110	1001	1001	0110	0110	1001	1001	0110
16	0000	0000	0000	0000	1111	1111	1111	1111	0000	0000	0000	0000	1111	1111	1111	1111
17	0101	0101	0101	0101	1010	1010	1010	1010	0101	0101	0101	0101	1010	1010	1010	1010
18	0011	0011	0011	0011	1100	1100	1100	1100	0011	0011	0011	0011	1100	1100	1100	1100
19	0110	0110	0110	0110	1001	1001	1001	1001	0110	0110	0110	0110	1001	1001	1001	1001
20	0000	1111	0000	1111	1111	0000	1111	0000	0000	1111	0000	1111	1111	0000	1111	0000
21	0101	1010	0101	1010	1010	0101	1010	0101	0101	1010	0101	1010	1010	0101	1010	0101
22	0011	1100	0011	1100	1100	0011	1100	0011	0011	1100	0011	1100	1100	0011	1100	0011
23	0110	1001	0110	1001	1001	0110	1001	0110	0110	1001	0110	1001	1001	0110	1001	0110
24	0000	0000	1111	1111	1111	1111	0000	0000	0000	0000	1111	1111	1111	1111	0000	0000
25	0101	0101	1010	1010	1010	1010	0101	0101	0101	0101	1010	1010	1010	1010	0101	0101
26	0011	0011	1100	1100	1100	1100	0011	0011	0011	0011	1100	1100	1100	1100	0011	0011
27	0110	0110	1001	1001	1001	1001	0110	0110	0110	0110	1001	1001	1001	1001	0110	0110
28	0000	1111	1111	0000	1111	0000	0000	1111	0000	1111	1111	0000	1111	0000	0000	1111
29	0101	1010	1010	0101	1010	0101	0101	1010	0101	1010	1010	0101	1010	0101	0101	1010
30	0011	1100	1100	0011	1100	0011	0011	1100	0011	1100	1100	0011	1100	0011	0011	1100
31	0110	1001	1001	0110	1001	0110	0110	1001	0110	1001	1001	0110	1001	0110	0110	1001
32	0000	0000	0000	0000	0000	0000	0000	0000	1111	1111	1111	1111	1111	1111	1111	1111
33	0101	0101	0101	0101	0101	0101	0101	0101	1010	1010	1010	1010	1010	1010	1010	1010
34	0011	0011	0011	0011	0011	0011	0011	0011	1100	1100	1100	1100	1100	1100	1100	1100
35	0110	0110	0110	0110	0110	0110	0110	0110	1001	1001	1001	1001	1001	1001	1001	1001
36	0000	1111	0000	1111	0000	1111	0000	1111	1111	0000	1111	0000	1111	0000	1111	0000
37	0101	1010	0101	1010	0101	1010	0101	1010	1010	0101	1010	0101	1010	0101	1010	0101
38	0011	1100	0011	1100	0011	1100	0011	1100	1100	0011	1100	0011	1100	0011	1100	0011
39	0110	1001	0110	1001	0110	1001	0110	1001	1001	0110	1001	0110	1001	0110	1001	0110
40	0000	0000	1111	1111	0000	0000	1111	1111	1111	1111	0000	0000	1111	1111	0000	0000
41	0101	0101	1010	1010	0101	0101	1010	1010	1010	1010	0101	0101	1010	1010	0101	0101
42	0011	0011	1100	1100	0011	0011	1100	1100	1100	1100	0011	0011	1100	1100	0011	0011
43	0110	0110	1001	1001	0110	0110	1001	1001	1001	1001	0110	0110	1001	1001	0110	0110

Table B.1 Walsh Chip within a Walsh Function[*] (Continued)

44	0000	1111	1111	0000	0000	1111	1111	0000	1111	0000	0000	1111	1111	0000	0000	1111
45	0101	1010	1010	0101	0101	1010	1010	0101	1010	0101	0101	1010	1010	0101	0101	1010
46	0011	1100	1100	0011	0011	1100	1100	0011	1100	0011	0011	1100	1100	0011	0011	1100
47	0110	1001	1001	0110	0110	1001	1001	0110	1001	0110	0110	1001	1001	0110	0110	1001
48	0000	0000	0000	0000	1111	1111	1111	1111	1111	1111	1111	1111	0000	0000	0000	0000
49	0101	0101	0101	0101	1010	1010	1010	1010	1010	1010	1010	1010	0101	0101	0101	0101
50	0011	0011	0011	0011	1100	1100	1100	1100	1100	1100	1100	1100	0011	0011	0011	0011
51	0110	0110	0110	0110	1001	1001	1001	1001	1001	1001	1001	1001	0110	0110	0110	0110
52	0000	1111	0000	1111	1111	0000	1111	0000	1111	0000	1111	0000	0000	1111	0000	1111
53	0101	1010	0101	1010	1010	0101	1010	0101	1010	0101	1010	0101	0101	1010	0101	1010
54	0011	1100	0011	1100	1100	0011	1100	0011	1100	0011	1100	0011	0011	1100	0011	1100
55	0110	1001	0110	1001	1001	0110	1001	0110	1001	0110	1001	0110	0110	1001	0110	1001
56	0000	0000	1111	1111	1111	1111	0000	0000	1111	1111	0000	0000	0000	0000	1111	1111
57	0101	0101	1010	1010	1010	1010	0101	0101	1010	1010	0101	0101	0101	0101	1010	1010
58	0011	0011	1100	1100	1100	1100	0011	0011	1100	1100	0011	0011	0011	0011	1100	1100
59	0110	0110	1001	1001	1001	1001	0110	0110	1001	1001	0110	0110	0110	0110	1001	1001
60	0000	1111	1111	0000	1111	0000	0000	1111	1111	0000	0000	1111	0000	1111	1111	0000
61	0101	1010	1010	0101	1010	0101	0101	1010	1010	0101	0101	1010	0101	1010	1010	0101
62	0011	1100	1100	0011	1100	0011	0011	1100	1100	0011	0011	1100	0011	1100	1100	0011
63	0110	1001	1001	0110	1001	0110	0110	1001	1001	0110	0110	1001	0110	1001	1001	0110

[*]Column entries are bit positions; row entries are Walsh codes.

Table B.2 Walsh Code Set for 64 kbs

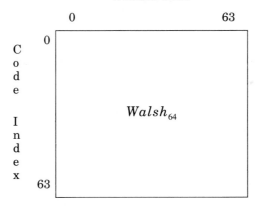

Table B.3 Walsh Code Set for 128 kbs

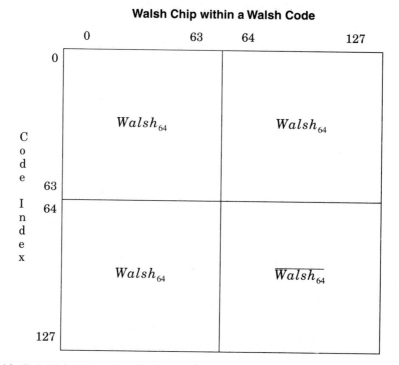

Table B.4 Walsh Code Set for 256 Codes

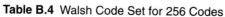

Table B.5 Walsh Code Set for 512 Codes

Walsh Chip within a Walsh Code

	0 63	64 127	128 191	192 255	256 319	320 383	384 447	448 511
0 – 63	$Walsh_{64}$	$Walsh_{64}$	$Walsh_{64}$	$Walsh_{64}$	$Walsh_{64}$	$Walsh_{64}$	$Walsh_{64}$	$Walsh_{64}$
64 – 127	$Walsh_{64}$	$\overline{Walsh_{64}}$	$Walsh_{64}$	$\overline{Walsh_{64}}$	$Walsh_{64}$	$\overline{Walsh_{64}}$	$Walsh_{64}$	$\overline{Walsh_{64}}$
128 – 191	$Walsh_{64}$	$Walsh_{64}$	$\overline{Walsh_{64}}$	$\overline{Walsh_{64}}$	$Walsh_{64}$	$Walsh_{64}$	$\overline{Walsh_{64}}$	$\overline{Walsh_{64}}$
192 – 255	$Walsh_{64}$	$\overline{Walsh_{64}}$	$\overline{Walsh_{64}}$	$Walsh_{64}$	$Walsh_{64}$	$\overline{Walsh_{64}}$	$\overline{Walsh_{64}}$	$Walsh_{64}$
256 – 319	$Walsh_{64}$	$Walsh_{64}$	$Walsh_{64}$	$Walsh_{64}$	$\overline{Walsh_{64}}$	$\overline{Walsh_{64}}$	$\overline{Walsh_{64}}$	$\overline{Walsh_{64}}$
320 – 383	$Walsh_{64}$	$\overline{Walsh_{64}}$	$Walsh_{64}$	$\overline{Walsh_{64}}$	$\overline{Walsh_{64}}$	$Walsh_{64}$	$\overline{Walsh_{64}}$	$Walsh_{64}$
384 – 447	$Walsh_{64}$	$Walsh_{64}$	$\overline{Walsh_{64}}$	$\overline{Walsh_{64}}$	$\overline{Walsh_{64}}$	$\overline{Walsh_{64}}$	$Walsh_{64}$	$Walsh_{64}$
448 – 511	$Walsh_{64}$	$\overline{Walsh_{64}}$	$\overline{Walsh_{64}}$	$Walsh_{64}$	$\overline{Walsh_{64}}$	$Walsh_{64}$	$Walsh_{64}$	$\overline{Walsh_{64}}$

Code Index

Table B.6 Hadamard Code Set for 96 codes

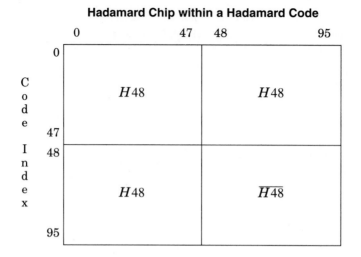

Table B.7 Hadamard Code Set for 192 codes

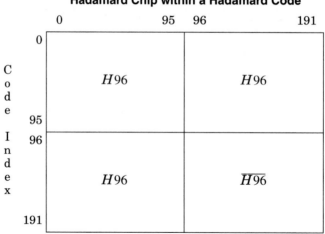

Table B.8 Hadamard Code Set for 384 Codes

Hadamard Chip within a Hadamard Code

	0	95	96	191	192	287	288	383
0	$H96$		$H96$		$H96$		$H96$	
95								
96	$H96$		$\overline{H96}$		$H96$		$\overline{H96}$	
191								
192	$H96$		$H96$		$\overline{H96}$		$\overline{H96}$	
287								
288	$H96$		$\overline{H96}$		$\overline{H96}$		$H96$	
383								

(Row label: C o d e   I n d e x)

Table B.9 Hadamard Code Set for 768 Codes

Hadamard Chip within a Hadamard Code

Code Index	0 – 95	96 – 191	192 – 287	288 – 383	384 – 479	480 – 575	576 – 671	672 – 767
0 – 95	$H96$	$H96$	$H96$	$H96$	$H96$	$H96$	$H96$	$H96$
96 – 191	$H96$	$\overline{H96}$	$H96$	$\overline{H96}$	$H96$	$\overline{H96}$	$H96$	$\overline{H96}$
192 – 287	$H96$	$H96$	$\overline{H96}$	$\overline{H96}$	$H96$	$H96$	$\overline{H96}$	$\overline{H96}$
288 – 383	$H96$	$\overline{H96}$	$\overline{H96}$	$H96$	$H96$	$\overline{H96}$	$\overline{H96}$	$H96$
384 – 479	$H96$	$H96$	$H96$	$H96$	$\overline{H96}$	$\overline{H96}$	$\overline{H96}$	$\overline{H96}$
480 – 575	$H96$	$\overline{H96}$	$H96$	$\overline{H96}$	$\overline{H96}$	$H96$	$\overline{H96}$	$H96$
576 – 671	$H96$	$H96$	$\overline{H96}$	$\overline{H96}$	$\overline{H96}$	$\overline{H96}$	$H96$	$H96$
672 – 767	$H96$	$\overline{H96}$	$\overline{H96}$	$H96$	$\overline{H96}$	$H96$	$H96$	$\overline{H96}$

Table B.10 Hadamard Code Chip of H48

Hadamard Chip within a Hadamard Function

Hadamard Function Index

			11	1111	1111	2222	2222	2233	3333	3333	4444	4444
	0123	4567	8901	2345	6789	0123	4567	8901	2345	6789	0123	4567
0	0000	0000	0000	0000	0000	0000	0000	0000	0000	0000	0000	0000
1	0000	1101	1011	0000	1101	1011	0000	1101	1011	0000	1101	1011
2	0001	1011	0110	0001	1011	0110	0001	1011	0110	0001	1011	0110
3	0101	0111	1000	0101	0111	1000	0101	0111	1000	0101	0111	1000
4	0011	0110	0011	0011	0110	0011	0011	0110	0011	0011	0110	0011
5	1001	0100	1110	1001	0100	1110	1001	0100	1110	1001	0100	1110
6	0100	0010	1111	0100	0010	1111	0100	0010	1111	0100	0010	1111
7	0010	1110	1100	0010	1110	1100	0010	1110	1100	0010	1110	1100
8	1001	1010	1001	1001	1010	1001	1001	1010	1001	1001	1010	1001
9	0101	1100	0101	0101	1100	0101	0101	1100	0101	0101	1100	0101
10	0011	0001	1101	0011	0001	1101	0011	0001	1101	0011	0001	1101
11	1000	0111	0101	1000	0111	0101	1000	0111	0101	1000	0111	0101
12	0000	0000	0000	1111	1111	1111	0000	0000	0000	1111	1111	1111
13	0000	1101	1011	1111	0010	0100	0000	1101	1011	1111	0010	0100
14	0001	1011	0110	1110	0100	1001	0001	1011	0110	1110	0100	1001
15	0101	0111	1000	1010	1000	0111	0101	0111	1000	1010	1000	0111
16	0011	0110	0011	1100	1001	1100	0011	0110	0011	1100	1001	1100
17	1001	0100	1110	0110	1011	0001	1001	0100	1110	0110	1011	0001
18	0100	0010	1111	1011	1101	0000	0100	0010	1111	1011	1101	0000
19	0010	1110	1100	1101	0001	0011	0010	1110	1100	1101	0001	0011
20	1001	1010	1001	0110	0101	0110	1001	1010	1001	0110	0101	0110
21	0101	1100	0101	1010	0011	1010	0101	1100	0101	1010	0011	1010
22	0011	0001	1101	1100	1110	0010	0011	0001	1101	1100	1110	0010
23	1000	0111	0101	0111	1000	1010	1000	0111	0101	0111	1000	1010
24	0000	0000	0000	0000	0000	0000	1111	1111	1111	1111	1111	1111
25	0000	1101	1011	0000	1101	1011	1111	0010	0100	1111	0010	0100
26	0001	1011	0110	0001	1011	0110	1110	0100	1001	1110	0100	1001
27	0101	0111	1000	0101	0111	1000	1010	1000	0111	1010	1000	0111
28	0011	0110	0011	0011	0110	0011	1100	1001	1100	1100	1001	1100
29	1001	0100	1110	1001	0100	1110	0110	1011	0001	0110	1011	0001
30	0100	0010	1111	0100	0010	1111	1011	1101	0000	1011	1101	0000
31	0010	1110	1100	0010	1110	1100	1101	0001	0011	1101	0001	0011
32	1001	1010	1001	1001	1010	1001	0110	0101	0110	0110	0101	0110
33	0101	1100	0101	0101	1100	0101	1010	0011	1010	1010	0011	1010
34	0011	0001	1101	0011	0001	1101	1100	1110	0010	1100	1110	0010
35	1000	0111	0101	1000	0111	0101	0111	1000	1010	0111	1000	1010
36	0000	0000	0000	1111	1111	1111	1111	1111	1111	0000	0000	0000
37	0000	1101	1011	1111	0010	0100	1111	0010	0100	0000	1101	1011
38	0001	1011	0110	1110	0100	1001	1110	0100	1001	0001	1011	0110
39	0101	0111	1000	1010	1000	0111	1010	1000	0111	0101	0111	1000
40	0011	0110	0011	1100	1001	1100	1100	1001	1100	0011	0110	0011
41	1001	0100	1110	0110	1011	0001	0110	1011	0001	1001	0100	1110
42	0100	0010	1111	1011	1101	0000	1011	1101	0000	0100	0010	1111
43	0010	1110	1100	1101	0001	0011	1101	0001	0011	0010	1110	1100
44	1001	1010	1001	0110	0101	0110	0110	0101	0110	1001	1010	1001
45	0101	1100	0101	1010	0011	1010	1010	0011	1010	0101	1100	0101
46	0011	0001	1101	1100	1110	0010	1100	1110	0010	0011	0001	1101
47	1000	0111	0101	0111	1000	1010	0111	1000	1010	1000	0111	0101

Traffic Tables

This appendix provides traffic tables (Tables C.1 through C.9) for a variety of blocking probabilities and channels.[*] The blocked-calls-cleared (Erlang-B) call model is used. In Erlang B, we assume that, when traffic arrives in the system, it either is served, with probability from the table, or is lost to the system. A customer attempting to place a call therefore either will see a call completion or will be blocked and abandon the call. This assumption is acceptable for low blocking probabilities. In some cases, the call will be placed again after a short period of time. If too many calls reappear in the system after a short delay, the Erlang-B model will no longer hold.

In Tables C.8 and C.9, where the number of channels is high (greater than 250 channels), linear interpolation between two table values is possible. We provide the deltas for one additional channel to assist in the interpolation.

---

[*]The data in the tables was supplied by V. H. MacDonald.

Table C.1 Offered Loads (In Erlangs) for Various Blocking Objectives: According to the Erlang-B Model—System Capacity from 1–20 Channels

P(B)= Trunks	0.01	0.015	0.02	0.03	0.05	0.07	0.1	0.2	0.5
1	0.010	0.015	0.020	0.031	0.053	0.075	0.111	0.250	1.000
2	0.153	0.190	0.223	0.282	0.381	0.471	0.595	1.000	2.732
3	0.455	0.536	0.603	0.715	0.899	1.057	1.271	1.930	4.591
4	0.870	0.992	1.092	1.259	1.526	1.748	2.045	2.944	6.501
5	1.361	1.524	1.657	1.877	2.219	2.504	2.881	4.010	8.437
6	1.913	2.114	2.277	2.544	2.961	3.305	3.758	5.108	10.389
7	2.503	2.743	2.936	3.250	3.738	4.139	4.666	6.229	12.351
8	3.129	3.405	3.627	3.987	4.543	4.999	5.597	7.369	14.318
9	3.783	4.095	4.345	4.748	5.370	5.879	6.546	8.521	16.293
10	4.462	4.808	5.084	5.529	6.216	6.776	7.511	9.684	18.271
11	5.160	5.539	5.842	6.328	7.076	7.687	8.487	10.857	20.253
12	5.876	6.287	6.615	7.141	7.950	8.610	9.477	12.036	22.237
13	6.607	7.049	7.402	7.967	8.835	9.543	10.472	13.222	24.223
14	7.352	7.824	8.200	8.803	9.730	10.485	11.475	14.412	26.211
15	8.108	8.610	9.010	9.650	10.633	11.437	12.485	15.608	28.200
16	8.875	9.406	9.828	10.505	11.544	12.393	13.501	16.807	30.190
17	9.652	10.211	10.656	11.368	12.465	13.355	14.523	18.010	32.181
18	10.450	11.024	11.491	12.245	13.389	14.323	15.549	19.215	34.173
19	11.241	11.854	12.341	13.120	14.318	15.296	16.580	20.424	36.166
20	12.041	12.680	13.188	14.002	15.252	16.273	17.614	21.635	38.159

Table C.2 Offered Loads (In Erlangs) for Various Blocking Objectives: According to the Erlang-B Model—System Capacity from 20–39 Channels

P(B)= Trunks	0.005	0.01	0.015	0.02	0.03	0.05	0.07	0.1
20	11.092	12.041	12.680	13.188	14.002	15.252	16.273	17.614
21	11.860	12.848	13.514	14.042	14.890	16.191	17.255	18.652
22	12.635	13.660	14.352	14.902	15.782	17.134	18.240	19.693
23	13.429	14.479	15.196	15.766	16.679	18.082	19.229	20.737
24	14.214	15.303	16.046	16.636	17.581	19.033	20.221	21.784
25	15.007	16.132	16.900	17.509	18.486	19.987	21.216	22.834
26	15.804	16.966	17.758	18.387	19.395	20.945	22.214	23.885
27	16.607	17.804	18.621	19.269	20.308	21.905	23.214	24.939
28	17.414	18.646	19.487	20.154	21.224	22.869	24.217	25.995
29	18.226	19.493	20.357	21.043	22.143	23.835	25.222	27.053
30	19.041	20.343	21.230	21.935	23.065	24.803	26.229	28.113
31	19.861	21.196	22.107	22.830	23.989	25.774	27.239	29.174
32	20.685	22.053	22.987	23.728	24.917	26.747	28.250	30.237
33	21.512	22.913	23.869	24.629	25.846	27.722	29.263	31.302
34	22.342	23.776	24.755	25.532	26.778	28.699	30.277	32.367
35	23.175	24.642	25.643	26.438	27.712	29.678	31.294	33.435
36	24.012	25.511	26.534	27.346	28.649	30.658	32.312	34.503
37	24.852	26.382	27.427	28.256	29.587	31.641	33.331	35.572
38	25.694	27.256	28.322	29.168	30.527	32.624	34.351	36.643
39	26.539	28.132	29.219	30.083	31.469	33.610	35.373	37.715

Table C.3 Offered Loads (In Erlangs) for Various Blocking Objectives: According to the Erlang-B Model—System Capacity from 40–60 Channels

P(B)= Trunks	0.005	0.01	0.015	0.02	0.03	0.05	0.07	0.1
40	27.387	29.011	30.119	30.999	32.413	34.597	36.397	38.788
41	28.237	29.891	31.021	31.918	33.359	35.585	37.421	39.861
42	29.089	30.774	31.924	32.838	34.306	36.575	38.447	40.936
43	29.944	31.659	32.830	33.760	35.255	37.565	39.473	42.012
44	30.801	32.546	33.737	34.683	36.205	38.558	40.501	43.088
45	31.660	33.435	34.646	35.609	37.156	39.551	41.530	44.165
46	32.521	34.325	35.556	36.535	38.109	40.545	42.559	45.243
47	33.385	35.217	36.468	37.463	39.063	41.541	43.590	46.322
48	34.250	36.111	37.382	38.393	40.019	42.537	44.621	47.401
49	35.116	37.007	38.297	39.324	40.976	43.535	45.654	48.481
50	35.985	37.904	39.214	40.257	41.934	44.534	46.687	49.562
51	36.856	38.802	40.132	41.190	42.893	45.533	47.721	50.644
52	37.728	39.702	41.052	42.125	43.853	46.533	48.756	51.726
53	38.601	40.604	41.972	43.061	44.814	47.535	49.791	52.808
54	39.477	41.507	42.894	43.999	45.777	48.537	50.827	53.891
55	40.354	42.411	43.817	44.937	46.740	49.540	51.864	54.975
56	41.232	43.317	44.742	45.877	47.704	50.544	52.902	56.059
57	42.112	44.224	45.667	46.817	48.669	51.548	53.940	57.144
58	42.993	45.132	46.594	47.759	49.636	52.553	54.979	58.229
59	43.875	46.041	47.522	48.701	50.603	53.559	56.018	59.315
60	44.759	46.951	48.451	49.645	51.570	54.566	57.058	60.401

Table C.4 Offered Loads (In Erlangs) for Various Blocking Objectives: According to the Erlang-B Model—System Capacity from 61–80 Channels

P(B)= Trunks	0.005	0.01	0.015	0.02	0.03	0.05	0.07	0.1
61	45.644	47.863	49.381	50.590	52.539	55.573	58.099	61.488
62	46.531	48.776	50.311	51.535	53.509	56.581	59.140	62.575
63	47.418	49.689	51.243	52.482	54.479	57.590	60.181	63.663
64	48.307	50.604	52.176	53.429	55.450	58.599	61.224	64.750
65	49.197	51.520	53.110	54.377	56.422	59.609	62.266	65.839
66	50.088	52.437	54.044	55.326	57.395	60.620	63.309	66.927
67	50.980	53.355	54.980	56.276	58.368	61.631	64.353	68.016
68	51.874	54.273	55.916	57.226	59.342	62.642	65.397	69.106
69	52.768	55.193	56.853	58.178	60.316	63.654	66.442	70.196
70	53.663	56.113	57.791	59.130	61.292	64.667	67.487	71.286
71	54.560	57.035	58.730	60.083	62.268	65.680	68.532	72.376
72	55.457	57.957	59.670	61.036	63.244	66.694	69.578	73.467
73	56.356	58.880	60.610	61.991	64.222	67.708	70.624	74.558
74	57.255	59.804	61.551	62.945	65.199	68.723	71.671	75.649
75	58.155	60.729	62.493	63.901	66.178	69.738	72.718	76.741
76	59.056	61.654	63.435	64.857	67.157	70.753	73.765	77.833
77	59.958	62.581	64.379	65.814	68.136	71.769	74.813	78.925
78	60.861	63.508	65.322	66.772	69.116	72.786	75.861	80.018
79	61.765	64.435	66.267	67.730	70.097	73.803	76.909	81.110
80	62.669	65.364	67.212	68.689	71.078	74.820	77.958	82.203

Table C.5 Offered Loads (In Erlangs) for Various Blocking Objectives: According to the Erlang-B Model—System Capacity from 81–100 Channels

P(B)= Trunks	0.005	0.01	0.015	0.02	0.03	0.05	0.07	0.1
81	63.574	66.293	68.158	69.648	72.059	75.838	79.007	83.297
82	64.481	67.223	69.104	70.608	73.042	76.856	80.057	84.390
83	65.387	68.153	70.051	71.568	74.024	77.874	81.107	85.484
84	66.295	69.085	70.999	72.529	75.007	78.893	82.157	86.578
85	67.204	70.016	71.947	73.491	75.991	79.912	83.207	87.672
86	68.113	70.949	72.896	74.453	76.975	80.932	84.258	88.767
87	69.023	71.882	73.846	75.416	77.959	81.952	85.309	89.861
88	69.933	72.816	74.796	76.379	78.944	82.972	86.360	90.956
89	70.844	73.750	75.746	77.342	79.929	83.993	87.411	92.051
90	71.756	74.685	76.697	78.306	80.915	85.014	88.463	93.146
91	72.669	75.621	77.649	79.271	81.901	86.035	89.515	94.242
92	73.582	76.557	78.601	80.236	82.888	87.057	90.568	95.338
93	74.496	77.493	79.553	81.202	83.875	88.079	91.620	96.434
94	75.411	78.431	80.506	82.167	84.862	89.101	92.673	97.530
95	76.326	79.368	81.460	83.134	85.850	90.123	93.726	98.626
96	77.242	80.307	82.414	84.101	86.838	91.146	94.779	99.722
97	78.158	81.245	83.368	85.068	87.827	92.169	95.833	100.819
98	79.075	82.185	84.323	86.036	88.815	93.193	96.887	101.916
99	79.993	83.125	85.279	87.004	89.805	94.217	97.941	103.013
100	80.911	84.065	86.235	87.972	90.794	95.240	98.995	104.110

Table C.6 Offered Loads (In Erlangs) for Various Blocking Objectives: According to the Erlang-B Model—System Capacity from 105–200 Channels

P(B)= Trunks	0.005	0.01	0.015	0.02	0.03	0.05	0.07	0.1
105	85.518	88.822	91.030	92.823	95.747	100.371	104.270	109.598
110	90.147	93.506	95.827	97.687	100.713	105.496	109.550	115.090
115	94.768	98.238	100.631	102.552	105.680	110.632	114.833	120.585
120	99.402	102.977	105.444	107.426	110.655	115.772	120.121	126.083
125	104.047	107.725	110.265	112.307	115.636	120.918	125.413	131.583
130	108.702	112.482	115.094	117.195	120.622	126.068	130.708	137.087
135	113.366	117.247	119.930	122.089	125.615	131.222	136.007	142.593
140	118.039	122.019	124.773	126.990	130.612	136.380	141.309	148.101
145	122.720	126.798	129.622	131.896	135.614	141.542	146.613	153.611
150	127.410	131.584	134.477	136.807	140.621	146.707	151.920	159.122
155	132.106	136.377	139.337	141.724	145.632	151.875	157.230	164.636
160	136.810	141.175	144.203	146.645	150.647	157.047	162.542	170.152
165	141.520	145.979	149.074	151.571	155.665	162.221	167.856	175.668
170	146.237	150.788	153.949	156.501	160.688	167.398	173.173	181.187
175	150.959	155.602	158.829	161.435	165.713	172.577	178.491	186.706
180	155.687	160.422	163.713	166.373	170.742	177.759	183.811	192.227
185	160.421	165.246	168.602	171.315	175.774	182.943	189.133	197.750
190	165.160	170.074	173.494	176.260	180.809	188.129	194.456	203.273
195	169.905	174.906	178.390	181.209	185.847	193.318	199.781	208.797
200	174.653	179.743	183.289	186.161	190.887	198.508	205.108	214.323

Table C.7 Offered Loads (In Erlangs) for Various Blocking Objectives: According to the Erlang-B Model—System Capacity from 205–245 Channels

P(B)= Trunks	0.005	0.01	0.015	0.02	0.03	0.05	0.07	0.1
205	179.407	184.584	188.192	191.116	195.930	203.700	210.436	219.849
210	184.165	189.428	193.099	196.073	200.976	208.894	215.765	225.376
215	188.927	194.276	198.008	201.034	206.023	214.089	221.096	230.904
220	193.694	199.127	202.920	205.997	211.073	219.287	226.427	236.433
225	198.464	203.981	207.836	210.963	216.125	224.485	231.760	241.963
230	203.238	208.839	212.754	215.932	221.180	229.686	237.094	247.494
235	208.016	213.700	217.675	220.902	226.236	234.887	242.430	253.025
240	212.797	218.564	222.598	225.876	231.294	240.090	247.766	258.557
245	217.582	223.430	227.524	230.851	236.354	245.295	253.103	264.089

Table C.8 Offered Loads (In Erlangs) for Various Blocking Objectives: According to the Erlang-B Model—System Capacity from 250–600 Channels

P(B)= Trunks	0.005	0.01	0.015	0.02	0.03	0.05	0.07	0.1
250	222.370	228.300	232.452	235.828	241.415	250.500	258.441	269.622
delta	0.961	0.977	0.988	0.998	1.015	1.042	1.069	1.107
300	270.410	277.144	281.853	285.707	292.142	302.617	311.866	324.961
delta	0.966	0.980	0.991	1.001	1.017	1.044	1.070	1.108
350	318.698	326.155	331.424	335.738	342.995	354.836	365.359	380.384
delta	0.969	0.984	0.994	1.005	1.018	1.045	1.071	1.109
400	367.163	375.334	381.128	385.963	393.895	407.096	418.890	435.813
delta	0.972	0.989	0.998	1.004	1.020	1.046	1.071	1.109
450	415.779	424.774	431.022	436.178	444.877	459.408	472.456	491.263
delta	0.975	0.987	0.997	1.006	1.021	1.047	1.072	1.109
500	464.518	474.130	480.890	486.480	495.919	511.759	526.049	546.730
delta	0.977	0.989	0.999	1.007	1.022	1.048	1.072	1.110
550	513.361	523.600	530.843	536.846	547.012	564.142	579.663	602.208
delta	0.979	0.991	1.000	1.008	1.023	1.048	1.073	1.110
600	562.292	573.142	580.859	587.267	598.145	616.552	633.295	657.697

Table C.9 Offered Loads (In Erlangs) for Various Blocking Objectives: According to the Erlang-B Model—System Capacity from 600–1050 Channels

P(B)= Trunks	0.005	0.01	0.015	0.02	0.03	0.05	0.07	0.1
600	562.292	573.142	580.859	587.267	598.145	616.552	633.295	657.697
delta	0.983	0.992	1.001	1.009	1.023	1.049	1.073	1.110
650	611.418	622.748	630.927	637.732	649.313	668.982	686.941	713.193
delta	0.981	0.993	1.002	1.010	1.024	1.049	1.073	1.110
700	660.462	672.410	681.042	688.238	700.511	721.432	740.598	768.697
delta	0.982	0.994	1.003	1.011	1.024	1.049	1.073	1.110
750	709.586	722.119	731.196	738.777	751.735	773.896	794.266	824.206
delta	0.984	0.995	1.004	1.011	1.025	1.050	1.074	1.110
800	758.762	771.872	781.386	789.346	802.981	826.375	847.943	879.719
delta	0.985	0.996	1.004	1.012	1.025	1.050	1.074	1.110
850	807.987	821.662	831.608	839.942	854.247	878.865	901.627	935.236
delta	0.985	0.996	1.005	1.012	1.026	1.050	1.074	1.110
900	857.256	871.487	881.857	890.561	905.530	931.365	955.317	990.757
delta	0.986	0.997	1.005	1.013	1.026	1.050	1.074	1.110
950	906.565	921.343	932.132	941.202	956.829	983.875	1009.013	1046.281
delta	0.987	0.998	1.006	1.013	1.026	1.050	1.074	1.111
1000	955.910	971.226	982.430	991.862	1008.142	1036.393	1062.715	1101.808
delta	0.988	0.998	1.006	1.014	1.027	1.050	1.074	1.111
1050	1005.289	1021.136	1032.748	1042.539	1059.468	1088.918	1116.420	1157.337

List of Abbreviations

A

Abs	Analysis by synthesis
ADPCM	Adaptive Differential Pulse Code Modulator
AGCH	Access Grant Channel
AIN	Advanced Intelligent Network
AM	Amplitude Modulation
AMPS	Advanced Mobile Phone System
ANM	Answer Message
APBS	Antipodal Baseband Signaling
APC	Automatic Power Control
ARC	Automatic Reverse Charge
ARQ	Automatic Request for retransmission
ATIS	Alliance for Telecommunications Industry Solutions
AUC	Authentication Center
AWGN	Additive White Gaussian Noise

B

BAI	Building Index
BCCH	Broadcast Control Channel
BCH	Broadcast Channel
BLD	Building Location Distribution
bps	bits per second
BPSK	Binary Phase Shift Keying
BS	Base Station
BSC	Base Station Controller
BSD	Building Size Distribution
BSIC	Base Station Identity Code
BSS	Base Station Subsystem
BTS	Base Transceiver System

C

CAC	Communication Access Channel
CAI	Common Air Interface
CBCH	Cell Broadcast Channel
CCCH	Communication Control Channel
CCF	Call Control Function
CCH	Control Channel
CCITT	Consultative Committee on Telephone and Telegraph
CDL	Coded Digital control channel Locator
CDMA	Code-Division Multiple Access
CDPD	Cellular Digital Packet Data
CDVCC	Coded Digital Verification Color Code
CEPT	Conference Europeenne des Postes et Telecommunications
CGSA	Cellular Geographical Service Area
CNIP	Calling Number Identification Presentation
CNIR	Calling Number Identification Restriction
CRC	Cyclic Redundancy Check
CSFP	Coded Super Frame Phase
CTIA	Cellular Telecommunications Industry Association
CT-1	Cordless Telephone-1
CT-2	Cordless Telephone-2
CT-3	Cordless Telephone-3

D

DCA	Dynamic Channel Allocation
DCC	Digital Color Code
DCCH	Dedicated Control Channel
DCT	Digital Cordless Telephone
DECT	Digital European Cordless Telecommunications
DMH	Data Message Handler
DQPSK	Differential Quadrature Phase Shift Keying
DS	Direct Sequence
DSSS	Direct Sequence Spread Spectrum

E

EEPROM	Electrically Erasable PROM
EIA	Electronic Industry Association
EIR	Equipment Identity Register
EIRP	Effective Isotropic Radiated Power
E-Mail	electronic mail
ESN	Electronic Serial Number
E-TDMA	Extended Time-Division Multiple Access

F

| FACCH | Fast Associated Control Channel |
| FC | Fast Channel |

FCC	Federal Communications Commission
FCCH	Frequency Correction Channel
FDD	Frequency-Division Duplex
FDMA	Frequency-Division Multiple Access
FE	Functional Element
FH	Frequency Hopping
FHMA	Frequency-Hopping Multiple Access
FHSS	Frequency Hop Spread Spectrum
FM	Frequency Modulation
FSK	Frequency Shift Keying

G

GFSK	Gaussian Frequency Shift Keying
GHz	gigahertz
GMSK	Gaussian Minimum Shift Keying
GOS	Grade of Service
GSM	Global System for Mobile communications

H

HAAT	Height Above Average Terrain
HLR	Home Location Register

I

IAM	Initial Address Message
ID	Identification
IEEE	Institute of Electrical and Electronic Engineers
IP	Internal Peripheral
IMEI	International Mobile station Equipment Identity
IMSI	International Mobile Station Identification
IMTS	Improved Mobile Telephone Service
ISDN	Integrated Service Digital Network
ITAR	International Traffic in Arms Regulation
IWF	Interworking Function

J

JDC	Japanese Digital Cellular

K

kHz	kilohertz

L

LAI	Location Area Identity
LAN	Local Area Network
LPC	Linear Predictive Coding

M

MAP	Mobile Application Part
MC	Multiple Carriers
Mcps	Million chips per second
MHz	Megahertz
MIN	Mobile Identification Number
MLPP	Multilevel Precedence and Preemption
MOS	Mean Opinion Score
MRI	Mobile Reported Interference
MS	Mobile Station
MSC	Mobile Switching Center
MT	Mobile Termination
MUF	Maximum Usable Frequency

N

NAM	Number Assignment Module
N-AMPS	Narrowband Advanced Mobile Phone Service
NMC	Network Management Center
NMT	Nordic Mobile Telephone system
NTT	Nippon Telephone and Telegraph
NSS	Network Switching Subsystem

O

OAM&P	Operation, Administration, Maintenance, and Provisioning
OMC	Operation Maintenance Center
OMSS	Operation and Maintenance Subsystem
OQPSK	Offset Quadrature Phase Shift Keying
OS	Operations System
OSS	Operational Subsystem

P

PACS	Personal Access Communications System
PAD	Packet Assembler/Disassembler
PBX	Private Branch Exchange
PCC	Power Control Channel
PCH	Paging Channel
PCM	Pulse Code Modulation
PCN	Personal Communications Network
PCP	Power Control Pulse
PCS	Personal Communications Services
PCSC	Personal Communications Switching Center
PIN	Personal Identification Number
PLMN	Public Land Mobile Network
PMC	Personal Mobility Controller
PMD	Personal Mobility Data store
PN	Pseudonoise

PS	Personal Station
PSC	PCS Switching Center
PS-CF	PS Call Forwarding
PSD	Power Spectral Density
PSK	Phase Shift Keying
PSPDN	Public Switched Packet Data Network
PSTN	Public Switching Telephone Network

Q

QAM	Quadrature Amplitude Modulation
QCELP	Quadrature Coded-Excited Linear Predictive
QPSK	Quadrature Phase Shift Keying

R

RACF	Radio Access Control Function
RACH	Random Access Channel
RAM	Random Access Memory
RASC	Radio Access System Controller
RCF	Radio Control Function
RELP	Residual Excited Linear Prediction
RES	Radio Equipment System
RF	radio frequency
RMS	Root Mean Square
RP	Radio Port
RPE-LTP	Regular Pulse Excited-Long-Term Predictive
RPI	Radio Port Intermediary
RPT	Radio Personal Terminal
RS	Radio System
RTF	Radio Terminal Function

S

SACCH	Slow Associated Control Channel
SAT	Supervisory Audio Tone
SC	Slow Channel
SCCH	Signaling Control Channel
SCCP	Signaling Connection Control Part
SCH	Synchronization Channel
SDCCH	Stand-alone Dedicated Control Channel
S/I	Signal-to-Interference ratio
SIM	Subscriber Identity Module
SMR	Specialized Mobile Radio
SMS	Short Message Service
SNR	Signal-to-Noise Ratio
SRF	Specialized Resource Function
SS	Spread Spectrum
SSB	Single Side Band
SSD	Shared Secret Data

SSF	Service Switching Function
SSP	Switching System Platform
SU	Subscriber Unit

T

TACS	Total Access Communications System
TCH	Traffic Channel
TCH/F	Traffic Channel/Full rate
TCH/H	Traffic Channel/Half rate
TDMA	Time-Division Multiple Access
TDD	Time-Division Duplex
TE	Terminal Equipment
TIA	Telecommunications Industry Association
TMC	Terminal Mobility Controller
TMD	Terminal Mobility Data store
TMSI	Temporary Mobile Station Identification
TVI	Terrain Varation Index

U

UHF	Ultra High Frequency
UIM	User Identity Module
UMTS	Universal Mobile Telecommunications System
UPCH	User Packet Channel
UPT	Universal Personal Telecommunications
USC	User Specific Channel

V

VI	Vegetation Index
VHF	Very High Frequency
VLR	Visitor Location Register
VLSI	Very Large Scale Integrated
VSELP	Vector Sum Excited Linear Prediction

W

WLAN	Wide Local Area Network
WPBX	Wireless Private Branch Exchange